Research Methods in Sport Management

2nd Edition

Research Methods in Sport Management

2nd Edition

Brenda Pitts
University of Georgia

Ming Li
Ohio University

Amy Kim
Florida State University

FiT PUBLISHING
A Division of the International Center for Performance Excellence
West Virginia University
375 Birch Street, WVU-CPASS
PO Box 6116
Morgantown, WV 26506-6116

Library of Congress Card Catalog Number: 2018936312

ISBN: 9781940067315

Cover Design: Wendy Lazzell, FiT Publishing
Cover Photos: (C) credits—ID: 553128571; By: Evgenii Matrosov; Shutterstock, Inc.
ID: 611813909; By: Evgenii Matrosov; Shutterstock, Inc.
ID: 619012244; By: violetkaipa; Shutterstock, Inc.
Production Editor: Eileen Harvey
Copyeditor: Kassi Roberts

10 9 8 7 6 5 4 3 2 1

FiT
PUBLISHING
A Division of the International Center for Performance Excellence
West Virginia University
375 Birch Street, WVU-CPASS
PO Box 6116
Morgantown, WV 26506-6116

800.477.4348 (toll free)
304.293.6888 (phone)
304.293.6658 (fax)

Email: fitcustomerservice@mail.wvu.edu
Website: www.fitpublishing.com

Contents

List of Exhibits

Dedication

This book is dedicated to my mother, Hae Young Kim, my brother, Byung Hoon Kim, and to the memory of my father, Young Jung Kim, with love.

—Dr. Amy Kim

I want to thank all of my students over the years—you keep me on my toes! I want to acknowledge the patience and support of my wife, Melita Reeves, and our "furry" family who provide the unconditional love and never-ending comic relief needed in life.

—Dr. Brenda Pitts

I want to delicate this book to my wife, Wan Chen, and two daughters, Madison and Holly, for their love and support.

—Dr. Ming Li

Lastly, the authors want to recognize and dedicate this edition of the book to our co-author on the first edition of this book, Dr. Jerome Quarterman. Jerome lost his battle with cancer just before we began work on the second edition. Jerome, you are loved and missed.

Acknowledgements

There are numerous people who work as a team to develop and bring a textbook to life. Writing or updating a textbook is a major undertaking. Although the authors spend a significant amount of time researching current trends to bring the most accurate information to the book, it is done with the patience, support, and help of several individuals.

We are particularly indebted to Richard Southall for his time and expertise in writing Chapter 4 Case Study Research as a contributing author.

We wish to acknowledge and thank the hard-working people at FiT Publishing located at West Virginia University in Morgantown, West Virginia. FiT was one of the first publishers to bring textbooks in sport management to the market. They have done an exceptional job, and we thank them for their work and especially for their patience with our endeavors.

Preface

It is with great pleasure we bring to you the second edition of *Research Methods in Sport Management.* The first words of the preface in the first edition, published in 2008, the first book about research methods in sport management were, "Over many years, we listened as numerous professors and students lamented that the field of sport management did not have its own research methods courses or book. Without such, our students were placed in courses on research methods in physical education, general education, or exercise science." Over the 10 years since the first edition, the field has made amazing strides forward in its progress, as other authors joined us with the goal of providing critically needed textbooks in the sport management field, and research methods in particular. And so we now present the second edition of the groundbreaking book *Research Methods in Sport Management.*

Expertise of the Authors

It is with great sadness that we dedicate this second edition to Dr. Jerome Quarterman, one of the three original authors, who passed away just as we were beginning work on this second edition. Jerome was a warm and kind person, a friend, and an amazing scholar in sport management who amassed a grand record and body of work in the field.

It is with great happiness that we present Dr. Amy Kim as our coauthor for this second edition. Dr. Kim is currently with the faculty of Florida State University and in her young career has already established a growing body of work in sport management.

The authors of this book have a vast knowledge and experience needed to produce a textbook on research methods suited to sport management. Each one of us has an established record and reputation as an expert researcher. All of us have published dozens of research papers in scholarly journals, gained numerous years of experience in service on editorial boards, as peer reviewers of scholarly research, as judges and mentors of national and international conferences, and as advisors of masters and doctoral student research in sport management.

Purpose of this Book

The Second Edition of *Research Methods in Sport Management* provides a detailed introduction to research itself, as well as the methods used in the field of sport business management. It presents, in a logical manner, an overview of many frequently used methods, their preparation, analysis, and presentation, as well as ethical considerations.

Audience

This textbook is written for everyone who seeks an introduction to and details about methods of research that apply to sport business management, including how to prepare, create, develop, and deliver research findings as a technical report for a sport business, or as scholarly research for submission to a peer reviewed journal for publication. With a direct focus on the sport management industry, this book does not provide an exhaustive inventory of all research methods available to scholars across subjects and fields of study. Rather, readers need not have any background in research methods, but will find that this textbook introduces and broadly overviews research methods best used within the sport management industry.

Instructors of courses including research methods in sport business management will be able to continue using this second edition without a major overhaul of course material and preparation. Instructors will find, however, new resources, details, information, and examples that are more in tune with contemporary sport business management practices.

Students will have in their possession a textbook to target research specifically to their chosen field of study: sport business management. Students will find that this book is full of content, examples, and methods of research approaches that are specific and applicable directly to sport business management.

Industry Professionals can use this textbook in their sport business to serve as a "how to" for conducting the basic research needed for their business. It offers information and guidelines on conducting and presenting research in a variety of situations as needed in business.

This Edition

As soon as one edition of a textbook is published, work toward the next edition begins. The purpose of a new edition is to bring the latest real-time, relatable information into the reader.

This edition builds upon and fully updates all of the information from the first edition. Where needed, current, timely examples are included. Additionally, the chapters have been reorganized to guide the reader through the process of preparation, research, analysis, and presentation. Lastly, since publishing the first edition about ten years ago, nearly all references and sources have been updated to reflect recent research methodology used in the field.

A Final Word from the Authors

When preparing the first edition of this book, we were profoundly aware of the challenges we faced in our attempt to provide the sport management field with its first book on research methods in sport management. We knew that this book has provided a starting place for introducing research methods to sport management students, and we are—in this second edition—still keenly aware its content can and should be enhanced through practical updates. We have incorporated your feedback over the years into this second edition, and we ask for your continued feedback and ideas to improve upon what has become a valuable resource in the field. We hope it exceeds your expectations.

Enjoy, and please, let us hear from you!

Brenda Pitts, Ming Li, and Amy Kim

Section I

Sport Management Research

1

Introduction to Sport Business Management Research

After completing the study of this chapter, you should be able to:

- Define research and sport management research.
- Understand the role of research for sport business.
- Understand the role of research for the academic field of sport management.
- Understand the value of research for the academic field of sport management.
- Understand the characteristics of sport business management research.
- Understand the purpose of sport business management research.
- Differentiate the philosophical approaches to sport management research.
- Differentiate applied sport business management research and basic sport business management research.
- Describe each of the three types of sport management research: exploratory research, descriptive research, and causal research.
- Understand the evolution of sport business management research.
- Discuss the future of sport business management research.

The purpose of this chapter is to provide an overview of sport business management research and its significance to the academic field of sport management and to the sport business industry. Specifically, the chapter will first define sport business management research and substantiate the importance of *research* to the academic field of sport management and to the sport business industry. The chapter will then discuss the characteristics of sport management research and the different types of research that sport business management researchers can use in solving their research problems. Lastly, the evolution and future of sport business management research will be presented.

Research for the Industry and Academic Field of Sport Business Management

At its most basic level, the purpose of research is to generate information. Research is carried out for a variety of reasons, in a variety of settings, and will be used for a variety of purposes. In a sport business setting, information is critical for making decisions and developing strategies in every part of the company. In an academic setting, information provides critical knowledge to be used in the classroom to educate students about the topic of sport business through several content identified areas by the accrediting organization, the Commission on Sport Management Accreditation (COSMA): sociocultural aspects of sport and sport management, sport business law, sport business marketing, sport management, sport communication, sport facility management, financial management in sport business, and ethics in sport business.

As an academic field of study, sport business management—also labeled with titles such as sport management, sport administration, and sport business—is relatively new compared to other disciplines, such as law, medicine, or education. Yet, Professor Brenda G. Pitts, an author of this textbook, believes that although sport business management is among the newest academic disciplines, it is most likely among the oldest professions. She argues that sport management is an activity that humans have been doing for thousands and thousands of years, or at least ever since humans started doing play, leisure, recreational, or sporting activities (Pitts, 2008; 2010; 2015; 2017; Pitts & Schwarz, 2013).

> *"The practice of sport management is one of the oldest professions . . . However . . . the study and teaching of Sport Management is one of the newest professions" (Pitts, 2008).*

That is, once humans started to play a game, someone was there to provide equipment, even though it might have been just a stick and a stone. Sooner or later, there was someone there to provide management, rules, officiating, facilities, training, and more. Dr. Pitts goes on to say that even though sport business management is something we have been doing forever, it is only recently that it has become a labeled, defined, and recognized academic discipline. In fact, the first named sport administration program has been traced to 1949, when Florida Southern University offered a curriculum titled Baseball Business Administration (Isaacs, 1964). Although the origins of the words that sports administration used to describe this field have not yet been traced in historical research, the earliest use of the label sports administration was found in a 1964 book entitled *Careers and Opportunities in Sports* (Isaacs, 1964). In this book, in which a variety of jobs in sport business are presented, there is a chapter titled "Sports Administration." This is a couple of years earlier than the use of the words sports administration seems to have been used in what is often touted as the first university academic program with the label—Ohio University in 1966.

In the sport business industry, research is critical for gathering and discovering information needed for making decisions and developing business strategies. In every part of the company, information is essential to developing products, determining pricing strategies, deciding about distribution, and creating promotional products for the company and its products. Therefore, research is the method of gathering or discovering new information to be used to investigate a question, find a solution for a problem, and create and develop the body of knowledge for the field of sport management.

Research in Our Daily Lives

Research touches us every day. Whether we reach for something in the fridge, get into our cars, turn on our cell phones, or put on our running shoes, each one of these items has been created, developed, manufactured, and marketed with the help of research.

Staying current with the knowledge of today is a demanding task. Information, old and new, is produced and flows at amazing speeds. When an incident happens on the other side of the world, we can read about it, hear about it, and see it within seconds. Additionally, information technology, the invention, production, and distribution of high technology equipment that allows us to access and manage information, improves daily at a blinding pace—we can surf the web, manage email, watch TV or movies, listen to music, and check our GPS for directions on a gadget that fits in the palm of a hand. And, oh, by the way, it's also a phone.

In the world of sports and in the sport business industry, research affects every part of the industry. In sporting apparel, the athlete's clothing is lighter, warmer, faster, safer, and basically whatever the athlete needs to maximize performance. In goods, advances in equipment have caused many sports to change rules to either restrict the technology, or to embrace it. In golf, for instance, golf course designers have had to lengthen courses because the technology of golf clubs and golf balls allows the average player to hit the ball farther.

In every sport business, information is critical. Sport business management professionals need information to make informed decisions and to generate strategies. The nature and scope of sport business management research is limited only by one's definition of business and of research. Research provides information needed about the consumer, the competitor, the company, and the climate—the 4 Cs of sport marketing the sport business must study constantly to stay informed and make good decisions and successful strategies (see Exhibit 1.1) (Pitts & Stotlar, 2013).

Sport Management Research Defined

In gaining an understanding of "what is research," it is first important to understand the academic discipline of sport business management and the industry of sport business. As an academic discipline, sport business management is the study and practice of all people, activities, businesses, or organizations involved in producing, facilitating, promoting, or organizing any sport-related business or product (Pitts & Stotlar, 2013). The field is known by other labels, some of which are sport management, sport business, sport business management, and sport(s) administration. As the field evolves and develops, a more accurate title may be "sport business management," because the purpose of the discipline is to prepare students who will be working in a sport business in a wide variety of management positions. Thus, the field focuses on the management of sport businesses. At universities, sport management degree programs are offered at the undergraduate, masters, and doctoral levels and may be found in such departments as a department of recreation management, a department of business, a department of management, a department of marketing, and a department of kinesiology and sport. Moreover, the discipline has grown to be large enough to have its own department at some universities, such as a department of sport management, department of sport, recreation, and tourism management, or department of sport business. Most of these departments are in a large unit called a college (or school) that houses several departments. Some of these are a college of business, school of hospitality and tourism, or college of education.

Exhibit 1.1. The 4 Cs of Sport Marketing

Consumer	Demographics Psychographics Lifestyle Geographics Purchase behavior
Competitor	The industry and marketplace Product differentiation Pricing strategies Financial strategies Positioning Promotion strategies
Company	Mission and objectives Strengths Production Product management Distribution strategies Promotion strategies Weaknesses
Climate	Economic Legal Social and cultural Political Ethical Trends Technology Education Community Corporate

Source. Pitts, B. & Stotlar, D. (2013). *Fundamentals of Sport Marketing* (4th ed.). Morgantown, WV: Fitness Information Technology.

A large portion of research conducted by faculty at universities is critical in building a field's body of knowledge—the body of literature. The purpose of the body of literature is to contain all of the information about a topic. An academic discipline is built upon a sound body of knowledge that defines the field. The body of literature is what is used in the academic field in college programs to teach students everything about their field of study, the industry in which they will work, and what skills and knowledge will be needed for their jobs and careers (Fielding, Pitts, & Miller, 1991; Kim, Chelladurai, & Kim, 2015; Pitts, 2016; Pitts, Danylchuk, & Quarterman, 2014). Research, at its most basic level, is the generation of information. Other common definitions of research are "to examine carefully to help build a scientific body of knowledge," "the systematic investigation of phenomena (behavior, events, people, things, etc.)," and "a diligent systematic inquiry or study to gain new knowledge or refine existing" (Research, 2017). One of the most comprehensive definitions of research is offered by Kerlinger and Lee (2000). They regard research as "a systematic, controlled, empirical, amoral, public, and critical investigation of natural phenomena. It is guided by theory and hypotheses about the presumed relations among such phenomena" (p. 14). McMillan and Schumacher (2006) defined research as "the systematic process of collecting and logically analyzing data for some purpose" (p. 9).

For the purpose of this book, sport business management research is defined as an orderly investigative process that involves purposeful and systematic collection, analysis, and interpretation of data to gain new knowledge or to verify existing knowledge primarily within the domain of managing human and material resources in entities of the sport business industry. The activities of sport business management research include the scientific, purposeful, systematic, and rigorous method of collecting, analyzing, and interpreting data objectively or subjectively about some characteristic (individual(s), group(s), organization(s), idea(s), demographic(s), concept(s), model(s), theory(ies), etc.) in order to gain new knowledge or add to the existing knowledge base of the field of sport organization management studies.

Research for the Body of Knowledge in Sport Business Management

A body of knowledge is the collective research and information known about a subject. This information is usually held as articles, papers, presentations, books, and journals. The full collection of information is known as the body of knowledge or the literature. Depending on the topic, a body of knowledge may be complete or incomplete. For an academic discipline, the body of knowledge is used for educational purposes, first and foremost to prepare students with the most current and full body of knowledge of information for their jobs and careers in the sport business industry. Thus, sport business management academicians, professors, educators, researchers, and scholars are responsible for developing the body of knowledge. Hence, this is the primary reason why professors are required to conduct research. This textbook is a direct result of sport business management professors conducting research to produce literature for the body of knowledge in sport management.

Sport business management is an emerging and expanding field of study as a science. Science involves systematic study for the development of new information or ways to validate existing information. Such systematic study leads to the creation of a scientific body of knowledge for sport administration education. As shown in academic journals in the field, such as *Journal of Sport Management*, *Sport Marketing Quarterly*, and *International Journal of Sport Management*, research topics and methods have grown such that they have made the study of sport management more scientific. There are now over 150 academic journals with a direct or indirect association to the discipline of sport business management in areas including sport management principles, psychological topics in sport, sport law, and other specific content areas. Directly related journals are those created and developed specifically by academic professionals who are most likely professors of sport management. Indirectly related journals are those created and developed by people outside of the field of sport management, but the topic of the journal is related to the field of sport management. Academics who study sport management can have sport management papers published in these journals as well (Shapiro & Pitts, 2014). Appendix 1-A provides a list of sport management journals by subject matter.

As an academic discipline, sport business management has grown tremendously over the past decade. There are now approximately a thousand undergraduate, masters, and doctoral programs at numerous colleges and universities around the USA and around the globe. Sport management professors have been hard at work developing the body of knowledge. For example, merely thirty years ago there were no textbooks about sport management. The first were published in the 1990s. Today, there are dozens of textbooks. However, there are more books in some of the content areas in sport management than others. Still, it has come a long way in a short period of time. Some of the very first books were published by Fitness Information Technology—now FiT Publishing—when a Sport Management Library Task Force consisting of six sport management professors worked for FiT to develop a list of needed textbooks, recruit the authors, and supervise the publication of the first full set of textbooks in sport management.

Value of Research for Students in Sport Business Management

Students in sport management are in a unique position to learn the value of research, why and when to conduct research, and how to conduct research. Knowledge and skill to conduct basic research are valuable for employees of organizations. When students gain the knowledge and skill while in college, they attain a major skill-set that future employers will value. The following outlines the value of research to the student in sport management:

- It enhances the development of critical thinking of students for study of supervisory management of firms (enterprises) of the sport industry.
- It helps students to become more authoritative in research procedures for the study of supervisory management in the sport industry.
- It helps students to become authorities in their designated fields or options of study, (i.e., sport promotion, professional sport management, sporting goods manufacture).
- It helps students to identify problems that need to be investigated in their designated fields or options of study.
- It helps students to masterfully apply research findings to the practice of the sport business.
- It help students to share more research findings with their colleagues and instructors about the study of sport business management in the sport industry.
- It motivates students to become involved either as principal investigators or as participants in research about the study of sport business management in the sport industry.

Major Purposes of Sport Business Management Research

Why is research important in sport business management? At its most fundamental base, the purpose of sport business management research is to generate information. Let's take this fundamental purpose and look at the detailed reasons for research in different settings.

The primary rationale for research is to explicate a scientific body of knowledge that will eventually define and describe sport business management as a unique field of study. The problem is that sport business management exists as a field of study and as a field of practice without a scientific heritage. Therefore, a scientific body of knowledge is needed to give autonomy and provide direction for the study and practice of sport business management in such areas as intercollegiate athletics administration, the sporting goods industry, and sport business marketing.

There is need to expand the knowledge base through scientific studies about management processes, management skills, management roles, organizational structures, and organizational behavior that will help those who are now studying in undergraduate and graduate programs, as well as those who now practice leadership and management in the various segments of sport management.

A second reason for research is to substantiate a body of knowledge that will lend support for the development of ideas, concepts, and theories unique to the field of sport management. Currently, much of sport management education functions from an untested theoretical (atheoretical) base. At the time of this writing, there are no known theories unique to the field of sport management. Most of the theories are borrowed from other fields and disciplines. The field is in need of original conceptual and theoretical frameworks unique to sport business management.

A third reason for research is to substantiate a body of knowledge that will lend support for teaching students in undergraduate and graduate degree programs. Too much of the literature utilized in classroom teaching is speculative and generally unsubstantiated by empirical evidence.

A fourth reason is to advance the emerging field of sport management as a credible practicing profession. As sport management strives to establish itself as a profession, it must develop, through empirical research, a unique body of knowledge—one of the primary criteria of a profession.

A fifth and final rationale for research is to ensure survival of sport business management as a part of the higher education enterprise. Currently, there are several hundred sport management programs at the undergraduate level; therefore, more faculty find themselves challenged and encouraged to conduct research not only because of the demands of the evolving field of sport management, but also because of the promotion/tenure requirements of colleges and universities. Research is one way to ensure survival of the sport management programs of undergraduate and graduate programs in higher education.

Philosophical Approaches to Sport Business Management Research

The philosophical nature of research is captured under what is known as a paradigm. According to Patton (1990), a paradigm is a world view, a general perspective, a way of breaking down the complexity of the real world. Guba (1990) also told us that a paradigm is an interpretative framework, which is guided by "a set of beliefs and feelings about the world and how it should be understood and studied" (p. 17). This perspective is used as a way to guide research and practice in academics for the study of sport business management.

There are two major research paradigms that guide the approach to inquiry in research: the positivist paradigm and the naturalistic paradigm (Creswell, 1994; 2005). It is important for sport management students to understand the assumptions of both paradigms. It is these two paradigms that provide the philosophical foundation and assumptions for the research process in the field of sport management. The two paradigms underlie psychological and sociological research and have different assumptions about the nature of reality, as well as different research objectives. The logic of these paradigms provides direction for sport management research. Exhibit 1.2 and Exhibit 1.3 provide detailed explanations on the differences of these two philosophical approaches to sport management research.

Positivist Paradigm

The *positivist paradigm* refers to a philosophical doctrine that asserts that scientific knowledge is logically embedded in the traditional scientific method, and because of this it is the only kind of factual knowledge. The positivist paradigm is rooted in deductive logic. It is the process of

Exhibit 1.2. Differences of the Positivist and Naturalistic Paradigms of Research

Positivist Paradigm	Naturalistic Paradigm
• the philosophical approach in assumption of quantitative research	• the philosophical approach in assumption of qualitative research
• reality is objective and apart from the research	• reality is subjective, as constructed by the study participants
• researcher is independent from what is being researched and does not influence the findings	• researcher interacts with what is being researched; the interaction process creates the finding
• researcher writes in a formal style, using quantitative terminology	• researcher write in a literary, informal style, using qualitative terminology
• definitions are set at the beginning of the study.	• definitions evolve during course of study
• deductive reasoning process	• inductive reasoning process
• focus on specific, discrete concepts isolated before the inquiry	• focus on the entire phenomena as they emerge during the inquiry
• emphasis is on fixed designs	• emphasis is on emerging designs; categories are identified during the inquiry process
• elicits primarily quantitative (numeric) information	• elicits primarily qualitative (narrative) information
• numeric data are analyzed by statistical procedures and techniques	• narrative data are organized into categories/ themes and analyzed for meaning
• most often associated with quantitative data	• most often associated with qualitative research

Sources. Creswell, J. (1998). Qualitative inquiry and research design. Thousand Oaks, CA: Sage Publications; Lincoln, Y. S., & Guba, E. G. (1995). Naturalistic inquiry. Thousand Oaks, CA: Sage Publications; and Morse, J. M. (Ed). (1994). Critical issues in qualitative research methods. Thousand Oaks, CA: Sage Publications.

using rigorous empirical tools and techniques to discover generalizable explanations. In other words, it takes the view that if something can't be proven in the laboratory or "field setting," it is not useful because it was not derived from scientific methods. Exhibit 1.4 shows deductive logic or reasoning.

The research that embraces this paradigm is often considered "hard science." The logic of the positivist paradigm underscores the premise that the quantity (numerical) of a measurable phenomenon gives rise to *quantitative research.* The positivist approach relies heavily on the quantitative research approach. The researcher is independent of what is being researched and does not influence the findings. Definitions are set at the beginning of the study, and deductive reasoning focuses on specific discrete concepts isolated before the inquiry.

Most research in sport business management has embraced the research paradigm of positivists. The paradigm provides objective reality that allows for researchers in sport management to claim and ascertain truth.

Naturalistic Paradigm

The *naturalistic paradigm* refers to the subjective nature of research. The naturalists believe that there are multiple interpretations of reality, and the goal of researchers working within this perspective

Exhibit 1.3. Positivist and Naturalistic Research Paradigms

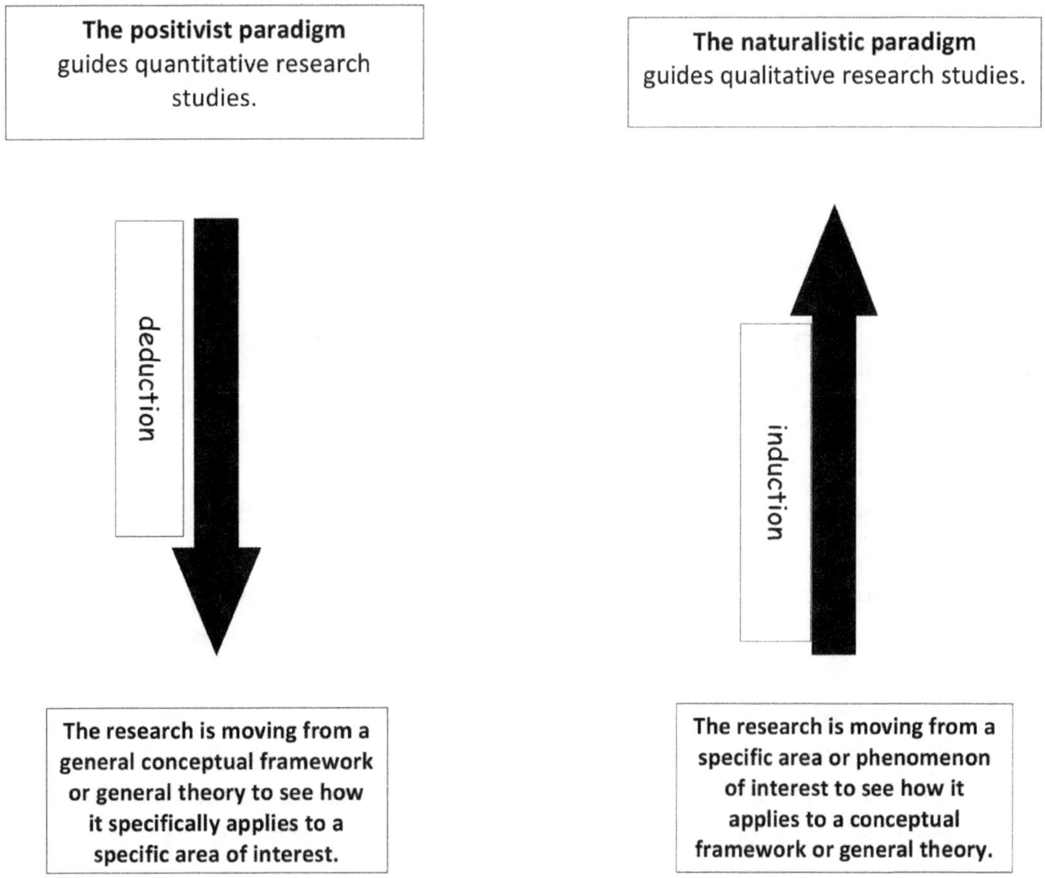

Exhibit 1.4. Deductive Logic or Reasoning

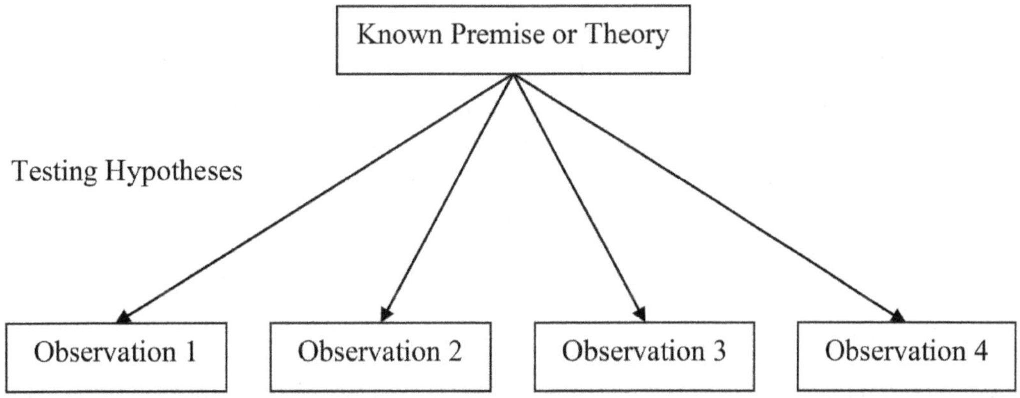

Exhibit 1.5. Inductive Logic or Reasoning

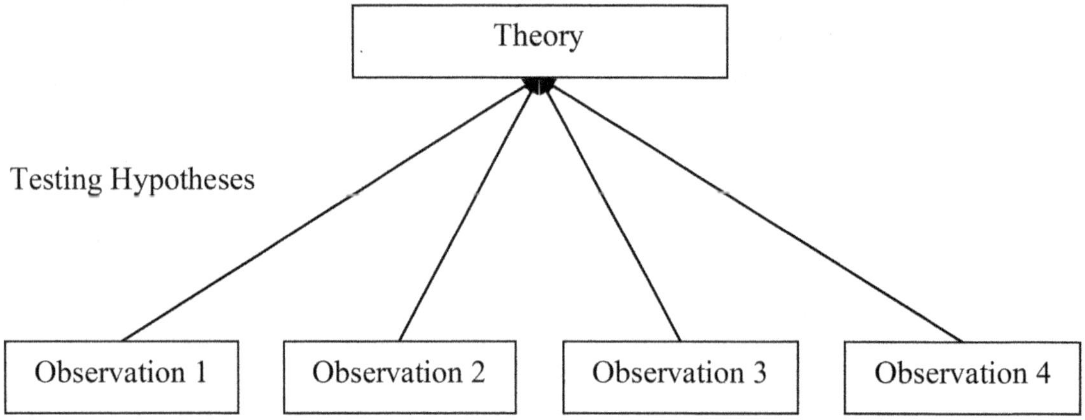

is to understand how individuals construct their own realities within their social contexts. It is logically embedded in a nontraditional approach to scientific inquiry and is primarily subjective in nature. The naturalist paradigm is primarily rooted in inductive logic. It is a complementary approach to positivism. It is often viewed as "soft science." Exhibit 1.5 demonstrates inductive logic or reasoning.

The naturalist paradigm underscores quality (nonnumerical) of phenomenon. As such, it relies heavily on the *qualitative research* approach. The researcher interacts with what is being researched; the interaction process creates the findings. Definitions evolve during the course of the study, and inductive reasoning focuses on the entire phenomena as they emerge during the inquiry.

Major Approaches to Sport Business Management Research

Approaches to and methodologies of research consist of several methods. Not all research is conducted in a laboratory with chemicals. Much research is conducted in numerous settings and involves a variety of methods. This section provides an overview of research methodologies.

Basic and Applied Research

Basic research (also known as pure research) is primarily theoretical in nature. Its primary purpose is the development of new knowledge or knowledge to validate existing knowledge for theory development. In order words, it aims at the "knowledge for the sake of knowledge." Basic research is the main vehicle to substantiate further research in sport management. The results of this type of research have no immediate practical applications, and in fact it may be years before the results of basic research are used in practical applications (Baumgartner, Strong, & Hensley, 2002).

Applied research is also known as practical research. It is primarily concerned with the development of new knowledge or with the validation of existing knowledge that can be applied immediately in practical settings in sport management. Unlike basic research, the primary purpose of applied research is the solution of an immediate practical problem. The results of applied research are less likely to be published in peer review or scholarly journals. Often the results are not published and a report is given to a committee or a board of directors. Unlike basic research, applied research is not conducted within an academic department of a college or university. It is usually conducted in the workplace, where immediate decisions can be made in terms of effective and efficient job satisfaction and work performance and production of goods and services. It is "knowledge for the sake of practical application." For example, Zhang, Pease, Hui, and Michaud (1995) conducted applied research to determine the variables affecting the spectators' decisions to attend NBA games. The major finding revealed that NBA teams should consider fans' socio-demographic backgrounds (age, economic, status, ethnicity, education, and occupation) when developing and implementing marketing strategies to lure spectators to the games.

Quantitative and Qualitative Research

Qualitative and quantitative are two distinctive and complementary approaches to research. Each has its own functions and characteristic activities. These approaches originate from different philosophical perspectives and use different methods for collection and analysis of data. Qualitative research is rooted in the fields of anthropology, sociology, and philosophy. Quantitative research is rooted in the fields of psychology, economics, and management. The differences between the two approaches are based on the extent to which the analysis can be done by converting observations into numbers or using narrative text to describe human experiences. It is best to think of the qualitative-quantitative distinction as dualistic in nature. Qualitative research is a method of research designed for discovery rather than verification; this type of research is used to explore little-known or ambiguous phenomena. Quantitative research uses numbers to measure variables and is a method to verify if there are differences or relationships about phenomena.

Sport business management research also can be classified as either qualitative research or quantitative research. Qualitative research is referred to as the type of research methodology that attempts to explain the phenomenon to be investigated without the use of quantitative measurements to describe the phenomenon (e.g., its characteristics). Quantitative research, on the other hand, is a type of research practice that employs measurements and statistical analyses to explain the phenomenon that is under investigation. The contents of this book are arranged along this classification of research methods in sport management.

Since the 1980s, quantitative research has been the predominant research methodology used in sport management research studies. Based on the positivist paradigm, this methodology emphasizes the search for the facts and causes of human behavior through objective, observable, and quantifiable data. The researcher is viewed as an objective scientist whose main tasks are to manipulate the external environment and observe the effects on the subjects (Stainback & Stainback, 1984).

Sport management researchers adopted the prevailing quantitative-positivist model as the primary base for research activities because they wanted to be accepted and respected by other academic fields in the higher education enterprise. Also, because almost all of such researchers were responsible for establishing the first doctoral programs in sport management, they instituted the quantitative-positivist model as the primary research model within these programs. In addition, it may have been that the primary sport management research publication outlets would accept only quantitative research reports for publication. Federal funding for research projects was awarded only to quantitative-research projects. Thus the quantitative approach is the most prominent research model within the field of sport management.

There is no debate of the fact that sport management researchers using quantitative research methodology have contributed greatly to the advancement of the field of sport management. However, in the past few years there has been increasing recognition in nursing and in other disciplines that relying solely on the quantitative approach to answer research questions has serious limitations. Sport business management researchers have been frustrated and disenchanted with trying to use only quantitative methods to gain a better understanding of the sport business management discipline and the sport industry. Leininger (1994) noted that individuals are not reducible and measurable objects that exist independently of their historical, cultural, and social contexts. To treat them as such, as quantitative research does, reduces them to machine-like figures that are only a sum of their parts. Sport management researchers are urged to think anew and to keep an open mind for new ways to collect, analyze, and interpret data for the betterment of the field of sport management.

Qualitative research is another major research method for investigating diverse phenomena. Chapter 3 of this book provides a detailed description of this type of research method. Qualitative research is a popular methodology and is being used with increasing frequency in sport management. The qualitative method, which has enjoyed a long and rich tradition in sociology and anthropology, is anchored theoretically in the naturalist paradigm (Creswell, 2005). Within the past two decades, researchers have been challenged to guide research and sport management and have supported the use of qualitative methodologies to take research beyond the traditional quantitative designs (Inglis, 1992; Pitts, 2016; Zhang, 2015). Qualitative studies are now appearing with increasing frequency in sport management research.

Exploratory, Descriptive, and Causal Research

Exploratory research involves preliminary activities to find answers or to refine the problem or question into a researchable form. The purpose of this type of research is to progressively shape the scope of the research into a well-defined project with specific, measurable, and appropriate objectives.

Descriptive research describes characteristics of a population or a phenomenon. A study conducted to acquire information about the season ticket holders in terms of their household income, education level, size of family, age, occupation, and so on is a typical example of descriptive research. Data collection usually includes the use of surveys, interviews, or observations (Gay, 1992; McMillian & Shumacher, 2004). Descriptive research is commonly used to collect detailed descriptions about existing demographic variables, such as age, sex, marital status, ethnicity, income, occupation, or religious status.

Descriptive research has a number of advantages. The major advantage is that it is used to describe the state of something. Often, descriptive research is needed as a base on which other research can be conducted. On the other hand, there are several noticeable disadvantages:

- independent variables cannot be manipulated the way they are in laboratory experiments;
- inappropriate wording or placement of questions within a questionnaire can bias results; and
- the potential problem of talking to the wrong people exists (e.g., telephone studies).

Causal research identifies cause-and-effect relationships among variables where the research problem has already been narrowly defined. A study conducted to examine ticket pricing and attendance or a study conducted to examine the effect of various marketing and promotion strategies on consumers' perceptions of an athletic product are good illustrations of causal research in sport business management. If organizational commitment scores for African-American female office workers of the Nike Corporation are significantly different than those of white female office workers, a relationship exists between the two groups of workers and their commitment to the Nike Corporation.

The Evolution and Future of Sport Business Management Research

Over the last twenty years, there has been tremendous growth of published research in the field of sport business management all around the world. The quality of published research has also increased with journals being able to reject 65% of submitted articles. In other words, researchers are contributing many more papers than can be accepted for publication in journals; editors and their respective editorial boards can reject two thirds or more of the manuscripts submitted.

To truly appreciate the growth of published research in the field of sport management, it is important to understand how fast it has evolved and grown over the past two decades. Many of the current academic conferences and publications for sharing sport management research are relatively new. For instance, although the first directly-associated sport management journal was started in 1987, most of the journals of today were started after 2000. Sport management research in academia began with graduate theses and dissertations in the 1960s (Pitts, 2016). Also, sport management–related manuscripts were initially published in academic journals primarily in the physical education field. On the other hand, trade conferences and publications for sport management have been around for much longer. For example, the NCAA began in 1910 and now hosts an annual convention for athletic directors; the *Sporting Goods Dealer* publication was first published in 1899.

The first academic organization for sport management started with the inception of the North American Society for Sport Management (NASSM) in 1985. During the 1985–1986 academic year, NASSM organized its first sport management conference, which was attended mainly by university and college faculty and students from the USA and Canada (Zeigler, 1987). In 1987, this organization launched its first academic publication, *Journal of Sport Management.* Since then, several journals published by other academic organizations have been established, including *Sport Management Review* (1990), *European Journal of Sport Management* (1992), and *Sport Marketing Quarterly* (1992). Today, as you can see in Exhibit 1.6, there are over one hundred academic journals directly or indirectly related to the topic of sport management—that is, some were developed primarily for other topics or fields but their content is related to sport management; and others were developed specifically by and for people in the new and developing field of sport management.

The academic discipline has grown around the globe. There is now a scholarly association for the six continents and two global associations:

- North American Society for Sport Management (NASSM)—started 1985
- European Association for Sport Management (EASM)—started 1993
- Sport Management Association of Australia and New Zealand (SMAANZ)—started 1995
- Asian Association for Sport Management (AASM)—started 1999
- Latin Association for Sport Management (ALGEDE)—started 2005
- African Sport Management Association (ASMA)—started 2010
- Global Sport Business Association (GSBA)—started 2012
- World Association for Sport Management (WASM)—started 2012

Additionally, there are now organizations in many individual countries, such as China, France, Japan, Korea, Italy, Germany, and the United Kingdom. A new movement in academic associations is the organization of topical associations. There are organizations for the study of sport law (Sports Lawyers Association) and for a focus on marketing (Sport Marketing Association.)

In the sport industry, there are numerous outlets for research, primarily in trade publications and conventions. Examples of sport business industry trade publications include *Bicycle Retailer & Industry, Golf Marketing, Fitness Product News, Golf Retailer, Sporting Goods Business,* and *Sports Executive Weekly.* Sport business associations:

Adventure Cycling Association
American Sportfishing Association
American Sportscasters Association
Golf Course Superintendents Association of America
Ice Skating Institute
National Association of Charterboat Operators
National Bicycle Dealers Association
National Collegiate Athletic Association
National Scholastic Surfing Association
Sporting Goods Manufacturers Association
Women's Basketball Coaches Association (WBCA)
Women's Sports Foundation

There are some research companies that specialize in conducting research for companies, such as Joyce Julius & Associates, Boulder Sports Research, and Goliath's Sports Research Corporation. These companies and more will continue to provide valuable research for companies that want to outsource the research activity.

The sport business industry is reported to be one of the largest and most varied industries in the world (Pitts & Stotlar, 2013). With industry segments such as participant sports (the largest segment of the industry), sporting goods (equipment, apparel, and shoes), sports television, sport sponsorship, sports facility design, construction and management, high school sports, college sports, professional sports, winter sports, water sports, sport licensing and merchandising, auto racing, extreme sports, Olympic and Paralympic sports, masters and senior sports, youth sports, and sports tourism, there will be no shortage of jobs and careers for students in sport business management for a long time.

Research is the key to success for a sport business. It is what discovers information, and information is critical for the development of a sport business and for the development and growth of the sport business management academic discipline.

Summary

The purpose of this chapter is to give students an introduction to sport management research. Sport business management research is defined as an orderly investigative process that involves purposeful and systematic collection, analysis, and interpretation of data to gain new knowledge or to verify existing knowledge primarily within the domain of managing human and material resources in entities of the sport business industry.

Characteristics of sport management research include it is interdisciplinary in nature; a wide variety of topical specialty areas are included in published reports; and most of the studies conducted by sport management researchers are quantitative in nature.

There is a need for research in sport management: (1) to explicate a scientific body of knowledge that defines and describes sport management as a unique field of study; (2) to substantiate a body of knowledge that will lend support for the development of ideas, concepts, and theories unique to the field of sport management; (3) to substantiate a body of knowledge that will lend support for teaching students in undergraduate and graduate degree programs; (4) to advance the emerging field of sport management as a credible practicing profession; and (5) to ensure survival of sport management as a part of the higher education enterprise.

The two major research paradigms that guide research inquiry in sport management are the positivist paradigm and the naturalistic paradigm. In general, sport management research can be classified in three different ways. First, it is either basic or applied research. Second, it is either quantitative or qualitative research. Third, it is exploratory, descriptive, or causal research.

Study Questions

1. What is the definition for each of the following two terms?
   a. Research
   b. Sport business management research
2. What is the role of research for sport business?
3. What is the role of research for the academic field of sport management?
4. What is the value of research for students in sport business management?
5. What are the three characteristics of sport business management research?
6. What are the differences of the two philosophical approaches to sport management research?
7. What are the differences between applied sport business management research and basic sport business management research?
8. What are the differences among the three types of sport business management research: exploratory research, descriptive research, and causal research?

Learning Activities

1. Conduct an online search for a data based research article that was published within the last year by one of the journals created specifically for sport business management or sport marketing and discuss, analyze, and critique how it was classified by type.
2. Conduct an online search for a qualitative research article that was published this year by one of the journals created specifically for sport business management or sport marketing and discuss, analyze, and critique how it was classified by use.
3. Conduct an online search for a qualitative research article that was published this year by one of the journals created specifically for sport business management or sport marketing and discuss, analyze, and critique how it was classified by design.

4. Conduct an online search for a qualitative research article that was published this year by one of the journals created specifically for sport business management or sport marketing and give the rationale as to why a qualitative study was conducted instead of a quantitative study.

References

Baumgartner, T. A., Strong, C. H., & Hensley, L. D. (2002). *Conducting and reading research in health and human performance* (3rd ed.). New York: McGraw-Hill.

Creswell, J. W. (1994). *Research design, qualitative and quantitative approaches*, London: Sage Publications.

Creswell, J. W. (2005). *Educational research, planning, conducting, evaluating, quantitative and qualitative research (*2nd ed.). Upper Saddle River, NJ: Pearson Merrill/Prentice Hall.

Fielding, L. W., Pitts, B. G., & Miller, L. K. (1991). Defining quality: Why should educators in sport management programs be concerned about accreditation? *Journal of Sport Management, 5,* 1–17.

Gay, L. (1992). *Educational research (4th ed.*). New York: Macmillan.

Guba, E. (1990). *The paradigm dialogue*. Newbury Park, CA: Sage Publications.

Guba, E. G., & Lincoln, Y. S. (1994). Competing paradigms in qualitative research. In N. K. Denzin & Y. S. Lincoln (Eds.), *Handbook of Qualitative Research* (pp. 105–117). Thousand Oaks, CA: Sage Publications.

Inglis, S. (1992). Focus groups as a useful qualitative methodology in sport management. *Journal of Sport Management, 6,* 173–178.

Isaac, S. (1964). *Careers and opportunities in sports*. New York: E. P. Dutton & Co.

Iso-Ahola, S. E. (1986). A theory of substitutability of leisure behavior. *Leisure Sciences, 8*(4), 367–389.

Kerlinger, F. N., & Lee, H. B. (2000). *Foundations of behavioral research* (4th ed.). Fort Worth, TX: Harcourt College Publishers.

Kim, A. C., Chelladurai, P., & Kim, Y. K. (2015). Scholarly thrusts in the *Journal of Sport Management*: Citation analysis. *Global Sport Business Journal, 3*(1), 1–19.

Leininger, M. (1994). Evaluation criteria and critique of qualitative research studies. In J. M. Morse (Eds.), *Critical issues in qualitative research methods* (pp. 95–115). Thousand Oaks, CA: Sage Publications.

Lincoln, Y. S., & Guba, E. G. (2000). Paradigmatic controversies, contradictions and emerging confluences. In N. K. Denzin & Y. S. Lincoln (Eds.), *Handbook of qualitative research* (2nd ed., pp. 163–188). Thousand Oaks, CA: Sage Publications.

McMillan, J. H., & Schumacher, S. (2006). *Research in education: Evidence-based inquiry* (6th ed.). Boston, MA: Pearson.

Morse, J. M. (1994). *Critical issues in qualitative research methods (Ed.).* Thousand Oaks, CA: Sage Publications.

Patton, M. Q. (1990). *Qualitative evaluation and research methods* (2nd ed.). Newbury Park, CA: Sage Publications.

Pitts, B. G. (2008). "Sport Business Management: Oldest Profession, Newest Education: Understanding the Critical Connection between Industry and Academia for Successful Sport Business Management." Keynote Address delivered at the International Conference on Sport Industry Management; National Taiwan Sport University, Taichung, Taiwan, June, 2008.

Pitts. B. G. (2010, December 11–12). *Sport Business Management – The Employment and Academic Preparation for the Sport Industry in the USA: Development and Trends*. Invited Keynote Speaker. ASIA Taipei International Sport Industry Forum 2010. December 11–12, 2010. Taipei, Taiwan.

Pitts, B. G. (2015). Thriving in a Globalized Sport Business Marketplace: Challenges, Trends, and Opportunities. Keynote Address delivered at the Latin American Association for Sport Management (ALGEDE: Asociacion Latinoamericana de Gerencia Deportiva) IV Conference. Sao Paulo, Brazil; March 18–22, 2015.

Pitts, B. G. (2016). Examining the sport management literature: Content analysis of the *International Journal of Sport Management. International Journal of Sport Management,* 17, 1–21.

Pitts, B. G. (2017). Thriving in a Globalized Sport Business Marketplace: Challenges, Trends, and Opportunities. Keynote address delivered for the V Congresso Nacional Gerencia y Derecho Deportivo Colombiano, Universidad Tecnologica de Pereira, Pereira, Colombia. May 25–27, 2017.

Pitts, B. G., Danylchuk, K., & Quarterman, J. (2014). Analysis of sport management literature: *European Sport Management Quarterly. Sport Management International Journal, 10*(2), 45–72.

Pitts, B. G., & Schwarz, E. (2013, June). A history of the development of sport management in the USA as a field of study: New findings for the classroom. Presentation delivered at the annual conference of the Sport Marketing Association Pedagogy Symposium, October 23–27, 2013, Albuquerque, New Mexico.

Pitts, B., & Stotlar, D. (2013). *Fundamentals of sport marketing* (4th ed.). Morgantown, WV: Fitness Information Technology.

Research (2017). In *Merriam-Webster Dictionary.* Retrieved from https://www.merriam-webster.com/dictionary/research

Shapiro, D. R., & Pitts, B. G. (2014). What little do we know: Content analysis of disability sport in sport management literature. *Journal of Sport Management, 28,* 657–671.

Stainback, W., & Stainback, S. (1984). A rationale for the merger of special and regular education. *Exceptional Children, 51,* 102–111.

Zhang, J. J. (2015). What to study? That is a question: A conscious thought analysis. *Journal of Sport Management, 29,* 1–10.

Zhang, J. J., Pease, D. G., Hui, S. C., & Michaud, T. J. (1995). Variables affecting the spectator decision to attend NBA games. *Sport Marketing Quarterly, 4*(4), 29–39.

Appendix 1-A. Sport Management Journals Identified by Subject Matter

Sport Management Principles

Academic Athletic Journal for N4A
Annals of Leisure Research
European Sport Management Quarterly
Global Sport Business Journal
Global Sport Management
ICHPERSD Journal of Research (Int. Council for HPERSD)
International Gambling Studies
International Journal of Applied Sports Sciences
International Journal of Asian Society for Physical Education, Sport and Dance
International Journal of Sport Management
International Journal of Sport Management and Marketing
International Journal of Sport Management, Recreation & Tourism
International Sports Studies
Japanese Journal of Sport Management
Journal of Applied Sport Management (Prev: Journal of Sport Administration & Supervision
Journal of Contemporary Athletics
Journal of Gambling Studies
Journal of Global Sport (SportsWorld)
Journal of Global Sport Management
Journal of ICHPER-SD
Journal of Intercollegiate Sport
Journal of Japan Society of Sports Industry
Journal of Leisure Research
Journal of Park & Recreation Administration
Journal of Physical Education and Sport Management
Journal of Physical Education and Sports Management
Journal of Physical Education, Recreation, & Dance
Journal of Quantitative Analysis in Sports
Journal of Shanghai University of Sport
Journal of Sport & Tourism
Journal of Sport Management
Journal of Youth Sports
Korean Journal of Physical Education
Korean Journal of Sport Management
Korean Journal of Sport Science
Leisure Information Quarterly
Leisure Sciences
Leisure Studies
Managing Leisure
National Intramural-Recreational Sports Association Journal
Olympika: The International Journal for Olympic Studies

Palaestra
Quest
Recreational Sports Journal
Research Quarterly for Exercise and Sport
Scandinavian Sports Studies Forum
Serbian Journal of Sports Sciences
SMART Online Journal (ceased publication)
Sport & Entertainment Review (Prev: Journal of Venue & Event Management)
Sport and EU Review
Sport, Business and Management: An International Journal
Sport Business International
Sport Journal
Sport Management International Journal - Choregia
Sport Management Review
Sports: An Online Journal
Sports Business Journal
Women in Sport and Physical Activity Journal
World Journal of Sport Sciences
World Leisure Journal

Sport Leadership

Applied Research in Coaching and Athletics Annual
Coach and Athletic Director
International Journal of Sports Science & Coaching
Moving Mountains: Journal of Sport for Development and Peace

Sport Operations – Facility Management

Journal of Venue & Event Management

Sport Governance

International Journal of Sport Policy and Politics
Journal of Issues in Intercollegiate Athletics
Journal of NCAA Compliance

Sociocultural Aspects of Sport and Sport Management

European Journal for Sport and Society
International Journal of Religion and Sport
International Journal of Sport and Society
International Journal of the History of Sport
International Journal of Religion and Sport
International Review for the Sociology of Sport
Journal of Issues in Intercollegiate Athletics
Journal of Olympic History
Journal of Sport and Social Issues
Journal of Sport Behavior
Journal of Sport for Development
Journal of Sport History
Journal of the Philosophy of Sport
NISR Journal of Sport Reform
Quest
Sociology of Sport Journal
Sport, Education, & Society
Sport History Review
Sport in History
Sport in Society (formerly Culture, Sport, Society)
Sporting Traditions (journal of the Australian Society for Sports history)

Psychological Topics in Sport

International Journal of Sport Psychology
International Review of Sport and Exercise Psychology
Journal of Applied Sport Psychology
Journal of Sport and Exercise Psychology
Psychology of Sport and Exercise
Sport Psychologist
The Sport Psychologist

Sport Marketing

International Journal of Leisure and Tourism Marketing
International Journal of Sports Marketing & Sponsorship
Journal of Prediction Markets
Journal of Sponsorship
Journal of Vacation Marketing
Sport Marketing Quarterly
Sports Marketing
The Cyber Journal of Sport Marketing (ceased publication)

Sport Communication

Communication and Sport
International Journal of Sport Communication
Journal of Sports Media
Modern Sport Communication

Sport Finance & Economics

International Journal of Sports Finance
Journal of Gambling & Business Economics
Journal of Sports Economics

Sport Law

Arizona State Sports and Entertainment Law Journal

Australian and New Zealand Sports Law Journal
Denver Sports & Entertainment Law Journal
DePaul Journal of Sports Law & Contemporary Problems
Entertainment and Sports Law Journal
Entertainment and Sports Lawyer
Entertainment, Arts and Sport Law Journal
Harvard Journal of Sports and Entertainment Law
International Sports Law Journal
International Sports Law Review
Journal of Legal Aspects of Sport
Legal Issues in College Athletics
Legal Issues in High School Athletics
Marquette Sports Law Review
Seton Hall Journal of Sport Law
Sport and the Law Journal (UK)
Sports and Entertainment Litigation Reporter
Sports Law eJournal (AUS)
Sports Lawyers Journal
Sports Litigation Alert
Sports, Park and Recreation Law Reporter
Sweet & Maxwell International Sports Law Review
Texas Review of Entertainment & Sports Law
The Sports Law Forum at Fordham University School of Law
University of Miami Entertainment and Sports Law Review
Villanova Sports & Entertainment Law Journal
Virginia Sports and Entertainment Law Journal
Willamette Sports Law Journal

Sport Management Education

Case Studies in Sport Management
Chronicle of Kinesiology and Physical Education in Higher Education
Journal for the Study of Sports and Athletes in Education
Journal of Hospitality, Leisure, Sport and Tourism Education
Journal of Physical Education, Recreation, & Dance
Physical Education & Sport Pedagogy
Physical Educator
Sport, Education, & Society
Sport Management Education Journal

Technology and Sport

International Journal of Computer Science in Sport
International Journal of Performance Analysis in Sport
International Journal of Sports Science and Engineering
Measurement in Physical Education & Exercise Science
Sports Engineering

2

Ethics in Sport Management Research and Publications

After completing the study of this chapter, you should be able to:

- Explain the three different approaches (i.e., the deontological approach, ethical skepticism, and utilitarianism) to research ethics.
- Trace the major historical development of ethical codes and guidelines for conducting research.
- List and explain three general ethical principles for conducting research on human subjects.
- Explain the major roles of the American Psychological Association (APA) in guiding ethics for research in sport management studies.
- Explain the major roles of institutional review boards in guiding ethics for research in sport management studies.
- Explain the major roles of the North American Society for Sport Management and the Sport Marketing Association in guiding ethics for research in sport management studies.
- Explain what is meant by informed consent.
- Identify the essential elements of an informed consent form.

The purpose of this chapter is to provide students with an understanding of ethical considerations for the study of humans in sport management research. The philosophical paradigms and historical events that have influenced the development of ethical research practices of human subjects will be discussed first, followed by the ethical principles underlying the basic rights involved when dealing with the protection of human subjects, historical events influencing the development of research practices of human subjects, and the development of ethical codes of conduct for conducting research on human subjects.

In the latter part of the chapter, the guidelines and procedures for protecting human rights when conducting research are presented. The chapter concludes with a discussion of the nature and scope of the processes of *informed consent* and the *institutional review boards* (IRBs). Examples of the aforementioned principles, guidelines, and procedures are then summarized from selected articles in the sport management literature.

The Nature and Scope of Research Ethics

What are *ethics*? Describing the nature of ethics in sport management is a difficult task given that many different perspectives on the field exist. Ethics are typically associated with morality, and both ethics and morality deal with matters of right and wrong. The term ethics has also been defined by a number of different scholars, including as "the principles and guidelines that help us uphold the things we value" (Johnson & Christensen, 2014, p. 126). These definitions either are too general or do not directly deal with the phenomenon of sport management. For our purposes, sport management research ethics refers to a set of consensually accepted principles and standards of behavior for sport management researchers to follow when conducting research with human subjects. These principles and guidelines are grounded in philosophical principles and theories.

Approaches to Ethical Issues in Research

Given the aforementioned operational definition, the question is how do we arrive at a comprehensive approach to making decisions to protect human rights in sport management research. This question is debated because different approaches can be adopted to make decisions about what is right and wrong (Schlenker & Forsyth, 1977). A starting point to consider for protecting human rights in sport management research is philosophical paradigms for interpreting ethical behaviors. This chapter will cover three predominant philosophies governing ethics: (1) the *deontological approach* of Immanuel Kant, (2) the *utilitarianism approach* of John Locke and John Stuart Mill, and (3) *ethical skepticism*.

The deontological approach refers to an approach that ethical issues must be assessed based on some universal code. Deontologic ethics are rule-based and give much weight to obeying principles and guidelines. When adhering to this mode of inquiry, sport management researchers must conduct their research on the basis of universal codes such as the Ethical Principles of Psychologists and Code of Conduct from the American Psychological Association (2003), which are commonly known as the APA Ethical Codes.

Utilitarianism is an approach in which judgments of the ethics of research rely on the consequences studies have for the research participants and the benefits that might result from the study (Johnson & Christensen, 2014). Hence, from this approach, we decide what is wrong based on balancing the action's consequences. In other words, ethical decisions are based on measuring the potential benefits against the potential costs. If the benefits are large enough relative to the costs, then the study is considered to be ethically acceptable. This approach has been the primary approach employed by the federal government, most professional organizations, and Institutional Review Boards, especially when research participants are at risk but the study also has the potential to contribute significant knowledge and benefit to humans.

Ethical skepticism is a radical ethical approach that posts that concrete and inviolate moral codes, such as those used by deontologists, cannot be developed. While this approach would not deny the significance of ethical principles, this approach says that ethical rules are a matter of individual conscience relative to one's culture and time (Johnson & Christensen, 2014). According to this approach, the researcher should do what the researcher thinks is right and avoid doing what the researcher thinks is wrong.

Ethical Concerns

When it comes to research ethics, researchers need to contemplate three critical areas: (1) the relationship between society and science, (2) professional issues, and (3) the treatment of research participants (Johnson & Christensen, 2014).

The ethical issue concerning the relationship between society and science reflects that societal and cultural concerns and values should lead the course of research. For instance, numerous funding agencies, including the federal government, spend millions of dollars annually on significant research that can impact our lives dramatically. Each funding agency has its own priority and criteria to select certain research projects to award research funds. For this, researchers should orient their research proposal toward these priorities and criteria, which indicates these funding agencies can dictate the type of research that is conducted.

Professional issues include the growing problem of research misconduct. Research misconduct is defined as "fabrication, falsification, or plagiarism (FFP) in proposing, performing, or reviewing research, or in reporting research results" (US Office of Science and Technology Policy, 2017). Here, fabrication refers to "making up data or results and recording or reporting them," falsification refers to "manipulating research materials, equipment, or processes, or changing or omitting data or results such that the research is not accurately represented in the research record," and plagiarism is defined as "the appropriation of another person's ideas, processes, results, or words without giving appropriate credit" (US Office of Science and Technology Policy, 2017). Whether the reason for research misconduct is due to personal factors (e.g., personality, value orientation) or nonpersonal factors (e.g., pressure to publish, competition for research funding, inadequate supervision of trainees), there is never any justification for engaging in such unethical behavior and the consequence of research misconduct is tremendous. One of the best ways to prevent professional issues is to create and nurture a proper institutional culture and translate it into actual action (Gunsalus, 1993). Johnson and Cristensen (2014) introduce a list of strategies for preventing research misconduct originally developed by Steinberg (2002). The list includes (1) have the researcher make it clear that the data collected are checked and verified, and then make sure that some of the data are checked, (2) ask some of the research participants who should have been contacted by each data collector if you may recontact them, (3) make sure there are no deviations from the approved study design, (4) watch for data collectors who complete data collection in record time, (5) teach ethical standards of conducting research in classes, including reviews of cases of misconduct, and (6) provide guidelines for handling cases of suspected misconduct (Johnson & Christensen, 2014, p. 129).

Historical Development of Ethical Research

Treatment of research participants is one of the most crucial issues that researchers encounter. Since the 1930s, there has been a history of clearly unethical research practices on human subjects, primarily in the medical field. Two of the most infamous examples were the Nuremberg Medical Studies and the Tuskegee Syphilis Study (Brandt, 1978). Although such practices were in medical research, they have influenced the formulation of guidelines in ethical decision making and ethical codes that give direction to the conduct of research in the psychological, sociological, historical, and political sciences, as well as in sport management.

Nuremberg Medical Studies

During the 1940s, prisoners in German concentration camps were deliberately infected with bacteria by German physicians. There were numerous types of experiments conducted on the prisoners. Three of the experiments: (1) freezing experiments (placing prisoners in ice water to determine how long it would take to lower body temperature to the point of death); (2) malaria experiments (exposing a prisoner to mosquitoes that carried the malaria virus to determine what drugs were most effective for treatment of malaria); and (3) high altitude experiments (exposing prisoners to high altitudes to determine how much pain one could take before dying).

Tuskegee Syphilis Study

Between 1932 and 1970, rural African-American males were used as subjects for a large-scale study of syphilis. The study was initiated in 1932 and conducted under the auspices of the US Public Health Services (Gamble, 1997). The purpose of the study was to test the effectiveness of mercury in treating syphilis. The participants were not made aware of the purpose of the study or the danger to which it exposed them. There was no attempt by the researchers or the government to explain the study to them. Participants were enticed with a variety of inducements, such as physical examinations, free rides to clinics, meals, and a $50 stipend as a burial fee when they died. The participants were also enticed to assure the physicians that they did not receive treatment from other physicians. This study violated nearly all of the guidelines of ethics for research with humans, from needing to inform to protecting participants from physical harm. The study did not become public knowledge until 40 years after it was started. President Clinton made a formal public apology to the eight remaining survivors of this study in the late 1990s (Gamble, 1997).

These types of practices supported the need for guidelines that subjects should be sufficiently informed and be allowed to volunteer for research studies. Such studies give rise to a need for ethical guidelines for the increase of human subjects in research for the current day academic enterprise. Many of the guidelines developed can be used to guide research practices in sport management.

The Development of Ethical Codes of Conduct for Research on Human Subjects

Since the 1940s, various codes for the responsible conduct of human experimentation in medical research have been adopted by different organizations. The best known of these are the Nuremberg Code of 1947, the Helsinki Declaration of 1964 (revised in 1975)—an extension of the Nuremberg Codes that was developed as principles for controlling unethical practices in the medical sciences—and the 1971 Guidelines (codified in Federal Regulations in 1974) issued by the US Department of Health, Education, and Welfare.

During the 1940s, the Nuremberg Codes (see Exhibit 2.1) were initiated because of the infamous Nuremberg medical studies that were conducted during the 1930s. The codes are recognized as the first international effort to establish codes of conduct for protecting human subjects when conducting research. The codes were embedded in both the philosophies of utilitarian and

Exhibit 2.1. The Nuremberg Code

1. The voluntary consent of the human subject is absolutely essential. This means that the person involved should have legal capacity to give consent; should be so situated as to be able to exercise free power of choice, without the intervention of any element of force, fraud, deceit, duress, over-reaching, or other ulterior form of constraint or coercion; and should have sufficient knowledge and comprehension of the elements of the subject matter involved, as to enable him to make an understanding and enlightened decision. This latter element requires that, before the acceptance of an affirmative decision by the experimental subject, there should be made known to him the nature, duration, and purpose of the experiment; the method and means by which it is to be conducted; all inconveniences and hazards reasonably to be expected; and the effects upon his health or person that may possibly come from his participation in the experiment.

   The duty and responsibility for ascertaining the quality of the consent rests upon each individual who initiates, directs, or engages in the experiment. It is a personal duty and responsibility that may not be delegated to another with impunity.

2. The experiment should be such as to yield fruitful results for the good of society, unprocurable by other methods or means of study, and not random and unnecessary in nature.

3. The experiment should be so designed and based on the results of animal experimentation and a knowledge of the natural history of the disease or other problem under study, that the anticipated results will justify the performance of the experiment.

4. The experiment should be so conducted as to avoid all unnecessary physical and mental suffering and injury.

5. No experiment should be conducted where there is an a priori reason to believe that death or disabling injury will occur; except, perhaps, in those experiments where the experimental physicians also serve as subjects.

6. The degree of risk to be taken should never exceed that determined by the humanitarian importance of the problem to be solved by the experiment.

7. Proper preparations should be made and adequate facilities provided to protect the experimental subject against even remote possibilities of injury, disability, or death.

8. The experiment should be conducted only by scientifically qualified persons. The highest degree of skill and care should be required through all stages of the experiment of those who conduct or engage in the experiment.

9. During the course of the experiment, the human subject should be at liberty to bring the experiment to an end, if he has reached the physical or mental state where continuation of the experiment seemed to him to be impossible.

10. During the course of the experiment, the scientist in charge must be prepared to terminate the experiment at any stage, if he has probable cause to believe, in the exercise of the good faith, superior skill, and careful judgement required of him, that a continuation of the experiment is likely to result in injury, disability, or death to the experimental subject.

Source: "Trials of War Criminals before the Nuremberg Military Tribunals under Control Council Law No. 10", Vol. 2, pp. 181–182. Washington, D.C.: U.S. Government Printing Office, 1949.

deontologic ethical theories as the balance of benefits outweigh risk and adequate protection of subjects from risk or harm was conducted (Katz, 1972; Katz, 1996).

The Tuskegee Syphilis Study and other nonconsensual experiments on human subjects led Congress to establish the National Research Act, which established the National Commission for the Protection of Human Subjects of Biomedical and Behavioral Research (1979). Out of this Commission the Belmont Report (National Institutes of Health, 2008) was formulated. The Belmont Report is recognized as the first effort to articulate ethical principles for conducting research in the US.

General ethical principles as opposed to ethical theories and international codes were found to be more inspirational in nature. Their intent is to guide and inspire researchers toward the very highest ethical ideals of the profession.

A Review of Ethical Principles

Several ethical principles grew out of the Belmont Report, as well as from scholarly work in ethics. According to the Belmont Report, there are three general ethical principles underlying the protection of human subjects: *respect for human dignity*, *beneficence*, and *justice*. Beauchamp and Childress (2001) have also identified four major principles: *autonomy*, *beneficence*, *nonmaleficence*, and *justice*. These can be used by sport management researchers in designing ethical guidelines for selecting subjects and acquiring written consents for proposed research investigations.

Autonomy is the principle of respect for people's rights and dignity requires researchers to treat research participants as autonomous individuals and to obtain their informed consent to participate in a research investigation (Hully & Cummings, 1988). This principle is embedded in utilitarian ethics and is articulated in the Belmont Report and the current APA Ethical Codes. It is composed of two ethical convictions: (1) assurance of the right to self-determination, and (2) assurance of the right to full disclosure. A research participant must be free to make independent decisions about participating in an investigation without the coercion of the researcher.

According to the principle of beneficence, the researcher must balance the benefits and risks of the subjects participating in a research investigation. The benefits are to outweigh the related categories of risks, including physical, psychological, social, and economic (Levine, 1988). The principle of generosity to ethical decisions in research is inherent in deontological and utilitarian ethics. It simply means that researchers must be generous to subjects who participate in their research.

According to the principle of nonmaleficence, sport management researchers must ensure that the research participants have the right to freedom from harm and exploitation during the research process. As with beneficence, this principle is inherent in deontological and utilitarian ethics.

The principle of justice is embedded in both deontological and utilitarian ethics. According to this principle, researchers are obligated to provide assurance of the right to fair treatment and assurance of the right to privacy of subjects who participate in research investigations.

Ways to Ensure that Research is Ethical

Following the Guidelines of Professional Associations

In addition to the Belmont Report, professional associations have formulated and articulated ethical principles and guidelines for conducting research activities with human subjects. In the

US, many professional associations in such disciplines as medicine, law, management, marketing, political science, and psychology have established their own codes of conduct in human subject research. The following are several examples of professional associations that exist for articulating standards of conduct in research on human subjects: American Medical Association, American Marketing Association, American Sociological Association, American Society of Criminology, American Political Science Association, American Statistical Association, American Historical Association, and American Psychological Association.

Since 1953, the American Psychological Association (APA) has published principles for research involving human subjects commonly known as the APA Ethical Codes. They are considered public statements of principles used to promote and maintain high standards of conduct guiding and inspiring researchers toward the very highest ethical ideals of a field or discipline. Codes for the conduct of social and behavioral research have adopted the format of the APA. The codes are inspirational in nature and available to guide the professional conduct of sport management researchers when ethical issues surface.

The American Psychological Association (2017) has published a set of ethical principles that govern psychologists' research. These principles are as follows:

- Using recognized standards of competence and ethics, psychologists must plan research so as to minimize the possibility of misleading results. Any ethical problems are resolved before research is started. The welfare and confidentiality of all participants are to be protected.
- Psychologists are responsible for the dignity and welfare of participants. Psychologists are also responsible for all research they perform or that is performed by others under their supervision.
- Psychologists must obey all state and federal laws and regulations, as well as professional standards governing research.
- Except for anonymous surveys, naturalistic observations, and similar research, psychologists must reach an agreement regarding the rights and responsibilities of both participants and researcher(s) before research is started.
- When consent is required, psychologists must obtain a signed, informed consent before starting any research with a participant.
- Deception is used only if no better alternative is available. Under no condition is there deception about (negative) aspects that might influence a participant's willingness to participate.
- Other issues covered include sharing and utilizing data, offering inducements, minimizing evasiveness, and providing participants with information about the study.

These codes are demonstrated in nearly all scholarly journals in the field of sport management. Flagship professional associations in sport management, such as the North American Society for Sport Management and the Sport Marketing Association, closely adhere to the APA Ethical Codes and the Belmont general ethical principles. It is the responsibility of sport management researchers to carefully examine each of their investigations with all the conscience and candor that they can summon to assure that the rights of their subjects are protected. The aforementioned guidelines underscore the ethical principles for conducting research on human subjects in sport management.

Obtaining Institutional Review Board Approval

The Institutional Review Board (IRB), also known as the Human Subjects Committee or Human Subjects Review Board, is considered the hub for upholding research standards and principles established by the federal government and the professional associations. The duty of the IRB is to make certain that all research proposals meet the federal government (National Institutes of Health, 2008) and professional association requirements for ethical research.

An IRB is made up of scholars and researchers across a wide range of disciplines of an institution. At least one member of the IRB must be a member of the community who is not associated with the institution in any way. The IRB system is decentralized. Therefore, each local IBR can interpret and implement both the federal and associations' procedures using its own forms and guidelines. Although the IRBs' criteria and procedures may vary slightly from one institution to another, the requirements of its membership is the same: a minimum of five persons make up the board, the members must come from varied backgrounds, they must be adequately competent to review research, and they must be diverse enough to promote the welfare of human subjects.

Federal regulations require that institutions (including all colleges and universities) must maintain an Institutional Review Board (IRB) to oversee all research involving human subjects. Further, any institution involved in research that receives significant federal funding or conducts drug or medical device research regulated by the Federal Drug Administration (FDA) must establish IRBs. The 2017 Code of Federal Regulations clearly described the activities under the review board's jurisdiction as follows: "any research, development, or related activities which depart from the application of those established and accepted methods necessary to meet the subject's needs or which increase the risk of daily life" (Federal Register, May 30, 1974). The following information should be provided by researchers who seek IRB approval:

- a statement that the study involves research, an explanation of the purposes of the research and the expected duration of the subject's participation, a description of the procedures to be followed, and identification of any procedures that are experimental;
- a description of any reasonably foreseeable risks or discomforts to the subject;
- a description of any benefits to the subject or to others that may reasonably be expected from the research;
- a disclosure of appropriate alternative procedures or courses of treatment, if any, that might be advantageous to the subject;
- a statement describing the extent, if any, to which confidentiality of records identifying the subject will be maintained;
- for research involving more than minimal risk, an explanation as to whether any compensation and an explanation as to whether any medical treatments are available if injury occurs and, if so, what they consist of, or where further information may be obtained;
- an explanation of whom to contact for answers to pertinent questions about the research and research subjects' rights, and whom to contact in the event of a research-related injury to the subject; and
- a statement that participation is voluntary, refusal to participate will involve no penalty or loss of benefits to which the subject is otherwise entitled, and the subject may discontinue participation at any time without penalty or loss of benefits to which the subject is otherwise entitled.

The IRB system places much emphasis on scrutinizing the process of informed consent to assure that the rights and welfare of subjects are protected. This group review also serves an important role in the protection of the rights and welfare of human research subjects. The board is responsible for ensuring that the procedures are not in any way harmful to the research participants and that the participants' privacy and anonymity are assured. An IRB can approve the proposed plan of study, require that it be modified, or disapprove the proposed plan.

The IRB is the primary agency for soliciting informed consent under the APA Ethical Codes. According to the Codes, when obtaining informed consent as required in "standard 3.10, Informed consent," research investigators must inform the participants of the following:

- the purpose of the research, expected duration, and procedures;
- their rights to decline to participate and to withdraw from the research after participation has begun;
- the foreseeable consequences of decline or withdrawing;
- reasonably foreseeable factors that may be expected to influence their willingness to participate, such as potential risks, discomfort, or adverse effects;
- any prospective research benefits;
- limits to confidentiality;
- incentives for participation; and,
- whom to contact for questions about the research and the research participant's rights.

It may not be necessary to cover all the aforementioned standards in order to inform a potential research participant; however, it is important to include what is appropriate for the study being conducted (Fischman, 2000).

Guidelines and Procedures for Protecting Human Rights

The majority of current guidelines for sport management research fall under the purview of the Belmont Report, the APA Ethical Codes, and the Institutional Review Board of the institution for which the research is being conducted. The guidelines are written in a more objective manner than the codes and principles. The procedures for protecting human rights in sport management research usually fall into one or more of the following guidelines through the process of informed consent: assurance of protecting the study participant's rights of confidentiality; clearly debriefing the participants about the purpose of the study; assurance of protecting the study participant's rights of anonymity; assurance of the study participant's rights to volunteer; assurance of protecting the study participants from harm; assurance of protecting the study participants' rights for self-determination; assurance of protecting the study participant's rights of full disclosure; assurance of protecting the study participant's rights to privacy; and assurance that the participants have the freedom to withdraw from a study at any time during the investigation.

Assurance of Protecting the Study Participant's Rights of Confidentiality

Anonymity and confidentiality are often confused. Confidentiality is usually maintained by use of code numbers. Confidentiality refers to the assurance given by the researcher not to reveal the identity of persons who provided research information (Ellis, 1998, p. 278). This guideline implies that sport management researchers must not disclose data or information that should be kept private in order to prevent harm to a research participant. If such data or information must be revealed about a subject, it must not be made public or available to others without the subject's consent.

Assurance of Protecting the Study Participant's Rights of Anonymity

As with full disclosure, anonymity is another way to guarantee privacy. This guideline requires researchers to not link or identify a specific participant to the information reported in a particular research investigation. Records cannot be linked to the participants' names. In this way, the identity of the participant is kept anonymous to everyone, including the researchers themselves in some studies. For instance, a study of Butts (2006) demonstrates how to protect the study participants' rights of anonymity. The study examined the overall and specific sports gambling activity among athletes and nonathletes enrolled in a Southern, regional NCAA II university. For this, both student-athletes and nonathlete students completed the survey instrument in the presence of research assistants. To assure the research participant's rights to anonymity, the author guaranteed complete anonymity by assuring names were not linked with the collected questionnaire and permission to conduct the study was granted by the IRB at the University of West Georgia.

Assurance of Protecting the Study Participant's Rights to Volunteer

A study of Frey, Czeck, Kent, and Johnson (2006) is a good illustration of the use of this principle in protecting the study participant's rights to volunteer. The participants were informed that their involvement was voluntary.

Assurance of Protecting Study Participants from Harm

This guideline implies that sport management researchers should take measures to ensure that research participants are protected from undue physical, emotional, legal, financial, or social harm. The risk of harm refers to how much a research participant is exposed to the possibility of injury when participating in a research investigation. For example, risks of injury in sport management research may involve emotional factors, such as exposure to stress or anxiety, or social factors, such as loss of privacy and confidentiality.

Assurance of Protecting the Study Participant's Rights of Self-Determination

This guideline is captured within the ethical principle of autonomy. It implies that sport management researchers must take measures to ensure that research participants feel free from constraints, coercion, or any undue influence during the research process. This guideline acknowledges that when the research participants, such as children or those in the prison system, lack the capacity for-self-determination, it is the researcher's duty to protect such rights of the participants.

Assurance of Protecting the Study Participant's Rights of Full Disclosure

Full disclosure is considered a basic right of human subjects and is one of the ways to protect the study participant's privacy. It can be done in a pre-study interview with each research participant prior to the investigation. This guideline acknowledges that the researcher must fully describe the nature of a study, ensuring that the research participant is provided with information about what participating in the study will involve, and not with false or misleading information.

Assurance of Protecting the Study Participant's Rights of Privacy

According to this guideline, the researcher must keep the participants' responses or performances strictly confidential. For example, the researcher must give each participant a code number and use it to label documents rather than use the person's name or any affiliation.

Folkman (2000) has identified six threats to privacy that are of concern for researchers in sport management: (1) subpoenas (legal actions), (2) access by third parties to data, (3) use of new technologies, (4) data sharing, (5) mailed questionnaires, and (6) publication or presentation of data (p. 52–56).

Assurance of Protecting the Study Participant's Rights of Withdrawal

Again, the study conducted by Frey et al. (2006) exemplifies how to protect the study participant's rights of withdrawal. All participants should be advised of the ability to terminate participation at any time.

Debriefing the Participants about the Results of a Study

This guideline implies that the research participant has the right to a post-study interview with the researcher. During this session, it is the responsibility of the researcher to protect the study participant's rights to full disclosure.

Obtaining Informed Consent

Informed consent refers to a legal principle that requires researchers to obtain voluntary consent from research participants. By law, the researcher must provide research subjects with all the facts about the study before the subjects participate. The information must be explained to each subject in a way that the subject clearly understands. Informed consent is closely related to voluntary participation. It ensures that potential subjects receive adequate information about the research, are able to comprehend the information, and have the option to choose freely to participate or not participate in the study (Ellis, 1998).

Before subjects participate in a study, they should be presented with consent forms and asked to give their consent to participate. According to Fischman (2000), a consent form is a written document attesting to the fact that participants are informed about the study for which they are volunteering. The consent form includes information about the purpose of the study, a brief identification of the primary researcher, an explanation of the risks or harm of participating in the study, an estimation of the time it will take to complete the survey, confidentiality, and the voluntary nature of participation in the study. It is the responsibility of the researcher to guarantee that consent is obtained before a volunteer participates in a study. Fischman (2000) has suggested that twelve components be included in a consent form. Researchers in sport management must make every effort to develop and implement a form that clearly communicates these components in order to properly inform those participating in a study. The following is the list of twelve components of a consent form introduced by Fischman (2000):

- *An invitation to participate.* An invitation to participate makes it clear that the participant is volunteering for the role of research participant.
- *Purpose of the research.* The purpose includes the overall reason for the research, including research goals at the individual and group level when appropriate.
- *Selection bias.* A clear statement of the reasons why the participant is appropriate for the study has the added advantage of allowing prospective participants to exclude themselves if they do not believe that they meet the criteria for inclusion.
- *Study procedures.* The study procedures should be clearly described to prospective study participants.
- *Descriptions of risks and discomforts.* Volunteers are unable to make informed decisions about whether to participate in a study if they are not adequately informed of possible risks and discomforts.
- *Description of benefits.* In general, benefits can be summarized under the general category of anticipated additions to a systematic body of knowledge.
- *Available alternatives.* The requirement to provide information about available alternatives is primarily for therapeutic studies in which nonvalidated interventions are being studied.
- *Assurance of confidentiality.* It is impossible to guarantee absolute confidentiality; however, the extent and limits to that guarantee should be described as part of the informed consent process.
- *Financial considerations.* Any costs of participating in the research should be clearly described. Economic advantages can include money, merchandise vouchers, food, access to improved facilities, therapy, physical exams, and subsidized transportation.
- *Offer to answer questions.* It is important for participants to know whom to contact for answers to any questions they might have, any problems they encounter as participants, and any inquiry they might incur while participating.
- *Noncoercive disclaimer.* Participation in research should be voluntary, and a decision not to participate or to discontinue should not result in any penalty or loss of benefits to the participants.
- *Incomplete disclosure.* Although the researcher should provide sufficient information for each potential participant to make an informed decision about whether to participate in a specific study, there are times when some information may be withheld. Although generally not the first choice of researchers, it is sometimes necessary to omit information in order to protect the validity of the data collected. Information necessary to make an informed decision should be presented and, if possible, a statement in the consent form that some information is being withheld is a reasonable approach to the issue (Fischman, 2000).

The American Psychological Association (2017) published a set of ethical guidelines that govern psychologists when obtaining informed consent to research. These guidelines note what must be disclosed to a study participant:

- the purpose of the research, expected duration, and procedures;
- the right to decline participation and to withdraw from the research once participation has begun;
- the foreseeable consequences of declining or withdrawing;
- reasonably foreseeable factors that may be expected to influence their willingness to participate, such as potential risks, discomfort, or adverse effects;
- any prospective research benefits;
- limits of confidentiality;
- incentives for participation; and
- whom to contact for questions about the research and research participants' rights.

There are two forms of informed consent: written informed consent and implied consent. When taking the written informed consent approach, the researcher essentially requests all the research participants to give their consent to participate in writing before executing the study. An example of the use of written informed consent in sport management is Mahony and Moorman's (2000) study to examine the "conditions impacting fans' preferences for watching their most disliked team and the favorite team" (p. 43). They used a two-stage model in data collection.

After receiving permission to collect data from the Human Subjects Committee at the University of Louisville, respondents signed an informed consent form before beginning the first stage of data collection. Each respondent was assigned a number that was placed at the top of the surveys for both stages of data collection. In order to assure confidentiality, the list of assigned numbers was kept separate from the surveys and was destroyed after all respondents completed the second stage of data collection" (p. 52).

Researchers rarely use written informed consent when they use self-administered questionnaires. Instead of the traditional written informed consent, they assume what is called implied consent. When using this type of consent, the researcher assumes that when a participant returns the questionnaire in its completed form, this action reflects voluntary response. More specifically, the return of the questionnaire in its completed form implies the consent to participate. Implied consent is not always warranted. For example, when conducting a study with a group of students in a class, the students may feel threatened that their final grade may be affected by not participating in a study.

An Example of IRB Approval in Sport Management Studies

Cianfrone and Zhang (2006) conducted a study to examine "the differential effectiveness of television commercials, athlete endorsements, venue signage, and combined promotions as assessed by Generation Y consumers" (p. 322). The following paragraph describes how the authors of the study obtained the IRB approval:

The institution's review board for the use of human subjects approved the study. The researchers contacted all 35 instructors who taught the general sport and fitness classes that enroll students from various academic departments and majors in the university. With the permission of the instructors, the researchers made an announcement in their classes to recruit participants who were within the upper age bracket of Gen-Y consumers. Those who volunteered for this study were randomly assigned into one of the eight experimental conditions...The study was conducted in a classroom setting. The participants in each group were informed that they were involved in a study that was related to action sports; however, the purpose of the study and the experimental conditions were not explained to them. After filling out an informed consent form, the study proceeded in the following sequence: (1) administration of background information questionnaire, (2) playing the videotape that was appropriate to the particular group, (3) administration of the unaided recall questionnaire, (4) administration of the aided recall questionnaire, and (5) administration of the recognition questionnaire. Each questionnaire was handed out when the previous round was completed. This process prevented subjects from obtaining information across the different sections of the survey. (p. 331)

Summary

The purpose of this chapter was to provide a common understanding of ethics in sport management. Ethics is typically associated with morality, and both ethics and morality deal with matters of right and wrong. For the purpose of sport management research, ethics is referred to as a set of consensually accepted principles and standards of behavior for sport management researchers to follow when conducting research with human subjects.

Utilitarian ethics and deontologic ethics are the two ethical theories that have been used in developing ethical principles and conduct codes in human subject research. The Nuremberg Codes, established on the basis of these two ethical theories, are recognized as the first international effort to establish codes of conduct for protecting human subjects when conducting research. The Belmont Report is recognized as the first effort to articulate ethical principles for conducting research in the US.

Four general ethical principles are relevant when conducting research on human subjects: respect for human dignity, beneficence, nonmaleficence, and justice. The principles of respect for people's rights and dignity require researchers to treat research participants as autonomous individuals and to obtain their informed consent to participate in a research investigation (Hully & Cummings, 1988). To follow the principle of beneficence, the researcher must balance the benefits and risks of the subjects participating in a research investigation. According to the nonmaleficence principle, sport management researchers must ensure that the research participants have the right to freedom from harm and exploitation during the research process. The principle of justice states that researchers are obligated to provide assurance of the right to fair treatment and assurance of the right to privacy of subjects who participate in research investigations.

The procedures for protecting human rights in sport management research fall into one or more of such guidelines as confidentiality, debriefing, institutional approval, harm, self-determination, full disclosure, privacy, and freedom to withdraw. There are two forms of informed consent: written informed consent and implied consent. The process of informed consent is to assure that the rights and welfare of subjects are protected.

Considered the hub for upholding research standards and principles established by the government and professional associations, Institutional Review Boards (IRBs) are made up of scholars and researchers across a wide range of disciplines of an institution.

Study Questions

1. What are the differences between a utilitarianism view and a deontological view of ethical philosophy?
2. Briefly illustrate the historical development of ethical codes and guidelines for conducting research.
3. What are the three general ethical principles for conducting research on human subjects?
4. What are the major roles of the American Psychological Association in guiding ethics for research in sport management studies?
5. What are the major roles of Institutional Review Boards in guiding ethics for research in sport management studies?

6. What are the major roles of the North American Society for Sport Management and the Sport Marketing Association in guiding ethics for research in sport management studies?
7. What is informed consent?
8. What are the essential elements of an informed consent form?

Learning Activities

1. Review an article that was published during this year in one of the journals in the field of sport management and document in writing how the process of informed consent was explained in the article.
2. Review an article that was published during this year in one of the journals in the field of sport management and discuss how one or more of the general ethical principles was explained in the article.
3. Visit the website for your institution and two others. Compare the IRB policies and procedures for each of the institutions and explain how those are similar and different across the institutions.
4. Review a research article that was published this year in one of the journals in the field of sport management and answer the following questions:
   - Was the study approved by an IRB or other group or committee?
   - Is there evidence that informed consent was obtained from all subjects or their representatives? How was it obtained?
   - Were the subjects protected from physical or emotional harm?
   - Is the research study designed to maximize the benefit(s) to human subjects and minimize the risks?
5. Review a research article that was published this year in one of the journals in the field of sport management and answer the following questions:
   - Were subjects coerced or unduly influenced to participate in this study?
   - Did they have the right to refuse to participate or withdraw without penalty?
   - Were vulnerable subjects used?
   - Were appropriate steps taken to safeguard subjects?
   - How have the data been kept anonymous and/or confidential?

References

American Psychological Association. (2017). *Ethical principles of psychologists and code of conduct.* Retrieved from http://www.apa.org/ethics/code/index.aspx

Beauchamp, T., & Childress, J. (2001). *Principles of biomedical ethics* (5th ed.). New York: Oxford University Press.

Brandt, A. M. (1978). Racism and research. The case of the Tuskegee syphilis study. *Hastings Center Report, 8*(6) 21–29.

Butts, F. (2006). A study of gambling activity in an NCAA division II institution. *The Sport Journal 9*(4). Retrieved from http://www.thesportjournal.org/article/study-gambling-activity-ncaa-division-ii-institution

Cianfrone, B., & Zhang, J. (2006). Differential effects of television commercials, athlete endorsement and venue signage during a televised action sports event. *Journal of Sport Management, 20*(3), 322–344.

Ellis, L. (1998). *Research methods in the social sciences*. New York: McGraw-Hill.

Federal Register. (1974). *Code of Federal Regulations*. Retrieved from: https://www.archives.gov/federal-register/cfr

Fischman, M. W. (2000). Informed consent. In B. D. Sales & S. Folkman, (Eds), *Ethics in research with human participants* (pp. 35–48). Washington, D.C: American Psychological Association.

Folkman, S. (2000). Privacy and confidentially. In B.D. Sales & S. Folkman (Eds). *Ethics in research with human participants* (p. 49–58). Washington, D.C: American Psychological Association.

Frey, M., Czech, D. R., Kent, R. G., & Johnson, M. (2006). An exploration of female athletes' experiences and perceptions of male and female coaches. *The Sport Journal*, (19)4. Retrieved from http://www.thesportjournal.org/article/exploration-female-athletes-experiences-and-perceptions-male-and-female-coaches.

Gamble, V. N. (1997). Under the shadows of Tuskegee: African Americans and health care. *American Journal of Public Health, 87*(1), 1773–1778.

Gunsalus, C. K. (1993). Institutional structure to ensure research integrity. *Academic Medicine, 68* (Sept. Suppl.), 533–538.

Hully, B., & Cummings, S. R. (1988). *Designing clinical research: An epidemiological approach*. Baltimore, MD: Williams & Wilkins.

Johnson, R. B., & Christensen, L. (2014). *Educational research: Quantitative, qualitative, and mixed approaches* (5th ed.). Thousand Oaks, CA: SAGE Publications, Inc.

Katz, J. J. (1972). *Semantic theory*. New York, NY: Harper & Row.

Katz, J. J. (1996). The unfinished Chomskyan revolution. *Mind and Language, 11*, 270–294.

Levine, R. J. (1988). *Ethics and regulation of clinical research*. New Haven, CT: Yale University Press.

Mahony, D. F., & Moorman, A. M. (2000). The relationship between the attitude of professional sport fans and their intentions to watch televised games. *Sport Marketing Quarterly, 9*(3), 131–139.

National Commission for the Protection of Human Subjects of Biomedical and Behavioral Research (1979). *The Belmont Report: Ethical principles and guidelines for the protection of human subjects of research*. Retrieved from http://ohsr.od.nih.gov/guidelines/belmont.html

National Institutes of Health. (2008). The Belmont Report: Ethical principles and guidelines for the protection of human subjects of research. Retrieved from http://ohsr.od.nih.gov/guidelines/belmont.html

Schlenker, B. R., & Forsyth, D. R. (1977). On the ethics of psychological research. *Journal of Experimental Social Psychology, 13*, 369–396.

Steinberg, J. A. (2002). Misconduct of others: Prevention techniques for researchers. *Observer, 15*(1), 40.

U.S. Office of Science and Technology Policy. (2017). Federal policy on research misconduct. Retrieved from https://ori.hhs.gov/chapter-2-research-misconduct-office-science-and-technology-policy

Section

II

Types of Sport Management Research

3

Interpretation of Qualitative Research in Sport Management Studies

After completing the study of this chapter, you should be able to:

- Define qualitative research.
- Explain the differences between qualitative and quantitative research strategies.
- Understand the circumstances under which qualitative research is chosen in sport management research.
- Use critiquing criteria to evaluate the qualitative research report.
- Determine the benefits of combining quantitative and qualitative research methods.

The purpose of this chapter is to provide an overview of *qualitative research* methods, techniques, and tools that can be used to assist the students in better appraising and interpreting sport management research.

The nature and scope of qualitative research will first be examined, followed by the discussion of the relationships between qualitative and *quantitative research* strategies, as well as the philosophical differences between these two types of strategies. The merits of combining qualitative and quantitative research as a research strategy will also be reviewed. Examples from the sport management literature are provided throughout the chapter to aid at facilitating the students' understanding of the information presented.

Qualitative Research Defined

The fundamental objective of qualitative research is to gain insight into and understand the meaning of a particular human experience. This may include human experiences as an individual, as a group, and as an organization. It is an in-depth, subjective analysis of the characteristics and significance of human experience. Unlike quantitative research, it is an interpretation of data based on the meaning that individuals give to their experiences, rather than verification. It is used to explore little-known or ambiguous phenomenon.

Qualitative researchers generally state the purpose for their research clearly and explicitly before beginning their research project. In most qualitative studies, some guiding research questions are first developed in order for the researcher to focus the data collection and analysis phase of the research. However, it is anticipated that these questions may change during the course of the study (Lincoln & Guba, 1985). Qualitative research designs are thus more fluid, flexible, and responsive to data than most quantitative designs.

Qualitative research questions are also more open-ended and exploratory. The generic question qualitative researchers often ask is "what's going on here and why?" Qualitative researchers attempt to approach their data without a priori assumptions, working to make "the familiar strange," to see events in a new way before interpreting what they see. They conduct their explorations in natural contexts and focus on looking at events and actions in a holistic, rather than a reductionist, manner.

Describing the nature of qualitative research is not an easy task because there are so many ways the concept has been defined by academic scholars. A content review of the definitions of qualitative research have been provided in a number of recent books and articles indicate that some authors view it as an interpretation of data based on the meaning of how individuals explain their experiences.

Creswell (2003) has defined qualitative research as an approach "in which the inquirer often makes knowledge claims based primarily on post-positivist claims for developing knowledge (i.e. cause and effect thinking, reduction to specific variables, hypotheses and questions, use of measurement and observation, and the test of theories), employs strategies of inquiry, such as experiments and surveys, and collects data on predetermined instruments that yield statistical data" (p. 18). Strauss and Corbin (1990) define it as "any kind of research that produces findings not arrived at by means of statistical procedures or other means of quantification" (p. 17). The authors have also maintained that the key contributions of qualitative research can be used to

1. assist scholars in better understanding phenomenon about which little is known;
2. gain new and different viewpoints on phenomena for which much is already known; and
3. gather in-depth information about that which may be hard to understand quantitatively.

In trying to condense these extended definitions to a useful description, the following comprehensive definition is offered: qualitative research method is a scientific, purposeful, systematic, and rigorous subjective method of collecting, analyzing, and interpreting data about some characteristic of individual(s), group(s), organization(s), concepts, and/or theories in order to gain new knowledge or add to the existing knowledge base of the field of sport organization management studies. Qualitative researchers tend to use inductive analysis of data to generate critical themes out of the data (Patton, 1990). Bogdan and Biklen (1991) generally define inductive qualitative data analysis as working with data, organizing it, breaking it into manageable units, synthesizing it, searching for patterns, discovering what is important and what is to be learned, and deciding what to tell others.

The key contribution of qualitative research is heuristic in nature, that is, its usefulness lies in its ability to explore new knowledge or add to existing knowledge for enhancement of theory or clinical practices in organizations of the sport industry. The study conducted by Lachowetz, McDonald, Sutton, and Hedrick (2003) is one example illustrating the use of qualitative research in sport management. They conducted in-depth interviews with the corporate sales managers for selected NBA teams to provide further insight regarding the factors that lead to high corporate sponsor retention. The collected data was analyzed using the inductive analysis method. As one of the well-recognized methods of coding qualitative data, inductive methodology allows categories to emerge both during and after analyzing the data. Results of the inductive analysis yielded three primary themes: added value, relationship building/developing, and customer education. According to the managers, these three areas were most responsible for their franchise retaining corporate purchasers over the three-season period of 1998–1999 to 2000–2001.

Qualitative Research Versus Quantitative Research

Qualitative research is one of two major research strategies for investigating diverse phenomena. Its counterpart is quantitative research. Qualitative and quantitative are two distinctive and complementary approaches to research. Each has its own functions and characteristic activities. These approaches originate from different philosophical perspectives and use different methods for data collection and analysis. Qualitative research is rooted in the fields of anthropology, sociology, and philosophy. Quantitative research is rooted in the fields of psychology, economics, and management. The differences between the two approaches are based on the extent to which the analysis can be done by converting observations into numbers or using narrative text to describe human experiences. It is best to think of the qualitative-quantitative distinction as dualistic in nature. Qualitative research is a method of research designed for discovery rather than verification. Thus, this type of research is used to explore little-known or ambiguous phenomena. Quantitative research uses numbers to measure variables. As such, it is a method to verify if there are differences or relationships about phenomena. Quantitative research uses numbers to measure variables. Qualitative research is in-depth, subjective analysis of the characteristics and significance of human experience. Exhibit 3.1 depicts the philosophical differences between quantitative and qualitative research modes of inquiry.

The Choice of a Qualitative Research Approach

Qualitative researchers prefer natural or real-world settings. They do not attempt to control variables, manipulate procedures, create research or comparison groups, or isolate a particular phenomenon. Rather, qualitative researchers immerse themselves in a naturally occurring setting to observe and understand it.

As mentioned earlier, qualitative and quantitative are two distinct and complementary approaches to research for the development of an empirical knowledge base in the field of sport management. The issue is not whether one type of research is better than the other. Each has its own functions and characteristic activities. Quantitative designs usually begin with a hypothesis, which, through measurement, generates data (usually expressed in numbers) and by deduction allows a conclusion to be drawn. In contrast, qualitative design begins with the intention of observing, describing, understanding, or interpreting experiences from the perspective of those experiencing it. Qualitative research is useful in the early stages of knowledge development. It is suggested that a qualitative research design is chosen under the following circumstances:

- A valid and reliable instrumentation is not available or readily developed.
- There is a need to personalize the evaluation process (i.e., meet with various individuals face-to-face to elicit their perceptions, experiences, and feelings).
- A single, in-depth case study approach is appropriate.

Exhibit 3.1. The Philosophical Differences Between Quantitative and Qualitative Research Modes of Inquiry

Factor	Quantitative	Qualitative
Paradigm	Positivist Reality is singular, objective	Naturalistic Reality is subjective
Relationship	Researcher is independent of what is being researched Researcher does not influence the findings Primarily rooted in psychology, philosophy, and sociology	Researcher interacts with what is being researched The interaction between the researcher and "researchee" create the findings Primarily rooted in history and social anthropology
Interpretation of data	Researcher writes in a formal style using quantitative terminology	Researcher writes in a literary style, using qualitative terminology
Research findings	Deductive reasoning (logic) through inferences from numerical data Researchers have specific questions and sites in mind and are usually more concerned with products Attempts to prove or disprove a hypothesis	Inductive reasoning (logic) through creativity and critical thinking Researchers use natural settings of study and are usually more concerned with outcomes or about meaning and understanding of phenomena Attempts to generate a hypothesis
Subjects	Known as participants or respondents	Known as informants
Data collected	Primarily numerical values	Primarily narrative descriptions
Data collection	Passive interaction through questionnaire and/or experimental design	Active interaction with sample population techniques (Observation by active participation)
Data analysis	Interpretive through statistical data Descriptive Inferential (parametric frequencies and nonparametric)	Interpretive through themes analysis Constant comparative procedures Descriptive (rankings percentages)
Sample population	Large sample size relative to concepts to be investigated	Small sample size relative to phenomenon or phenomena to be investigated
Data collection	Survey (non-experimental)	Observation and interview
Research variables	Small number Before/after treatment (experimental)	Large number
Sample population	Random sampling	Purposive sampling
Instrument tools	Questionnaires, scales, equipment, etc.	Tape recorder, transcriber, computer

- Sport management professionals are interested in detailed descriptive information.
- There is a need for information that cannot be tapped by quantification.
- Decision-making personnel need subjective information about the strengths and weaknesses of a program.
- An inductive approach is desired.
- No preconceived notions about an idea, concept, theory, issue, opinion, attitude, or program have been proposed.

The Choice of a Quantitative Research Approach

A quantitative research design is often chosen under the following circumstances:

- A standardized instrument is available that is reliable and valid.
- Use of a standardized instrument is easily administered, cost-effective, and safe.
- Comparisons of the findings are to be made at various points in time.
- The evaluation question asks, "how much?"

Data Collection Techniques of Qualitative Research

Data collection for qualitative research typically is extensive and is in the form of words rather than numbers. The primary concern is with observation and opinion, as they are collected from a subject's environment. Quantitative research designs usually begin with a hypothesis, which through measurement, generates data (usually expressed in an amount or count) and by deduction allows for conclusions to be drawn. In contrast, qualitative research design begins with the intention of observing, describing, understanding, or interpreting experiences from the perspective of those experiencing an event. The most common data collection techniques of qualitative research are case studies, interviews, focus groups, and observation. Each of these four qualitative research techniques will be discussed in detail in the following chapters.

Trustworthiness of Qualitative Research Rigor

Four criteria used to measure the rigor of qualitative research include creditability, transferability (fittingness), dependability, and confirmability (Guba & Lincoln, 1989; Miles & Huberman, 1984; Lincoln & Guba, 1985). According to the aforementioned authors, in qualitative research, trustworthiness is analogous to the psychometric measurements of validity and reliability in quantitative research. Credibility of qualitative research is analogous to the validity of quantitative research. Transferability of qualitative research is analogous to generalizability of quantitative research. Dependability of qualitative research is analogous to reliability of quantitative research. Confirmability of qualitative research is analogous to objectivity of quantitative research. For a summary of the criteria, see Exhibit 3.2.

Combining Qualitative and Quantitative Research

One relative advantage of qualitative research is that it can be a source of rich descriptions and explanations of lived experiences. Although information gained from purely qualitative research

Exhibit 3.2. A Comparison of Accuracy and Precision in Qualitative and Quantitative Research

Quantitative concept	Equivalent in qualitative research
Validity Addresses the question: Are we measuring what we think should be measured?	*Credibility* Addresses accuracy of describing the precise parameters of the study (who, where, when). It addresses the ability of the study participants to distinguish the researcher's interpretation of the data collected as being correct. It is ensured by having the participants review the researcher's conclusions to determine if the interpretation was correct. Techniques to support credibility: • prolonged engagement • triangulation • peer debriefing • member checking (Lincoln & Guba, 1985) • data saturation (Sandelowski, 1994)
Reliability Reliability addresses the question: Are there consistency and stability of the scores of the measurement scale(s) that are being utilized?	*Dependability* Addresses trustworthiness in qualitative research and is accomplished through securing credibility of the findings
Transferability Extent to which the results of a study can be applied to groups other than the one studied	*Generalizability* Addresses the ability to generalize the findings of a study to the population at large (Lincoln & Guba, 1985)
Objectivity Addresses the question is the result of the research applicable outside the research situation	*Confirmability* (Dempsey & Dempsey, 2000) Addresses the trustworthiness of the findings as an indicator of the extent to which the researcher conducted the investigation in a rigorous manner. It refers to an assessment of the overall quality of the findings of a study. For example, by using an expert reviewer or by verifying the interpretations from a study with the original participants.

may be useful, combining qualitative and quantitative approaches can help the researcher benefit from the relative advantages of each method (Teddlie & Tashakkori, 2003). Accordingly, researchers recognize that mixed method research—combining qualitative and quantitative methods—can lead to stronger inferences and enhance overall knowledge of the research issues (Rohm, Milne, & McDonald, 2006).

There are two ways of combining qualitative and quantitative research simultaneously: qualitative plus quantitative and quantitative plus qualitative. Such procedures are also called *mixed method designs*. When qualitative methods initiate the inquiry and are then followed by quantitative techniques, the problem implies an exploratory design. When data are collected sequentially—quantitative first, qualitative second—the quantitative phase provides general results that are then explained with qualitative data. When data are collected sequentially, an explanatory design is implied (McMillan &

Schumacher, 2006). For qualitative plus quantitative, there is a qualitative foundation and quantitative methods are used to provide complementary information. This is a mixed method design. For a quantitative plus qualitative method, there is a quantitative foundation and qualitative methods are used to provide the complementary information (Creswell, 2002; McMillan & Schumacher, 2006). Following are examples of recent mixed studies in sport management research using a qualitative plus quantitative method. Martin (1990) identified and analyzed "mentoring and networking among selected male and female administrators employed by National Collegiate Athletic Association (NCAA) institutions." In his study, a mail questionnaire was followed with a semi-structured interview for approximately 30 minutes at one of the institutions or by telephone. Young (1990) identified and analyzed mentoring and networking among selected male and female administrators employed by National Collegiate Athletic Association (NCAA) institutions. She used frequencies, contingency tables, and chi-square statistics in addition to semi-structured interview.

In the Martin (1990) and Young (1990) examples, data were collected sequentially through a mail questionnaire and follow-up interview. In this example, quantitative data were collected first, then qualitative data were collected second. When data are collected sequentially in this manner, an explanatory design is implied (McMillan & Schumacher, 2006).

Future Directions of Qualitative Research

During recent years, there has been a heightened interest in qualitative research, as recorded by an increasing numbers of reports of qualitative research in sport management and sport marketing peer reviewed journals (Quarterman, Pitts, Jackson, Han, Haueng & Ahn, 2007, unpublished work). This has been illustrated by an increasing number of reports of qualitative research investigations in sport management journals. A review of three journals, *Journal of Sport Management*, *International Journal of Sport Management*, and *Sport Marketing Quarterly*, indicates that between 1987 (or the journal's first issue date if prior to 1987) and 2004, these types of reports have more than tripled (see Exhibit 3.3 and Exhibit 3.4 for details). Exhibit 3.5 demonstrates the results of a study conducted by Peetz and Reams (2011) that found the research methods of over one third of the investigations published in Sport Marketing Quarterly in the first 20 years of existence since its inception in 1992.

It must be noted that this Exhibit does not conclusively reflect the total amount of qualitative research conducted in sport management. It is merely offered as an example of an upward trend in qualitative research in sport management publication outlets. The upward trend implies that researchers have come to recognize the value of qualitative research in sport management studies.

Many fields have affirmed the value and impact that qualitative inquiry can have on professional practice. In fact, many journals have committed to publishing qualitative research projects. Examples relative to sport management and sport marketing include *Journal of Sport Management* (1987), *European Sport Management Quarterly* (1994), *The Sport Journal* (1995), *Sport Management Review* (1998), *International Sport Journal* (started 1999; ended 2004), *International Journal of Sport Management* (2000), *The SMART-Journal* (2005), *Sport Management Education Journal* (2007), *Sport Marketing Quarterly* (1992), *Cyber Journal of Sport Marketing* (started 1997; ended 2000), *International Journal of Sport Marketing and Sponsorship* (2005), and *International Journal of Sport Management and Marketing* (2006).

Arguments about whether quantitative or qualitative research has more merit have raged for years and have produced a number of debates and propositions. Sport management research must

Exhibit 3.3. Percentage of Qualitative Research Articles Among All Articles Published in Three Sport Management Related Journals

Type of Article	Operational Definition	JSM 1987–2006 No	JSM 1987–2006 %	IJSM 2000–2006 No	IJSM 2000–2006 %	SMQ 1992–2006 No	SMQ 1992–2006 %	Row Total No	Row Total %
Quantitative research articles	Articles in which the writer(s) presented an essay in the form of research of a significant idea, concept, theory, issue, opinion, attitude, or program that was primarily represented by an amount or count.	186	53.45%	106	70.67%	174	62.37%	466	59.97%
Qualitative research articles	Articles in which the writer(s) presented research in the form of data of a significant idea, concept, theory, issue, opinion, attitude, or program that was primarily represented by words and phrases.	34	9.77%	14	9.33%	17	6.09%	65	8.37%
Mixed research articles	Articles in which the writer(s) presented research that employed both quantitative and qualitative values.	11	3.16%	5	3.33%	8	2.87%	24	3.09%
Conceptual articles	Articles in which the writer(s) presented an essay in the form of a critical analysis of some significant idea, concept, theory, issue, opinion, or program without employing data analysis.	117	33.62%	25	16.67%	80	28.67%	222	28.57%
Column Total		348	100%	150	100%	279	100%	777	100%

Source. Quarterman, J., Pitts, B., Jackson, N., Han, K., Haung, J., & Ahn, T. (2007). Statistical data analysis techniques employed in the *Journal of Sport Management, Sport Marketing Quarterly*, & *International Journal of Management*: 1987 to 2006. Unpublished manuscript.

recognize that these two research approaches answer different types of questions, and both can facilitate an understanding of the field of sport management. In many instances, a study can use both quantitative and qualitative methods in a mixed-methods approach.

Qualitative research has come to be recognized as a separate field with its own growing body of literature providing guidance on various aspects of the conduct of qualitative research. A number of doctoral programs in sport management now recommend their students to take a course in qualitative research.

Exhibit 3.4. Trends in Qualitative Research in *JSM, SMQ*, and *IJSM*: 1987–2006

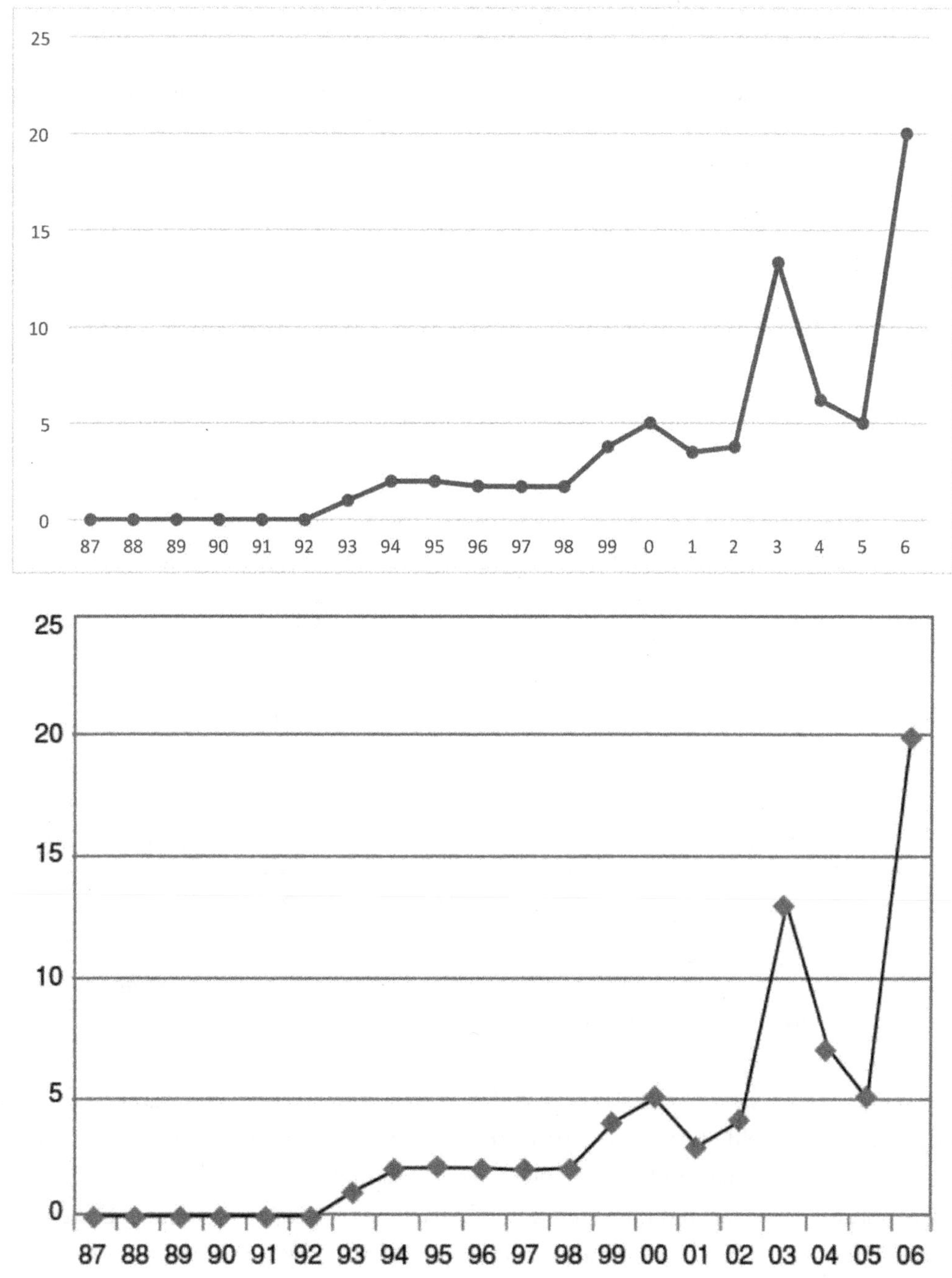

Summary

Qualitative research has been recognized as an important strategy for inquiry in the field of sport management. It is highly valued as a strategy for articulating the meaning of human experience in sport industry organizations. While the tradition of qualitative research in the field is relatively short, an increasing number of qualitative investigations are expanding our understanding of the human experience in the relatively new and emerging field of sport management.

Qualitative research is a scientific, purposeful, systematic, and rigorous subjective method of collecting, analyzing, and interpreting data about some characteristic of individual(s), group(s),

Exhibit 3.5. Research Methods Used in Sport Marketing Quarterly Between 1992 and 2011

Research Methods		Frequency	% of Total	% of Type of Research
Quantitative		212	61.3%	26.4%
	Descriptive	56	16.2%	16.5%
	Factor Analysis	35	10.1%	14.6%
	t-test and ANOVA	31	9.0%	12.3%
	Regression	26	7.5%	10.8%
	MANOVA	23	6.6%	8.0%
	Chi-Square	17	4.9%	7.5%
	Other	16	4.6%	2.8%
	Correlation	6	1.7%	0.9%
	ANCOVA	2	0.6%	59.7%
Qualitative		124	35.8%	20.2%
	Descriptive	74	21.4%	7.3%
	Theoretical	25	7.2%	6.4%
	Interview	9	2.6%	3.2%
	Historical	8	2.3%	3.2%
	Focus	4	1.2%	26.4%
	Observational	4	1.2%	16.5%
Mixed Method		10	2.9%	14.6%
Total		346	61.3%	12.3%

Note: Percentages may not equal 100% due to rounding.

Source. Peetz, T. B., & Reams, L. (2011). A content analysis of *Sport Marketing Quarterly*: 1992–2011. *Sport Marketing Quarterly*, 20, 209–218.

organization(s), concepts, and/or theories in order to gain new knowledge or add to the existing knowledge base of the field of sport management. It is social science-based, designed for discovery rather than verification. The fundamental objective of qualitative research is to gain insight into and understand the meaning and interpretation of a particular human experience. Data collection for qualitative research typically is extensive and is in the form of words rather than numbers.

There are four criteria by which rigor is measured in qualitative research: creditability, fittingness, auditability, and confirmability. Case studies, interviews, focus groups, and observation are qualitative research methods that are commonly used by sport management researchers. Both qualitative and quantitative research methods are viable means of inquiry in sport management research. When the answer to a research question or hypothesis requires accuracy, quantitative methods are preferred. However, when the answer to a research question or hypothesis involves capturing complexity in a situation or understanding meaning, qualitative methods are preferred.

Study Questions

1. What is qualitative research?
2. Explain the differences between qualitative and quantitative research strategies.
3. What are the circumstances under which qualitative research is chosen in sport management research?
4. What are the criteria that can be used to evaluate a qualitative research report?
5. What are the benefits of combining quantitative and qualitative research methods?

Learning Activities

1. Review a qualitative research article that was published this year in one of the journals created specifically for sport management or sport marketing and argue for or against the type of qualitative design that was used in the article.
2. Identify a topic of interest in sport management and design a study using a qualitative research technique as a means of data collection.

References

Bogdan, R.C., & Biklen, S. K. (1991). *Qualitative research for education: An introduction to theory and methods.* Boston, MA: Allyn and Bacon.

Creswell, J. W. (2002). *Educational research: Planning, conducting, and evaluating quantitative and qualitative research.* Upper Saddle River, NJ: Merrill/ Prentice Hall.

Creswell, J. W. (2003). *Research design: Qualitative, quantitative, and mixed methods approaches* (2nd ed.). Thousand Oaks, CA: Sage Publications, Inc.

Guba, E. G., & Lincoln, Y. S. (1989). *Fourth generation evaluation.* Newbury Park, CA: Sage.

Lachowetz, T., McDonald, M., Sutton, W. A., & Hedrick, D. G. (2003). Corporate sales activities and the retention of sponsors in the National Basketball Association (NBA). *Sport Marketing Quarterly, 12*(1), 18–26.

Lincoln, Y. S., & Guba, E. G. (1985). *Naturalistic inquiry.* Beverly Hills: Sage Publications, Inc.

Martin, C. L. (1990). The employee/customer interface: An empirical investigation of employee behaviors and customer perceptions. *Journal of Sport Management, 4*(1), 1–20.

McMillan, J. H., & Schumacher, S. (2006). *Research in education: Evidence-based inquiry* (6th ed.). Boston, MA: Pearson.

Miles, M., & Huberman, A. (1984). *Qualitative data analysis.* Beverly Hills, CA: Sage Publications, Inc.

Patton, M. Q. (1990). *Qualitative evaluation and research methods* (2nd ed.). Newbury Park, CA: Sage Publications, Inc.

Peetz, T. B., & Reams, L. (2011). A content analysis of sport marketing quarterly: 1992–2011. *Sport Marketing Quarterly, 20*, 209–218.

Quarterman, J., Pitts, B. G., Jackson, E. N., Han, K., Haung, J., & Ahn, T. (2007). Statistical data analysis techniques employed in the *Journal of Sport Management, Sport Marketing Quarterly, & International Journal of Management*: 1987 to 2006. Unpublished manuscript.

Rohm, A. J., Milne, G. R., & McDonald, M. A. (2006). A mixed-method approach for developing market segmentation typologies in the sports industry. *Sports Marketing Quarterly, 15*, 29–30.

Strauss, A., & Corbin, J. (1990). *Basics of qualitative research: Grounded theory procedures and techniques.* Newbury Park, CA: Sage Publications, Inc.

Teddlie, C., & Tashakkori, A. (2003). Major issues and controversies in the used of mixed methods in the social and behavioral sciences. In A. Tashakkori & C. Teddie (Eds.), *Handbook of mixed methods in the social & behavioral research.* Thousand Oaks, CA: Sage Publications, Inc.

Young, D. (1990). Mentoring and networking: Perceptions by athletic administrators. *Journal of Sport Management*, *4*(1), 71–79.

4

Case Study Research

Richard M. Southall, University of South Carolina
Crystal Southall, Western State Colorado University

After completing the study of this chapter, you should be able to:

- Identify, define, and discuss various presented terms.
- Describe basic types of case studies.
- Discuss five steps in designing a case study.
- Explain some useful standards for evaluating case study data collection techniques.
- Delineate the concept of triangulation/crystallization.
- Categorize at least four types of triangulation/crystallization.
- Detail several strengths, as well as some limitations, of case study research.
- Outline some basic strategies for analyzing case studies.

Conducting case study research is a challenge for sport management researchers, because it requires using a variety of techniques to obtain evidence from diverse data collection points. In addition, since many sport management researchers have limited social-science research training, they are either not sure what constitutes case study research, or they may view it as hierarchically inferior to other forms of research: an exploratory step-child to surveys and histories, or a distant and impoverished cousin of causal/experimental inquiries. Such perspectives inherently relegate case study research to nothing more than a preliminary strategy, not sufficiently rigorous to describe or test propositions.

However, this interpretation overlooks that social-science research, by definition, not only occurs within a social setting, but also involves investigating individual, group, organizational, or institutional (i.e., social) phenomena. Therefore, from an interpretive perspective, a qualitative researcher seeks to understand how and why such social phenomena occur. Case study research, as will be detailed in this chapter, fosters such understanding by answering focused questions regarding short-term (e.g., a few weeks to a year), contemporary social events. Since case-studies involve relatively short time-periods, researchers do not seek to control events. However, despite these social limitations, properly conducted case study research still has sufficient rigor for a researcher to answer how and why research questions.

Case study research's limitations reflect the social world's complexity. To better understand the complexity of case study research, it is helpful to first discuss its historical origins. After presenting a historic backdrop, the remainder of this chapter will answer some practical questions: (1) What are some types of case studies, (2) how can I design a case study, (3) how can I conduct a case study, (4) what is involved in analyzing case study evidence, and (5) what are some suggested ways to report case study results?

Case Study Defined

In an attempt to define what constitutes a case study, Hamel, Dufour, and Fortin (1993) posed two basic questions: "Is the case study a method? Or is it an approach?" (p. 1). Responding to these fundamental questions, Yin (2003b) clearly delineated case study research as one of several ways of doing social science research and referred to it as a qualitative research strategy distinct from other social science research methods (i.e., experiments, surveys, histories, and archival analysis). Even though case study research has been widely viewed as a distinct strategy, some scholars also subsume case studies within ethnographies or participant-observations, thereby reducing case study research to a data collection technique utilized in conducting fieldwork. Stake (2003) posited an inclusive criterion for what constitutes case study research by not defining it as a specific methodology, but rather asserting that a case study is defined by the study's uniqueness and specificity. Amis (2005) was not as inclusive in his discussion of what constitutes case study research, but argued for case study research that goes beyond focusing on a single case and also advocated for multi-level data analysis, including societal, governmental, organizational, subunit, and/or individual.

In 1990, the United States General Accounting Office developed a widely cited definition of case study as "a method for learning about a complex instance, based upon a comprehensive understanding of that instance obtained by extensive description and analysis of that instance taken as a whole and in its context" (U.S. General Accounting Office, 1990, p. 14). This definition assists in framing a discussion of case study research, because it highlights the contextual complexity in which case study research occurs and the need for researchers to consider not only particular actors (or groups of actors), but also the interactions between actors. This definition also suggests the multiperspective nature of case study research.

Researchers have consistently viewed case study research as a method or strategy that utilizes an inductive approach in which the definition of the subject of investigation is not a matter of conjecture or happenstance. The subject's definition emerges or asserts itself during the course of investigation. By establishing increasingly close ties, an investigator uncovers actors' personal experiences, which allows the social reality to emerge and provides answers to two questions:

1. How does the individual, group, organization, or society generate/reflect the problem or phenomenon being investigated?
2. Why has this problem or phenomenon been deemed as such by the individual, group, organization, or society?

Eisenhardt (1989) defined a case study as "a research strategy which focuses on understanding the dynamics present within single settings" (p. 534). Stake (2003) defined a case as a "specific one," while allowing that a case may be simple or complex. Yin (2003a, 2003b) also utilized the scope of a case study in his definition, but also noted the desire to investigate the case's contextual conditions as a central definitional element. Yin (2003b), recognizing the importance of allowing uncontrolled contextual variables and phenomenon to remain entangled, noted: "A case study is an empirical inquiry that investigates a contemporary phenomenon within its real-life context, especially when the boundaries between phenomenon and context are not clearly evident" (p. 13). Because a case study is a real-life, nonexperimental setting, utilized data collection and analysis strategies are helpful in delineating it from other forms of inquiry. While scholars may not all agree whether case study research is a method or a research design (Mertens, 2005), there is seeming consensus among researchers that case study inquiry occurs in situations where there are more variables of interest than data points. As a result, case study data collection should be guided by theoretical propositions, rely on multiple sources of evidence, and data analysis should converge in a triangulating fashion.

Exhibit 4.1 is a case study involving a single tragic car accident. It illustrates many of the elements regarding what constitutes a case study discussed above. The specific case illustrates the dynamic nature of case study research and how a single incident can illuminate a much larger societal issue.

Case Study as Research Design in Sport Management

The sport business industry is a goal-oriented and results-driven enterprise. From the field or court to the front office, success is measured by wins and losses, playoff successes, gate receipts, attendance, and sponsorship packages. Most sport management professionals are concerned with research's practical applicability. Because case study research utilizes inductive, open-ended strategies, it offers practical results for sport management professionals. One advantage of case study research is that the results and generated theories are understandable and credible. In addition, since a primary goal of case study research is to improve existing practices, it appeals to sport management professionals looking to improve their organization's bottom line. Case study research allows researchers and practitioners to jointly analyze contexts, processes, and meanings. In some instances, a case study can be the basis for developing a survey that can then be used to assess the generalizability of the case study's results. Based upon a case study's results, researchers may also perform secondary analysis and utilize other research methodologies to reach broader conclusions and implement policy changes.

Case study research forces researchers and participants to develop various ways to describe the events, situations, and actions studied. Such formulated understandings can be beneficial to an organization's understanding of itself. Participants' perspectives on events and situations are part of the social reality the researcher hopes to understand. In addition, since case study research typically involves a unique situation, small samples, and aggregations of data from many samples, the researcher is able to concentrate on the contextual uniqueness of a particular case and more fully explore it. This focus allows researchers to report data and managers to concentrate on their own organization and learn from a relatively removed and more objective analysis of their own situation.

Exhibit 4.1. Case Study Example 1—The National Football League and Its "Culture of Intoxication": A Negligent Marketing Analysis of *Verni v. Lanzaro*

On October 24, 1999, the parking lots at Giants Stadium, which opened four hours before kickoff, witnessed a Sunday game-day ritual played out at professional football stadiums across the United States: "pregame and post-game tailgate parties at which persons consume alcoholic beverages" (Sixth amended complaint and jury demand, *Verni v. Lanzaro*, 2003, p. 7). As the afternoon turned to early evening, the New York Giants had easily beaten the New Orleans Saints 31–3, to run their record to 4–3 on the season (NFL 1999 Season Archives, n.d.). Celebratory Giants fans, among them 30-year-old Daniel R. Lanzaro of Cresskill, New Jersey, poured out of Giants Stadium to begin their journey home or to continue their celebrations at one of the local bars, such as Shakers or The Gallery, in the borough of Hasbrouck Heights in Bergen County (Sixth amended complaint, 2003).

Returning home from a family outing to pick up pumpkins for Halloween, Ronald Verni, accompanied by his wife, Fazila, and their two-year-old daughter, Antonia, passed through Hasbrouck Heights, about five minutes from Giants Stadium, via Terrace Avenue (Gaudino, 2002; Cable News Network, 2003). After purchasing and consuming at least 14 beers and an undisclosed amount of marijuana (Sixth amended complaint, 2003) while participating in pregame tailgating and viewing the day's game at Giants Stadium, Lanzaro lost control of his 1994 Ford pickup, hit one vehicle, and then collided, head-on, with the Verni family's 1999 Toyota (Coffey, 2005).

The results of the accident were horrific. Antonia, who was resuscitated by Hasbrouck Heights emergency medical technicians, suffered a broken neck and spent the next 11 months in the hospital and in rehabilitation. The accident left the child "a quadriplegic in need of round-the-clock care.* Her mother went into a coma, needed reconstructive surgery on her face, and had a rod inserted into her leg" (Coffey, 2005, para. 6). Antonia's father, Ronald, was unhurt in the accident.

According to a newspaper report of the accident, Hasbrouck Heights police officer Corey Lange found Lanzaro and a passenger sitting on the curb near the accident site. Both Lanzaro and the passenger had trouble standing up. When asked how much he had been drinking, Lanzaro replied, "Too much" (Gaudiano, 2002, para. 2). A security guard at the hospital reported finding a marijuana "joint" in Lanzaro's pocket, and Lange later said that Lanzaro admitted to drinking before and during the game, and also to smoking some "pot" (Gaudiano, para. 2). According to police, Lanzaro's blood alcohol level was "0.266, 2 ½ times the legal limit of 0.10" (Para. 2). In August 2003, Lanzaro pled guilty to vehicular assault and was sentenced to five years in Riverfront State Prison (Coffey, 2005).

In all likelihood, this tragic occurrence would have remained just another tragedy in just another newspaper on just another weekend, soon forgotten by the general public, if the Vernis' had simply chosen to sue Mr. Lanzaro. However, the Vernis' chose to include several other parties in their complaint, including the New Jersey Sports and Exposition Authority, the New York Giants, Giants Stadium, Aramark, two local bars [Shakers, The Gallery], the National Football League (NFL), and NFL commissioner Paul Tagliabue (Sixth amended complaint, *Verni v. Lanzaro*, 2003, p. 1).** Suddenly, this tragic accident became much more than an isolated incident. It evolved into a lawsuit with major implications for arena/stadium concessions management policies and procedures.

* It should be noted that Antonia Verni (age 2) was wearing a seatbelt, but was not in a car seat at the time of the accident (Dvorchak, 2005).

** Since Lanzaro's insurance policy had a $200,000.00 liability limit, the list of defendants was consistent with a complaint based on allegations that the Giants and the NFL had the authority and ability to exercise control over Aramark as the concessionaire at Giants Stadium.

Source. Southall, R.M., & Sharp, L.A., originally published in *Journal of Legal Aspects of Sport*, © 2006. Reprinted with permission.

Building Theory from Case Study Research

Various qualitative research strategies or methods have often been criticized for being "unscientific," since they do not allow for manipulation of variables in explanatory or causal inquiries (Shavelson & Townes, 2002). In addition, some researchers recognize the appropriateness of various qualitative research techniques, but only for specific types of research (i.e., case studies mainly for an exploratory purpose, and surveys and histories for seeking descriptive information). If case studies are only a viable preliminary research methodology, it has been said they cannot be used to describe or test theoretical propositions. This attitude may lead to the question, if a case study is a specific one, how can case study research help us either test or build a generalizable *theory*? In the following sections, what constitutes a theory, the role of theory in designing and conducting case studies, and how case study research can be utilized to build or develop theories will be discussed.

What Is a Theory?

In the context of case study research, the term theory encompasses more than the common medical definition of a working probability-based hypothesis based on experimental evidence or factual or conceptual analysis (Theory, 2002). Any discussion meant to define theory has difficulties. Merton (1967) recognized this difficulty:

> *Like so many words that are bandied about, the word theory threatens to become meaningless. Because its referents are so diverse—including everything from minor working hypotheses, through comprehensive but vague and unordered speculations, to axiomatic systems of thought—use of the word often obscures rather than creates understanding. (p. 39)*

Sutton and Staw (1995) and Kaplan (1998) agreed with Merton (1967) that theories encompass answers to queries of why, and involve not just connections among phenomena, but are stories about why acts, events, structure, and thoughts occur. Building theory is a process. As such, it is full of contradictions and internal conflicts. In addition, while theory may often emphasize the nature of causal relationships and allows for identification of what comes first, as well as the timing of events, in case study research it also involves the design of research steps in accordance with a relationship to extant literature, policy issues, or substantive sources (Yin, 2003a).

Types of Theories

In addition to understanding what a theory is and the elements of theory-building, a researcher should develop an awareness of the various types of relevant theories. Through the process of reviewing the literature and questioning what is to be studied, why it is important to study it, and what the researcher hopes to learn from a study, a researcher can more fully explore pertinent theories and choose what type is most relevant to his/her study. Available theories can be categorized in accordance with units of analysis. Yin (2003a) was one of the first case study theorists to explicitly detail theory types and developed four theory types: (1) individual, (2) group, (3) organizational, and (4) societal. An individual theory is related to individual development; individual personality, perception, and learning; and individual interpersonal relationships. Group theories are concerned with explaining family, informal group, team, and interpersonal networks. Organizational theories explain bureaucracies, and organizational structures, performance. Societal theories predict societal relationships, such as international behavior, market functions, and emerging technologies.

Theory Building

In order to build theory from case study research it may be helpful to discuss some necessary theory-building elements. Whetten (1989) delineated four essential elements as the building blocks of theory development. The first one is found by determining what factors (i.e., constructs, variables) should be included in the explanation. While such a judgment should necessarily be guided by evaluating the comprehensiveness and parsimony of the proposed theory, it is preferable to err on the side of initially including too many variables rather than too few. A second element involves determining how the included factors are connected. This delineation explicitly involves causality and seeks to preliminarily identify relationships among and between variables. Third, a theorist should justify the selection of variables and proposed causal relationships by providing a logical rationale for the adoption of the proposed theory that necessarily includes an overarching view of human nature or organizational/societal processes. Finally, limitations on the propositions generated by the theoretical model should also be identified in order to answer questions of who, where, and when. This type of mental exercise seeks to test the generalizability and contextual limits of a theory's core propositions.

In addition to identifying several theoretical building blocks, it is also helpful for researchers to eliminate from consideration research components that may be critical to a successful research project, but should not be confused with theoretical elements (Sutton & Staw, 1995). An initial research-project component that is not theory, is performing a solid review of previous research (e.g., building a reference list). Referring to theories developed in previous research is a critical component in developing a context for proposing or developing a theory, but simply listing previous theories does not involve the third building block of theory building—explaining the casual logic contained in the listed theories. In addition, while organizational/social theory is based on data, data only describe observable patterns. Data do not explain why empirical patterns were observed. Data are also not predictive; they cannot explain what patterns are expected to be observed in the future. Data should not be confused with the inspired inferences. Just as data are important parts of theory, but are not the theory itself, so too variables and constructs are important to building theory, but do not alone constitute theory. Theory, according to Weick (1989), must explain why variables or constructs exist and how they are connected. Finally, hypotheses are also not theory. While they can serve as connections between theory and data, they do not contain logical arguments detailing why empirical relationships are predicted to occur. Hypotheses detail what is expected to occur, not why it is expected.

As can be seen from this brief discussion, theory building is not confined to quantitative research. For several decades, social scientists, including Glaser and Strauss (1967) and Strauss and Corbin (1990), have defended qualitative research's ability to build *grounded theories* that can explain collected data. According to grounded theory, researchers can generate theories from unanticipated contextual influences identified by either researcher or participants. Many grounded theorists have adopted a positivist paradigm that assumes an objective, external reality in which the researcher allows for the existence of subjective realities among subjects, while the researcher adopts the role of a neutral observer who is able to discover data, develop analytic interpretations of data, and utilize these interpretations to focus further data collection (Charmaz, 2003). In addition to focusing data collection, developed theoretical frameworks allow for further refinement of theoretical analysis.

The *positivist paradigm* has been questioned by grounded theorists who have adopted a *constructivist view* that assumes multiple subjective social realities created by the researcher and research subjects (Schwandt, 2000). A constructivist approach to grounded theory according to Charmaz (2003) is not rigid or prescriptive, furthers interpretive understanding, and can be adopted without adhering to the strictures of positivism. This reformulation of grounded theory is much more in alignment with Yin's (2003b) contention that a preliminary theory related to the topic of study should be constructed prior to conducting data collection. Yin's discussion of preliminary theoretical development reflects, in many ways, Eisenhardt's (1989) contention that an a priori specification of constructs (i.e., conflict, power) can help shape theory-building research, since such specificity permits more accurate measurement of constructs if and when they emerge during data collection. If an emergent theoretical construct proves critically important, it possesses an increased empirical grounding based on the utilization of triangulated data collection methods.

From this perspective, preliminary theory development (specification of constructs) is a research-design component, regardless of whether the research's ultimate goal is to develop or test a proposed theory. The development of preliminary constructs places a case study in its appropriate context within the extant research literature. This type of theory development assists the researcher in more clearly articulating and addressing the study's questions, propositions, and unit(s) of analysis. By defining the unit of analysis (What is the actual case?), the researcher can more easily screen and select potential candidates. This screening may also suggest possible relevant variables and data that should be included in a case study. In addition, the theoretical framework, which emerges from a specified construct, forms the basis for development of a logical method to link developed propositions to collected data and interpreting the study's findings. These research design steps will be discussed more fully later in the chapter.

Exhibit 4.2 is an example illustrating the use of a theoretical construct to place a case study involving the Women's United Soccer Association (WUSA)—a short-lived women's professional soccer league founded after the 1999 FIFA Women's World Cup—in its appropriate context. The theories utilized assisted in clearly articulating the study's questions and propositions. In addition, several variables and important data that needed to be collected as part of the case study emerged from the theoretical frameworks. The theories also greatly assisted in the process of interpreting the study's findings. Even many years later, the theories frame the case study's conclusions, allowing a new generation of readers to benefit from the research.

Types of Case Studies

Building from various theorists' (Eisenhardt, 1989; Stake, 2003; Yin, 2003b) case study definitions and utilizing our understanding of case study methodologies, we can develop a case study matrix that describes cases by unit of analysis, purpose, and focus. This typology of case studies includes cases involving single or multiple units of analysis; cases with exploratory, descriptive, or explanatory purposes; and cases with either an intrinsic, instrumental, or collective focus.

The first type of case we can discuss is a single case. In addition, researchers also utilize case study research to study multiple cases. In defining case study research, Eisenhardt (1989) described it as an examination of the dynamics present within a single setting, while Stake (2003) defined a case as a

Exhibit 4.2. Case Study Example 2—Build It and They Will Come? The Women's United Soccer Association: Who Built It, Who Came, and Why It Failed

Founded in 2001, the Women's United Soccer Association (WUSA) basked in the reflected glory of the unprecedented media coverage (11.4 Nielsen rating for final) and attendance (90,000 at the Rose Bowl for the Women's World Cup Finals) that accompanied the United States national team's 1999 World Cup shootout victory over China (Women's United Soccer Association, 2000a). The league was hailed as a historic culmination of several trends in women's sports in the United States: (1) Title IX's nearly 30-year influence on the culture and opportunities for women in collegiate sport, (2) the resulting, but specific, development of National Collegiate Athletic Association (NCAA) women's soccer, and (3) the 20-plus-year boom in soccer participation among America's youth.

For any sport league to be successful it must generate sufficient revenue to cover operating expenses. There are several revenue sources that can support a league (e.g., ticket sales, broadcast rights, sponsorship, and merchandise). All of these revenue sources depend upon a sound marketing strategy for success. Although many components of a league contribute to its success or failure, this paper will primarily focus on the inability of the WUSA to either identify and communicate to fans, broadcast partners, or sponsors a single coherent marketing strategy, or to convince these stakeholders to adequately support the league.

Marketing Theories

Exchange Theory

In any successful business transaction, all parties must feel there has been an exchange of value that satisfies the participants (Howard & Crompton, 2004). The desired outcomes are achieved only when the parties are willing to act in the best interest of all stakeholders or participants (Blalock & Wilken, 1979). Exchange theory is based upon three principles: rationality, marginal utility, and fairness (McCarville & Copeland, 1994). Rationality in marketing or sponsorship agreements focuses on the elucidation of the goals of all parties involved and the achievement of these stated goals. Blau (1967) and Homans (1974) contended that rationality was a response to presented agreement stimuli. A past favorable outcome for participants increases the likelihood of future agreements (McCarville & Copeland). Conversely, if a previous agreement has not fulfilled participants' expectations, the likelihood of any future agreement being finalized is diminished.

Marginal utility encapsulates the ability to obtain rewards and the value assigned to such rewards (McCarville & Copeland, 1994). Homans (1974) noted the existence of an inverse relationship between the ease and availability of compensation and its perceived value. Initial goals achieved during an initial contract period are not likely to be highly valued during subsequent negotiations. As a result, offering organizations must continually upgrade the available benefits included in their advertising or sponsorship packages. If the benefits remain constant, the law of diminishing returns often occurs. As a result, corporate partners may become dissatisfied with the status quo and demand new, more-valued benefits as part of any new agreement.

The fairness principle of exchange theory involves the perceived equitable distribution of rewards (McCarville & Copeland, 1994). Organizations entering into agreements seek specific and identifiable benefits (Kuzma, Shanklin, & McCally, 1993; Stotlar, 2001). If these benefits can be met through other more cost-effective means, the participating organization will be unlikely to agree to the initial agreement or renew an existing one. To ensure that corporate partners are receiving fair and equitable treatment, long-term contracts should have official (or unofficial) renegotiation clauses that allow both parties to propose altering the agreement's terms if the financial climate has changed since the agreement's inception (McCarville & Copeland).

(*Continued*)

Exhibit 4.2. (*Continued*) Case Study Example 2—Build It and They Will Come? The Women's United Soccer Association: Who Built It, Who Came, and Why It Failed

Cause Marketing

Cause-related marketing (CRM) is most often defined as the public association of a for-profit company with a nonprofit organization, and is intended to promote the company's product or service and to raise money for the nonprofit (Polonsky & Macdonald, 2000). Pringle and Thompson (2001) described CRM as a strategic positioning and marketing tool that links a company or brand to a relevant social cause or issue, for mutual benefit. An example of a cause-related marketing campaign was American Express's 1983 involvement in the restoration of the Statue of Liberty. Cause-related marketing is not *social marketing*, which is the use by nonprofit and public organizations of marketing techniques in an attempt to impact societal behavior (e.g., stop smoking, don't pollute, don't use drugs, don't drive drunk). Cause-related marketing attempts to link an organization's product(s) directly to a social cause through the implementation of a strategic marketing plan (Pringle & Thompson, 2001). Generally, an organization prefers to support causes that are of interest to its target market. Although there may be a philanthropic motive to cause-related marketing, the efforts of a cause-related marketing campaign tend to produce relatively short-term, product-related outcomes (LeClair & Ferrell, 2000).

Strategic Philanthropy

Strategic philanthropy evolved from nonprofit organizations recognizing the need for a new giving paradigm that combined elements of benevolent philanthropy (corporate giving and related activities not purposely aligned with the strategic goals and resources of the corporation) with a more synergistic use of an organization's core competencies and resources to address key stakeholders' interests and to achieve both organizational and social benefits (LeClair & Ferrell, 2000). It involves a company's long-term investment in a cause that provides societal benefits while also enhancing the company's reputation (Stotlar, 2001). Such strategic philanthropy may involve the participating organization's endurance of short-term business losses for the good of the cause and for the fulfillment of the organization's social responsibilities and long-term gain. As a result, strategic philanthropy requires support from a corporation's top management and shareholders, and coordination of corporate giving and employee volunteer programs with the overall corporate mission.

This redefinition of philanthropy recognizes that although businesses should be good corporate citizens, they must not forget their fundamental obligation to their shareholders and employees, and to the company's profit-and-loss statement. As Carroll (1979, 1991) noted, there are four components to the corporate responsibility model: economic, legal, ethical, and discretionary responsibilities. As a result of this need to synthesize organizational and societal needs, neither traditional benevolent philanthropy, nor strictly business objectives have primacy; both work in concert to benefit the other. However, it is often difficult for the various stakeholders in this process to discern where one element ends and another begins.

"specific one," while allowing that a case may be simple or complex. A single case study is appropriate when the chosen case represents a critical instance that allows for testing a well-formulated theory. In addition, selecting a single case may be beneficial if it is either an extreme or unique case that may rarely occur, if it is a typical case that provides a representative example that is commonplace, or if it will reveal a situation or context that has not previously been explored. Although a single case may be unique, it may also reoccur over time, so a single case study may become a longitudinal case. In this situation, a researcher revisits the same case at various periods in time. This allows for the evaluation of a theory that posits how certain conditions change over time.

An often-leveled criticism of case study research is a lack of generalizability. However, it is also appropriate to view evidence from multiple cases as providing more than just additional data. From this alternative perspective, data from multiple cases can be seen as analogous to multiple experiments, not as multiple respondents to a single survey or multiple subjects within a single experiment. Consequently, a researcher should conduct multiple cases to replicate findings from an initial case study. In replicating the initial case, the researcher may also attempt to duplicate the exact conditions present in the first case. Such cases will be deemed literal replications that will predict results similar to the initial case. In other instances (theoretical replications), altered conditions can be considered and contrasting, but predictable, results may be theorized (Yin, 2003b). As can be seen, the use of multiple case studies should be based on rich theoretical frameworks that can guide analysis of conditions in which a particular phenomenon will or will not occur.

Case studies can also be categorized by their purpose. A given case study's initial purpose may be to explore a particular organization or situation in order to develop hypotheses and propositions to be used in developing a theoretical framework to guide further research that involves new cases. By its very nature an *exploratory case study* involves fieldwork and data collection that is out in front of or a precursor to theoretical development. Such a case study is often viewed as a pilot study, which is useful in designing a subsequent distinct case study. If an exploratory case study is not viewed in this manner, a researcher may mistakenly utilize the collected data in a subsequent case study. Since the exploratory study or phase did not involve research questions, hypotheses, or even data collection methods, once it is completed, it should be considered complete and utilized as a resource or reference.

A *descriptive case study* utilizes a descriptive theory that delineates the scope and depth of the case being described. A descriptive case study does not simply involve random data collection, nor does it occur prior to the development of a descriptive theory. A descriptive case study must have a design (based on a descriptive theory) that allows for an examination (description) of the similarities and differences between previously developed theoretical patterns. A descriptive case study requires the development of idealized rival scenarios representing best-case alternatives. These best-case scenarios should be developed and reviewed by experts with no stake in the data collection. Data collection protocols can then be developed from these scenarios. It is important to remember that unanticipated findings are not precluded by the developed scenarios; such scenarios provide structure for the data collection.

A third case study purpose is to be explanatory in nature. An explanatory case study explains how events happen and so is much more aligned with an experimental approach. *Explanatory case studies* most often utilize either *factor theory* (utilizing correlation) or a causally-linked explanatory theory to explain the case. Both types of explanatory case studies involve the identification of various independent variables. Some explanatory case studies simply discuss the degree of correlation between the independent variables and the dependent variable. Other explanatory case studies go further and posit a causal relationship between variables. Because explanatory case studies propose a relationship between variables, correlation analysis, factor analysis, regression analysis, and analysis of variance are often utilized to account for interactions among independent variables. For many case study theorists, an explanatory case study's strength is dependent on the strength of the underlying theories (Yin, 2003b).

In addition to classifying case studies by their purpose, they may also be categorized by their focus. If a case study's focus is to better understand a particular case, without seeking to make any generalizations about other cases, the case study is termed an *intrinsic case study*. If, on the other hand, the focus of the case is to provide insight into a general issue or develop—or redevelop—a generalization drawn from the individual case, or a series of cases, the case study is an *instrumental case study*. With this focus, a particular case is of secondary interest. What is important is the developed generalization. Finally, a *collective case study* is actually multiple instrumental cases combined to allow for development of theories that can explain a large number of cases and/or generalizations (Stake, 2003).

Designing a Case Study

Now that we have discussed what a case study is, what a case study can do, how we can build theory utilizing a case study, and various types of case studies, we are ready to begin designing a case study. Fundamentally, designing a case study is no different from designing any research project. Research involves asking questions and searching for answers to those questions. Before you begin the process of designing a case study, you should already have asked other preliminary questions:

- Why do I want to do a case study on a certain topic?
- Should I do an experiment, survey, or archival search instead?
- Am I interested in conducting a specific type of case study? (See above discussion of types of case studies.)
- Who has conducted research on my topic of interest?

Based on these questions, a preliminary review of the literature can be conducted. Such a review should not be seen as just a step that has to be completed—a box to be checked off on the way to getting the real research started. A literature review should also not be seen as an end in itself. It is not just a search for answers—a catalogue of what is known about a topic—but a method for sharpening the questions to be asked about a topic. Look for what is missing. A thorough literature review should raise as many questions as it answers. After this initial questioning has occurred, the following guidelines will help you design a case study research project.

The first step in the design process is to develop specific research questions. Case study research questions guide the search for how and why answers. Because case studies are preferred in examining contemporary events in situations in which the examined behaviors cannot be manipulated, the questions developed should recognize case study's research limitations. For example, in developing a research question, you should ask yourself if there are any aspects of the question that cannot be answered by utilizing case study methodologies. In addition, if you can answer the developed research questions by successfully conducting a case study, how would you justify the significance of your findings to a reviewer? Will you have uncovered something significant, rare, or impressive? As can be seen, this initial questioning process should add precision to the proposed study. Research questions should have clarity and focus.

After research questions have been formulated, the next design step is identifying the study's purpose or identifying its propositions. An exploratory case study may not have specified propositions; nevertheless, the study should still have a purpose. Identifying and formulating propositions forces you to state and describe the theoretical underpinnings that will guide your exploration for relevant evidence. If the case study is exploratory and there are no theoretical underpinnings, you still need to describe the criteria by which you will deem the study a success. Propositions, therefore, should describe the theoretical framework(s), rationale, or direction of your research. At the very least, initial assumptions should be outlined. If there is an important theoretical issue to be examined, it should be detailed.

The third step in case study design is to specify the unit of analysis, or more specifically defining just what constitutes your case. Is the case an individual, several individuals, a decision or a series of decisions, programs, organizational change, or an implementation process? The research questions or propositions formulated are useful in guiding your selection of your case's unit of analysis. If you have accurately delineated your research questions, this will help in determining your unit of analysis. If you encounter difficulty in determining the unit of analysis, then perhaps your research questions are too vague or too numerous. By clearly defining your case, you can continue to clarify your unit of analysis and distinguish those persons or organizations to be included as part of the immediate topic of the case study. This focusing process demonstrates that the process of designing a case study is not necessarily a clear-cut linear process, but rather one that may involve revisiting previously listed steps.

After you have clearly developed your research questions and determined the appropriate unit of analysis, the next step is to logically link the propositions to the data collected. By having an idea of how the evidence is to be analyzed, you can more easily prioritize your analytic strategy by defining and testing rival explanations. Being aware of rival explanations prior to data collection allows you to determine if there is evidence about other possible influences. Considering alternative patterns and investigating possible alternative explanations is a crucial part of the research-design process. It allows you to pretest your research questions and decrease the likelihood you are only looking for data to support your original hypothesis. Patton (1990) describes this logical linkage process as decreasing a researcher's susceptibility to stacking the deck by pursuing other influences or logical explanations.

The final step is to clearly articulate the criteria that will be utilized in interpreting the study's findings. What criteria will be used in comparing the rival propositions developed and determining which proposition best matches or explains the collected data? As will be detailed in the next section, this process of evaluating findings involves all of the previously discussed steps. As Yin (2003b) details, it is important for you, as a researcher, to develop these criteria a priori, since both your research design and findings will be subject to scrutiny by others. If you have not taken the time to evaluate your study's efficacy and develop tactics to answer the logical tests used to evaluate it, you put at risk the trustworthiness of your study's findings.

Evaluating Case Study Research Designs

When designing a case study, the researcher should consider some basic questions about how the case's design will be evaluated. By asking these questions a priori, the result should be a better study. So, what are some of these basic questions?

1. What are the logical tests appropriately utilized to judge the quality of a case study's design and data collection?
2. What are some case study tactics I can utilize to pass these tests?
3. When do I need to perform these tactics?

Researchers or those more familiar with quantitative research are familiar with the basic logical tests utilized to evaluate a research study's design and methodology: *reliability* and *validity*. However, one of the issues in qualitative research, of which case study research is a part, is the confusion caused by attempting to impose positivist or post positivist definitions of reliability and validity onto research studies that have adopted constructivist ontology. As a result, many qualitative researchers (Guba & Lincoln, 1989; Mertens, 2005) have proposed replacing quantitative standards, such as reliability and validity, with more alternative qualitative referents. With this in mind, in our discussion of standards for evaluating the quality of case study data collection, we will utilize alternative concepts: *dependability* (in lieu of reliability) and *credibility* (instead of validity).

Dependability

Since change is a constant in an interpretive paradigm, it makes little sense to demand stability over time, but it is still critical to minimize errors and biases in a case study. Another researcher should be able to repeat an earlier case study by following documented procedures. To address this need for operational dependability through documentation, Mertens (2005) suggested the performance of dependability audits by members of the community of interpreters to "attest to the quality and appropriateness of the inquiry process" (p. 351). Yin (2003b) offered more specific tactics to overcome external reviewers' suspicions about a case study's dependability, including the development and use of a case study protocol and a case study database.

Yin's (2003b) proposed case study protocol should have the following components:

- An overview of the project, including objectives, issues, and relevant readings.
- Field procedures (credentials, case study site access, sources of information, and procedural guidelines).
- Specific case study questions to guide data collection, table shells for specific data arrays, and potential information sources for answering each question. (A table shell is a table outline in which the rows and columns are clearly defined, but no actual data has been entered.)
- Case study report guide: Outline data format, use and presentation of documents, and reference material. (p. 69)

Because any case study report should contain enough data to allow readers to draw their own conclusions about the case study, the development of a case study database is an important element to increase the study's dependability. The four database elements described by Yin (2003b) are:

1. Case study notes from investigator's interviews, observations, or document analysis.
2. Relevant case study documents arranged in primary and secondary files with annotated bibliographies to assist in easy retrieval for later inspection.
3. Tabular materials, including any surveys, observational counts, archival data, or other quantitative data.
4. Narratives and open-ended answers to questions in the case study protocol. These answers should be answered by referring to relevant evidence (i.e., interviews, documents, observations, or archival evidence). (pp. 102–104)

Credibility

A discussion of a case study's credibility involves attempting to answer questions regarding the correspondence of the social constructs of case study subjects to the manner portrayed in a case study.

In other words, are the researcher's inferences that a particular event resulted from some earlier situation as accurate as possible if the previous event was not observed? Prolonged engagement and persistent observation are tactics that can increase a study's credibility. In addition, peer debriefing and discussions, and triangulation or crystallization of data sources, investigators, theories, and methods greatly enhance credibility. The use of multiple sources of evidence (e.g., field observations with diaries, numerous and extensive interviews, public records, personal files, and news articles) by multiple investigators may be a challenging proposition, but such triangulation/crystallization is invaluable in representing a case study's depth and establishing a study's credibility.

The third example of a case study is shown in Exhibit 4.3. It highlights complex legal, political, and public policy issues and utilizes triangulation/crystallization, peer debriefing and discussions, and the use of multiple sources of evidence. The case study's depth and use of multiple data sources, investigators, legal theories, and methodologies greatly enhances its credibility. While the specific case has largely been forgotten, the multiple perspective approach provides a template from which much can still be learned about the use of multiple sources and methodologies to provide depth and breadth, and insure triangulation/crystallization.

Conducting a Case Study

Now that we have discussed the tests used to determine a study's quality, it is important to address practical considerations and suggestions for actually conducting a case study. As with any project, proper preparation (including investigator skill assessment and improvement, and case study protocol development) is crucial. Part of our preparation process is making arrangements for conducting the study. Finally, guided by our discussions regarding dependability and credibility, it is important to outline and follow specified, but flexible, data collection procedures.

Preparing for Data Collection

Any researcher serious about conducting quality case study research should possess certain skills. Unfortunately, the requisite skills are not easily attainable, and no set method for their development is possible. In addition, it is difficult to assess a researcher's skill level, since there are no objective tests for case study researchers to pass before they can conduct case studies. However, a good case study researcher needs to possess and constantly work to improve certain skills. One of the most important skills is the ability to ask good questions and interpret answers. Consequently, a case study researcher must be an attentive and open-minded listener, and not allow theories, ideologies, or preconceptions limit what is heard. An open-minded attitude must allow for contradictory evidence. In addition, because each situation is unique, an adaptable and flexible researcher can turn a threatening situation into an opportunity for great data collection. Because reducing events and information to manageable proportions is critically necessary in case study research, an investigator should have an authoritative grasp of the issues being examined. If a researcher does not possess these skills, such skills can be developed. However, before these skills can be developed or enhanced, an honest assessment must take place.

Before any case study is undertaken, whether by a single investigator or an investigative team, background reading on the case's subject matter, dialogue regarding theoretical issues that may affect the study's design, and delineation of the study's data-collection methods and tactics must

Exhibit 4.3. Case Study Example 3—*The University of Minnesota v. Haskins*: The University of Minnesota Men's Basketball Academic Fraud Scandal

Analysis of Haskins' Employment Agreement with the University of Minnesota

Haskins and the University entered into a written contract effective July 1, 1992, for a term of 10 years. The contract was in full force and effect at the time the scandal arose. According to the terms of the agreement, Haskins was to "diligently and conscientiously devote his full time, attention and best efforts" to being head coach at the University of Minnesota. The agreement also required him to comply with the "laws, policies, rules, and regulations" of the University, as well as the "constitution, bylaws, and rules and regulations" of the NCAA and Big Ten Conference. Further, he agreed to "attempt to have all assistant men's basketball coaches and any other University employees for whom Haskins was administratively responsible comply with" the same laws, policies, rules, and regulations (Complaint...Exhibit A, 2000).

The compensation terms of the contract included a base salary and provisions for annual raises, to be recommended by the athletic director and approved by the University president. The annual raises would "equal or exceed" the prior year's base salary, plus "the University's average percentage increase in the salaries of its tenured faculty." Further, Haskins was eligible for competition and administrative performance bonuses, media compensation (guaranteed to be at least $85,000 per year), and normal University benefits. Finally, Haskins was allowed to accept commercial endorsements, including operating a summer basketball camp (Complaint...Exhibit A, 2000, section 1.4).

In addition to the compensation provisions, the contract also contained extensive provisions for termination of the contract prior to the expiration of its primary term, including termination both for just cause and without cause. The contractual conditions allowing termination for just cause included:

- a major violation of a NCAA or Big Ten rule involving Haskins,
- a major violation of such a rule by an assistant coach with knowledge,
- two separate major violations of such a rule by an assistant coach that Haskins "should have known about,"
- commission and conviction of a felony by Haskins,
- substantial failure to perform the duties included in the contract, and
- a subsequent secondary rules violation by Haskins, after the NCAA or Big Ten had warned the University that Haskins' conduct evidenced a lack of institutional control, based on his commission of multiple prior rules violations. (Complaint...Exhibit A, 2000)

Included in the for-just-cause termination section (Section 2.2 Deferred Compensation) was a detailed description of deferred compensation in the event that Haskins (1) was terminated by the University with or without cause, (2) died, or (3) became disabled. If the University chose to terminate Haskins' employment for just cause, Haskins would have been owed deferred compensation of $348,722.00 (prior to 7/1/1999) or $423,021.00 (after 7/1/1999).

The University also had the right to terminate Haskins' employment without cause by giving Haskins 90 days' notice of the termination. In the event that the University terminated him without cause, the University was required to pay Haskins his base salary, deferred compensation, media compensation, and University benefits for the remaining years left on the contract (Complaint...Exhibit A, 2000).

(*Continued*)

Exhibit 4.3. (*Continued*) Case Study Example 3—*The University of Minnesota v. Haskins*: The University of Minnesota Men's Basketball Academic Fraud Scandal

Legal Options and Remedies Available to the Board of Regents of the University of Minnesota

Faced with newspaper/website headlines such as "Academic Scandal Makes Haskins' Philosophy a Fraud" (Ledbetter, 1999) and calls by local and national sports editors/columnists for Haskins' firing or resignation, the Board of Regents was in a precarious legal position. At this point, there was no legally sufficient evidence to support a just cause termination under any of the contractual provisions for early termination. If the University relied entirely on media accounts, rumors, and hearsay to terminate Haskins for just cause, and did so without reliable proof, Haskins might file a breach of contract suit against the University. Further, from a practical standpoint, failure to prove the required conditions for a just cause termination would leave the University open to negative publicity, including the dampening effect on the hiring of a new coach, together with possible charges of racism.

On the other hand, if the University decided to terminate Haskins without cause, then it would be liable for the balance of the agreed compensation for the remaining years of the contract. Moreover, the University faced the same potential negative publicity. The University was between the proverbial rock and hard place; either apparent legal option was perilous.

June 24–25, 1999: Haskins Is Bought Out

Apparently recognizing their tenuous position, the Board of Regents approached Haskins, seeking a mutual face-saving resolution to the dilemma. Rather than terminating Haskins through either alternative, the Board of Regents negotiated a settlement with Haskins, and he stepped down from his coaching duties on June 30, 1999 ("U of M reaches," 1999). The University "for the good of the program ... [and] to restore public confidence" announced that a new men's basketball coach would be hired ("Mark G. Yudof statement," 1999, para. 2). Under the terms of the negotiated settlement, Haskins agreed to resign, in return for a lump-sum payment of $1,500,537.00 from the University. He also retained his faculty retirement plan and optional 403(b) retirement plan, received various transition services and expenses as needed for his departure, and received four (4) tickets to all University of Minnesota Men's home basketball and Big Ten Men's basketball tournament games (*University of Minnesota v. Haskins*, 2000).

In addition, the University and the Board of Regents agreed to the following: (1) release of all claims against Haskins in a nonadmission of fault disclaimer, (2) issuance of public statement(s) thanking Haskins for his years of service, and (3) assurances that any statements regarding Haskins' resignation simply reflect the University's desire to go in a different direction, not an admission of any wrongdoing on Haskins' part ("Official Public Statement," 1999). University administrators, bound by the Agreement and Release clauses, were not permitted to publicly proclaim that Haskins was fired. As a result, President Mark Yudof's public statements could not raise the possibility of Haskin's direct involvement in the scandal. Interestingly, President Yudof not only refrained from making any derogatory statements about Haskins' possible culpability in the scandal, but proclaimed:

> *He [Haskins] and his teams have provided many moments of entertainment and value for the university community and the citizens of the state of Minnesota... The university is grateful for the 13 years of service Coach Haskins has provided and we wish him a future of success and good health. ("Mark G. Yudof statement," 1999, para 3)*

(*Continued*)

Exhibit 4.3. (*Continued*) Case Study Example 3—*The University of Minnesota v. Haskins*: The University of Minnesota Men's Basketball Academic Fraud Scandal

In a further show of institutional support and recognition of Haskins' "parental influence upon his athletes," University of Minnesota Board of Regents member William Hogan noted, "Coach Haskins' and Yevette Haskins' contributions to the community and church in general will long be remembered as the family approach that tried so hard to provide a sense of love and discipline to the players, and at the same time build their spiritual lives" ("William E. Hogan statement," 1999, para. 2).

Source. From Southall, R. M., Nagel, M. S., Batista, P. J., & Reese, J. T. (2003). Originally published in *Journal of Legal Aspects of Sport*. Used with permission.

take place. Before data collection takes place, everyone involved in the study must know why the study is being conducted, what sort of evidence is being sought, and what will constitute supportive or contradictory evidence. This preparatory immersion allows team members to not only increase their expertise, but provides team members who may not have participated in the initial question-defining or research-design phases an opportunity to become familiar with the proposed research project. It may also uncover problems with the case study plan or the research team's abilities, reveal incompatibilities among team members, or expose unrealistic deadlines or expectations. However, such negative possibilities can most often be overcome by confronting them early in a project. In addition, focused training may uncover positive rapport among team members that suggests team pairings during data collection and allow for the development of investigation-team norms for data collection. Above all, such preparation will insure the next steps of making arrangements for the study and actual data collection are more easily accomplished.

Arrangements for the Study

The designation of teams arises out of the preparatory phase discussed above. After teams have been selected (based upon the strengths/weaknesses, abilities, and skills determined during the preparatory phase), all team members should have a solid understanding of the study's purpose. One way to determine each member's level of understanding is to have each person draft a brief project summary, from which the overview section for the eventual case study report may be developed. Having each team member draft a project overview prior to data collection serves to check the team's understanding of the purpose of the case study.

The eventual official study overview that emerges from these member-drafts should be part of an introductory packet sent to all prospective case subjects (e.g., individuals or organizations). This introductory packet should include a formal letter that serves to introduce team member(s), briefly describes the nature of the study, and solicits participation and/or support for the proposed case study.

Case Selection and Conducting a Pilot Study

Case study research is a scientific endeavor. The research's purpose is to gain an understanding of the chosen situation or phenomena. For this reason, the specific case or cases chosen are expected to be representative of a population of cases. Due to the nature of case study research and its small sample sizes, random sampling is not warranted. Instead, purposive sampling is an appropriate methodology. A researcher may choose to study an extreme or unique case, or may choose from several possible cases. Cases may be selected based upon the site's willingness to participate and the richness of the available evidence. In addition, various documentary sources may provide background information that may lead to a chosen case.

In some instances, a pilot case study may be conducted to provide an opportunity to refine collection plans. The pilot study may be a convenient site, either geographically proximal or possessing an abundance of documents or data. The pilot study is not a pretest, but is formative in nature; it allows the researcher to develop relevant questions and refine the research design. The pilot study will likely be much broader and may not be nearly as focused as the actual case study. However, a pilot study allows insight into conceptual and methodological issues, as well as the issues being studied.

Collecting the Evidence

Although everything that has preceded the actual collection of evidence is crucial to a successful case, no design will guarantee success. The case study data collection process is a complex one. This complexity requires that a researcher possesses flexibility and versatility, but there is also a need for methodological consistency and formalized procedures to insure dependability and credibility during data collection.

In order to collect case study evidence, it is important that a researcher be aware of the basic sources of such evidence. Although the richness and variety of evidence is a characteristic of case study research, there are several easily identified sources of evidence (Yin, 2003b). These sources:

1. Documents: letters, memos, agendas, minutes of meetings, proposals, formal studies, newspapers, mass-media articles, Internet blogs, etc.
2. Archival records: service records, organizational records, maps and charts, databases, survey data, and personal records.
3. Interviews: open-ended, but guided, conversations with key individuals.
4. Direct observation: formal to casual data collection opportunities during site visits.
5. Participant observation: evidence collected by researcher while the researcher fulfills a designated real-life, on-site role.
6. Physical or cultural artifacts: tools, instruments, works of art, computer printouts, or technological devices.

Principles of Case Study Data Collection

In order to make the most effective use of the data collected from the identified evidence sources and to increase the data's dependability and credibility, there are a few basic principles that should be followed. These principles are applicable to other qualitative research methods, but our discussion will focus on their applicability to case study research. The data-collection principles:

- using multiple sources of evidence (i.e., triangulation or crystallization),
- creation of a case study database, and
- maintenance of a chain-of-evidence, which allows for a confirmability audit. (Denzin, 1989; Janesick; 2000; Mertens, 2005; Yin, 2003b)

Triangulation/Crystallization

By its very nature, one of the strengths of case study research is the opportunity to access and uncover a wide variety of evidence sources. If multiple sources of evidence are not utilized in developing a case, the effectiveness of the case is often dramatically diminished. Because each investigator should also be viewed as a source of evidence, the use of investigative teams is also desirable. The concepts of triangulation and the more recently developed crystallization offer a methodological rationale for multiple sources of evidence. Denzin (1978, 1989) identified four types of triangulation:

1. *Data triangulation.* Utilizing a variety of data sources;
2. *Investigator triangulation.* Using several different researchers;
3. *Theory triangulation.* Employing multiple perspectives to interpret a single set of data; and
4. *Methodological triangulation.* Drawing upon a variety of methods to investigate a problem.

Richardson (2000) agreed with the need for a multiplicity of data sources and investigative perspectives, but she proposed that the triangle, "a rigid, fixed, two-dimensional object," (p. 934) inadequately imagines a fixed point or object to be investigated and triangulated. Richardson contended that there are many more than three sides to every story. A crystal grows, changes, and alters, but is not an amorphous entity. It is a prism "that reflects externalities and refracts within themselves...What we see depends upon our angle of repose...[W]e have moved from plane geometry to light theory, where light can be both waves and particles" (p. 934) at the same time and for two different observers.

Both triangulation and crystallization deconstruct the notion of validity and, within the postmodernist/constructivist paradigm, propose a complex, varied, and thorough data-collection process. However, within this paradigm, our understanding of the case, though deep, will still be partial. Although we will know more, we will doubt what we have learned. In addition, we will recognize there is much more to be learned.

Case Study Database

In drafting a case study report, the researcher presents a synopsis of the collected data, analyses of the data, and conclusions drawn from the analyses. In contrast to experimental or survey research, it is quite common and acceptable for a case study researcher to perform data collection and data analysis at the same time. It is possible for the researcher to present the collected data as part of a developed narrative in the case report. However, in order to allow readers to draw their own conclusions based upon an inspection of the collected data, a researcher should develop a distinct case study database. Developing and maintaining a formal, presentable database allows other investigators, theoretically, to review the collected evidence directly and not be limited to the case report.

Yin (2003b) outlined the four minimum components of a case study database: field notes, documents, tables, and narratives. Field notes may include interviews, observations, or content/document analysis. They may be handwritten, typed, or audio/video tapes. Retaining all documents collected during a case study may be problematic, but with modern electronic storage capabilities it is much more practical today. If it is determined that some documents will not be retained, an annotated bibliography of all documents should be created. Any tabular material collected on-site or created by the research team that was not included in the case study report should be organized and retained for later use. Finally, narratives, both questions and answers, should be considered part of the formal case study database and retained. By developing and maintaining such a database, the study's credibility, as well as the ability to conduct future projects, will be greatly enhanced.

Chain of Evidence

Maintaining a chain of evidence increases a case study's informational dependability by allowing an observer (i.e., the reader of the case study) to follow the entire research process, from the development of research questions to case study conclusions. The concept is analogous to the care of evidence during a police investigation. Detectives and other forensic personnel collect evidence, and it is placed under police control in an evidence room until being produced at trial. In other words, the case study report must contain the same evidence collected during the investigation. In addition, the report must contain sufficient citation to the specific elements of the developed case study database (i.e., documents, interviews). Moving back through the chain of evidence, if the database should ever be examined by an external auditor, it should contain the actual evidence. Finally, the database should also contain a description of the time and location of data collection, and this should be consistent with what is reported in the case report. If a consistent chain of evidence is maintained, an external reviewer will be able to ascertain and follow the entire case study process, and see the connection between procedures and collection of evidence.

Case Study Analysis

After case study evidence has been gathered, it is important for a case study researcher to be able to analyze the collected data. Yin (2003b) outlines three basic and general strategies (listed from most to least preferable) for analyzing case studies: (1) relying on general propositions, (2) using rival explanations to develop an alternative framework from the one initially developed, and (3) developing a descriptive framework for organizing the data. During the initial phase of designing the case study, theoretical propositions—reflecting the study's research questions, review of literature, and newly developed hypotheses—are developed. These propositions guide data collection and, therefore, should be used to prioritize the analytic strategies. Theoretical propositions help the researcher determine which data to focus upon and which to ignore. The strategy of using rival explanations utilizes rival hypotheses that might have been included in original theoretical propositions. Developing rival explanations should have been part of the case study design process and rival explanations should have been vigorously pursued during data collection and utilized again in the analysis stage. (Remember, we previously noted that data collection and analysis may often occur simultaneously in case study research.) A final analysis strategy is to develop a case description that can be used to organize the case study. Such a descriptive approach may help identify the causal links to be analyzed. Utilizing one or all of these analytic strategies helps guide the case study's analysis and help define and prioritize what to analyze and why to analyze it.

After a strategy (or combination of strategies) has been chosen by which a case will be analyzed, there are several analytic techniques that case study researchers have developed. These techniques are not simply statistical techniques that can be performed on data; they are introspective processes that will take time and practice to master. Once mastered, however, they can be used to perform powerful analyses resulting in relevant and forceful case studies.

Campbell (1975) and Yin (2003b) proposed the use of time-series or other pattern matching logics to compare an empirically based pattern with a predicted pattern or series of patterns. The time series pattern can be simple, involving only one dependent and one independent variable, or it can be more complex, with two different patterns of events having been hypothesized and compared to the collected evidence. Other pattern matching techniques may utilize rival explanations involving patterns of mutually exclusive independent variables. In other words, if one explanation is valid, the other ones cannot be.

A specific type of pattern matching proposed by Yin (2003b) is explanation building. The goal of this technique is to build an explanation of the case by stipulating a presumed set of causal links and a resulting series of iterations.

- Making an initial theoretical statement or initial proposition about policy or social behavior.
- Comparing the findings of an initial case against a statement or proposition.
- Revising the statement or proposition.
- Comparing the details of the case against the revision.
- Comparing the revision to the facts of a second, third, or more cases.
- Repeating the process as many times as needed. (pp. 121–122)

Another technique that is often identified as a type of time-series analysis is compiling a chronology of events. Such a compilation of events is not simply a descriptive technique. Arraying events in a chronological fashion can be used to investigate presumed causal events. The purpose of this technique is to compare the empirical chronology with that predicted by a proposed theory. A chronology can be utilized in analysis, if the explanatory theory has specified one or more of the following conditions:

- Some events must always occur before other events, with the reverse sequence being impossible.
- Some events must always be followed by other events, on a contingency basis.
- Some events can only follow other events after a prespecified interval of time.
- Certain time periods in a case study may be marked by classes of events that differ substantially from those of other time periods. (Yin, 2003b, p. 126)

Case analysis is the most difficult and complex phase in case study research. Analysis techniques cannot be robotically performed; they require practice and are fraught with challenges. If a researcher can begin by designing and conducting simple cases, more complex cases can be tackled as more experience and insight is gained.

Presenting a Case Study

Presenting a case's results is the most critical phase of case study research. Results must be organized so not only is the underlying theory either confirmed or rejected, but the reader is able to find answers to questions of concern. Fundamentally, a case study presents a series of answers to such questions. Questions may be organized thematically or chronologically, and may reflect the interests and concerns of various parties, as well as the researcher.

A case study's organization should allow for inclusion of important contextual and explanatory information and descriptions of the organization(s) and/or individual(s). Since presenting results in a mechanical fashion may preclude in-depth descriptions and a more fluid and creative reporting structure, a researcher must balance the ease of writing a case following a template with the need to provide a richer analysis. In addition, a researcher who does not believe in the expressed dichotomy between qualitative and quantitative research can utilize both types of data in presenting the case study.

The case study examples in this chapter demonstrate various methods of organizing the presentation of a case study. However, while each example's presentation protocol may vary, it is important to note the consistency within each case. Such consistency should extend to the use of parallel procedures and methods within each case study, especially a multi-case study with multiple investigators, and is critical to design and conduct a case study. The protocol should be readily apparent in the study's presentation.

In writing or presenting a case study's results, the researcher should refer to the study's protocol, which itemizes or details questions or issues that should have been addressed by the case study investigator(s) and the procedures that were followed throughout the field investigations. As the researcher's role changes from investigator to author/presenter, the researcher can obtain data to illustrate the protocol's questions from a variety of sources: documents, direct observations, interviews, etc. As a result, the topics or questions in a well-designed protocol can serve as an outline for the case study report.

Stylistically, determining how to present a case study may involve asking several questions. Stake (2003) suggested several questions to be asked before developing a case study report:

- Should the report be a story? How much of a story should it be?
- Is it effective/relevant to compare this case with others?
- Should readers be left to draw their own generalizations, or should they be formalized in the presentation?
- How much description of organization(s) or individual(s) should be included in the report?
- How personal or anonymous should the report be?

As Stake (2003) noted, "Single or few cases are poor representation of a population of cases and questionable grounds for advancing grand generalization" (p. 156). However, he also correctly pointed out, "The purpose of a case report is not to represent the world, but to represent the case" (p. 156). Therefore, in presenting a case, the researcher must do their best to present a cogent and particularized experience for readers.

Evaluating the Quality of Case Study Reports

Although researchers and theorists may vary in their preferences for utilizing case study research, there are some basic guidelines that can be utilized for evaluating the efficacy of a case study report. Whether the case study method is the only research design utilized or is part of a multi-method research approach, there are certain areas that can be reviewed. These include the report's design, the data collection, the database management and analysis techniques, the reporting itself, and the level of impartiality and generalizability (Datta, 1997).

Many of the guidelines for evaluating a case study have already been covered in sections in this chapter. However, it is a great idea to develop a checklist that contains several questions for each area as a guide for evaluating other researchers' case study reports, as well as a tool for self-evaluation of one's own ongoing research.

Based upon several sources (Datta, 1997; Janesick 2000; Stake, 2003; and Yin, 2003b), some possible questions for such a checklist are provided.

- Are the evaluation questions clearly and explicitly stated?
- How was the case selected? Is the basis for this selection presented?
- Are the data collection methods discussed fully?
- Are the data sources clearly and completely described?
- Are the data collection methods appropriate for the case and its purpose?
- If there are multiple investigators, are they all properly trained, and is such training clearly delineated?
- Were there any differences in evaluators' interpretations? If so, were these differences resolved?
- Is adequate information provided to allow the reader to judge the generalizability of the conclusions?
- Does the report discuss the limitations of the study?

Concerns with Case Study Research

Because a case is, by definition, a particular instance, expressed concerns associated with the case study method involve not only the degree to which a particular case can be used to make generalizations about existence, but also the extent to which case study research exposes both researchers and those researched to more risk than other research methodologies. The elements of a particular case that make it a distinct occurrence must be balanced against the case's ability to allow generalizations to be drawn from it. Many academic researchers (Denzin, 1989; Stake, 2003; Yin 2003b) have noted concerns that case study research may be more appropriate as an exploratory means to lay the groundwork for studies that offer more generalizability. Because many case study researchers are intrinsically interested in the subject(s) of the study, it is critical that what is important about the case within its own world can be effectively translated or communicated to others.

Case study researchers face difficult decisions throughout the research process. They must decide how much and for how long a complex case should be studied. The case must be sufficiently described so that readers can experience and understand the situation and draw their own conclusions; however, not everything about a case can or must be studied. It is possible for a researcher to become either too focused on uncovering everything about the case being studied and lose sight of possible generalizable elements, or too committed to generalizing or theorizing that his/her attention is drawn away from the case's uniqueness.

Summary

This balancing act is part of what makes case study research so challenging. The various contexts in which the researcher may be placed and the varieties of evidence derived from diverse data collection points and techniques with which the researcher may be forced to grapple make case study research unique. In addition, since many sport management researchers or students have limited social-science training, and are therefore unfamiliar with just what constitutes a case study, they are sometimes intimidated and reluctant to utilize the methodology in their research. In addition, while other researchers may not be timid about utilizing case study methods, they may view it as simply exploratory and subservient to other forms of inquiry.

It is important to remember that investigating individuals, groups, or organizations, as well as other social and political phenomena, occurs within a social context. In order to understand such entities, it is necessary to uncover not only what constitutes them, but also how and why such social interaction occurs within these socially constructed realities. As has been detailed in this chapter, case study research encourages our understanding of how and why this takes place.

In addition, case study research has the ability to serve an important refining or tempering role. Although a generalizable theory may be more easily realized through utilizing another research methodology, case studies have the ability to reveal the limitations of such generalizability by highlighting the need for further theoretical refinement and/or additional in-depth, specific investigation. Perhaps case studies serve their highest purpose as reflections of the human experience and their ability to speak for and to those whose personal cases may have gotten lost in a statistical shuffle. While not discounting the usefulness of a test for significance, it is also imperative to remember that such a test is "intended to assess the evidence provided for data against a null hypothesis H_0 in favor of an alternative hypothesis H_a" (Moore & McCabe, 1989, p. 478). Significance, in a statistical sense, does not mean important. It is up to each of us individually, and all of us collectively, to determine the importance of our research.

Study Questions

1. There were several definitions of a case study offered in this chapter. List the similarities and differences of the various definitions.
2. Can you describe each of the following types of case studies?
   - Exploratory
   - Descriptive
   - Explanatory
   - Intrinsic
   - Instrumental
   - Collective
3. What are the five steps in designing a case study? Explain.
4. What are the two standards useful for evaluating the quality of case study data collection? How do these two concepts differ from the logical tests (reliability and validity) utilized to evaluate quantitative research?
5. Why is the concept of triangulation/crystallization so crucial to a case's research effectiveness? Identify at least four types of triangulation/crystallization.
6. What are several of the strengths, as well as some of the limitations, of case study research?
7. What are three basic strategies useful for analyzing case studies? Which strategy do you find preferable? Why?

Learning Activities

1. Utilizing an Internet search engine, type in case study and case studies in sport management. Browse the results and evaluate any appropriate case studies using the strategies discussed in this chapter.
2. Obtain a recent newspaper, magazine, or trade journal and find a recent sport-industry story that involves a managerial, political or policy issue. After identifying the issue from this one news account, find at least five additional accounts from different sources/authors. In addition, obtain additional information from the sports organization itself (website, personal communication, press release) regarding the situation. Using this information, develop a course of action to address the situation that you would suggest if the organization were to retain you as a consultant.

References

Amis, J. (2005). Interviewing for case study research. *Qualitative methods in sports studies*, 104–138.

Blalock, H. M., & Wilken, P. H. (1979). *Intergroup processes: A micro-macro perspective.* New York: Free Press.

Cable News Network. (2003, October 13). Lawsuit targets NFL. CNN.com. Retrieved from http://cnnstudentnews.cnn.com/ TRANSCRIPTS/0310/13/ltm. 14.html.

Campbell, D. T. (1975). Degrees of freedom and case studies. *Comparative Political Studies, 8*, 178–193.

Carroll, A. B. (1979). A three-dimensional conceptual model of corporate performance. *Academy of management review*, *4*(4), 497–505.

Carroll, A. B. (1991). The pyramid of corporate social responsibility: Toward the moral management of organizational stakeholders. *Business Horizons*, *34*(4), 39–48.

Charmaz, K. (2003). Grounded theory: Objectivist and constructivist methods. In N. K. Denzin & Y. S. Lincoln (Eds.), *Strategies of qualitative inquiry* (2nd ed., pp. 249–291). Thousand Oaks, CA: Sage Publications.

Coffey, W. (2005, January 29). Wasted innocence. New York Daily News.com. Retrieved from http://www.nydailynews.com/sports/

Complaint, Board of Regents of the University of Minnesota vs. Clem Smith Haskins, Case No _ , Exhibit A: Employment Agreement between Clem Smith Haskins and Regents of the University of Minnesota, dated February 11, 1994 (D.C. Minn. 2000) (hereinafter "Employment Agreement").

Datta, L. (1997). Multimethod evaluations: Using case studies together with other methods. In E. Chelimsky & W. R. Shadish (Eds.), *Evaluation for the 21st century: A handbook* (pp. 344–359). Thousand Oaks, CA: Sage Publications.

Denzin, N. K. (1978). *The research act: A theoretical introduction to sociological methods.* New York, NY: McGraw-Hill.

Denzin, N. K. (1989). *Interpretive interactionism.* Newbury Park, CA: Sage Publications.

Dvorchak, R. (2005, April 3). $135 million jury award forces new look at high costs of sports and drinking. Post-gazette.com Retrieved from http://www.post-gazette.com/pg/05093/481516.stm

Eisenhardt, K. M. (1989). Building theories from case study research. *Academy of Management Review*, *14*(4), 532–550.

Gaudiano, N. (2002). Crusade against DWI: Young girl inspires Heights to tackle big problem. *The Record (Bergen County, NJ).* A01.

Glaser, B. G., & Strauss, A. L. (1967). *The discovery of grounded theory: Strategies for qualitative research.* Chicago: Aldine.

Guba, E.G., & Lincoln, Y. S. (1989). *Fourth generation evaluation*. Newbury Park, CA: Sage Publications.

Hamel, J., Dufour, S., & Fortin, D. (1993). *Case study methods*. Qualitative research methods series: Vol. 32. Newbury Park, CA: Sage Publications.

Howard, D., & Crompton, J. (2004). Financing sport (2nd ed.). Morgantown, WV: Fitness Information Technology, Inc.

Janesick, V. J. (2000). The choreography of qualitative research design: Minuets, improvisations, and crystallization. In N. K. Denzin & Y. S. Lincoln (Eds.), *Handbook of qualitative research* (2nd ed., pp. 379–399). Thousand Oaks, CA: Sage Publications.

Kaplan, A. (1998). *The conduct of inquiry: Methodology for behavioral science*. New York: Oxford University Press.

Kuzma, J. R., Shanklin, W. L., & McCally, J. F. (1993). Number one principle for sporting events seeking corporate sponsors: Meet benefactor's objectives. *Sport Marketing Quarterly*, *2*(3), 27–32.

LeClair, D. T., & Ferrell, L. (2000). The role of strategic philanthropy in marketing strategy. Retrieved from http://www.ebusinessethics.com/lf/strategic.html

Ledbetter, D. 0. (1999, March 28). Academic scandal makes Haskins' philosophy a fraud. Milwaukee Journal Sentinel. Retrieved from http://www.jsonline.com/sports/sday2/led32899.asp

McCarville, R. E., & Copeland, R. P. (1994). Understanding sport sponsorship through exchange theory. *Journal of Sport Management*, *8*(2), 102–114.

Mertens, D. M. (2005). *Research and evaluation in education and psychology: Integrating diversity with quantitative, qualitative, and mixed methods* (2nd ed.). Thousand Oaks, CA: Sage Publications.

Merton, R. K. (1967). *On theoretical sociology*. New York: Free Press.

Moore, D. S., & McCabe, G. P. (1989). *Introduction to the practice of statistics*. New York: W. H. Freeman and Company.

NFL 1999 Season Archives (n.d.). 1999 NFL scores and schedules. Retrieved September 28, 2004, from http://home.earthlink.net/-oblgui/nflsbar/ nflsch99.htm

Patton, M. Q. (1990). *Qualitative evaluation and research methods* (2nd ed.). Newbury Park, CA: Sage Publications.

Polonsky, M. J., & Macdonald, E. K. (2000). Exploring the link between cause-related marketing and brand building. *International Journal of Nonprofit and Voluntary Sector Marketing*, *5*, 46–57.

Pringle, H., & Thompson, W. (2001). *Brand spirit: How cause related marketing builds brands*. Indianapolis, IN: Wiley Publishing, Inc.

Richardson, L. (2000). Writing: A method of inquiry. In N. K. Denzin & Y. S. Lincoln (Eds.), *Handbook of qualitative research* (2nd ed., pp. 923–948). Thousand Oaks, CA: Sage Publications.

Schwandt, T. A. (2000). Three epistemological stances for qualitative inquiry: Interpretivism, hermeneutics, and social constructivism. In N. K. Denzin & Y. S. Lincoln (Eds.), *Handbook of qualitative research* (2nd ed., pp. 189–214). Thousand Oaks, CA: Sage Publications.

Shavelson, R., & Townes, L. (2002). *Scientific research in education*. Washington, D.C.: National Academies Press.

Sixth amended complaint and jury demand, Verni v. Lanzaro, Docket No. BERL-10488-00 (Superior Ct. N.J. Law Div. Bergen County, Oct. 16, 2003).

Southall, R. M., Nagel, M. S., Batista, P. J., & Reese, J. T. (2003). The Board of Regents of the University of Minnesota v. Haskins: The University of Minnesota men's basketball academic fraud scandal: A case study. *Journal of Legal Aspects of Sport*, *13*(2), 121–144.

Southall, R. M., Nagel, M. S., & LeGrande, D. (2005). Build it and they will come? The Women's United Soccer Association: A collision of exchange theory and strategic philanthropy. *Sport Marketing Quarterly*, *14*(2), 158–167.

Southall, R. M., & Sharp, L. A. (2006). The National Football League and its "culture of intoxication:" A negligent marketing analysis of Verni v. Lanzaro. *Journal of Legal Aspects of Sport*, *16*(1), 101–127.

Stake, R. E. (2003). Case studies. In N. K. Denzin & Y. S. Lincoln (Eds.), *Strategies of qualitative inquiry* (2nd ed., pp. 134–164). Thousand Oaks, CA: Sage Publications.

Statement: Mark G. Yudof, President, University of Minnesota. (1999, June 25). University News Service. Retrieved from http://www 1 .umn.edu/urelate/newsservice/newsreleases/99_7yudofstate.html (hereinafter" Mark G. Yudof Statement").

Statement: Official Public Statement of the University of Minnesota June 25, 1999. (1999, June 25). University News Service. Retrieved from http://wwwl.umn.edu/urelate/newsservice/newsreleases/99_7publicstate.html (hereinafter "Official Public Statement").

Statement: William E. Hogan II, Regent, University of Minnesota. (1999, June 25). University News Service. Retrieved from http://wwwl .umn.edu/urelate/newsservice/newsreleases/99_7hoganstate.htm 1 (hereinafter" William E. Hogan Statement").

Strauss, A. L., & Corbin, J. (1990). *Basics of qualitative research: Grounded theory procedures and techniques.* Newbury Park, CA: Sage Publications.

Stotlar, D. K. (2001). *Developing successful sport sponsorship plans.* Morgantown, WV: Fitness Information Technology, Inc.

Sutton, R. I., & Staw, B. W. (1995). What theory is not. *Administrative Science Quarterly*, *40*, 371–384.

Theory. (2002). *Merriam-Webster's medical desk dictionary* (rev. ed.). Springfield, MA: Merriam-Webster, Incorporated.

U of M reaches contract agreement with men's basketball coach Clem Haskins. (1999, June 24). University News Service. Retrieved from http://wwwl.umn.edu/urelate/datebook/1999/99_7haskins.html

U. S. General Accounting Office (GAO). (1990, November). *Case study evaluations.* Transfer Paper 10.1.9. Retrieved from http://archive.gao.gov/f0202/143145.pdf

Weick, K. E. (1989). Theory construction as disciplined imagination. *Academy of Management Review*, *14*(4), 516–531.

Whetten, D. A. (1989). What constitutes a theoretical contribution? *Academy of Management Review*, *14*(4), 490–495.

Yin, R. K. (2003a). *Applications of case study research* (2nd ed.). Applied social research methods series: Vol. 34. Thousand Oaks, CA: Sage Publications.

Yin, R. K. (2003b). *Case study research: Design and methods* (3rd ed.). Applied social research methods series: Vol. 5. Thousand Oaks, CA: Sage Publications.

5

Interview Research

After completing the study of this chapter, you should be able to:

- Understand the research objectives of interviews as a qualitative technique.
- Appreciate the factors affecting the use of interviews in problem solving.
- Understand the differences in types and formats of interviews.
- Identify an appropriate setting to conduct interviews based on the research objectives.
- Understand the roles of interviewers.
- Understand how to train interviewers.
- Follow the proper procedure to conduct interviews.
- Appreciate the advantages and disadvantages of interviews.

As one of the qualitative research methods, *interviews* frequently have been used by sport management researchers to provide in-depth information about a particular research issue or question through gathering the needed information from a small number of participants. For example, in a study to examine "corporate social responsibility" (CSR) as an important issue in sport businesses, Djaballah (2017) conducted interviews with nine organizers and stakeholders of the Quiksilver Pro France event—a professional surfing tournament and one round of the World Surfing Championship (WSC). Held every year since 2002, this contest involves 48 of the world's top surfers. Because the Quiksilver Pro France event and organization have committed to corporate social responsibility activities since the mid-2000s, the purpose of this study was to explore the organizers and stakeholders thinking, reasons, and perceptions of the development of CSR and its successes and failures since its inception.

The purpose of this chapter is to introduce how sport management professionals can use interviews as a valuable technique in solving their research problems. Specifically, a number of issues related to the use of the interview technique, such as the objectives of interviews and the types and formats of interviews will be discussed first. In the latter part of the chapter, other aspects of the interview technique, including the place to conduct interviews, the roles of interviewers, the training of interviewers, and the procedure to conduct interviews, will also be reviewed. The advantages and disadvantages of interviews will be assessed in the last part of the chapter.

The Nature of the Interview as a Qualitative Research Technique

The interview is a qualitative research method that involves a planned question and answer discussion session with an individual to collect information. It can be a challenging method, but it is a rewarding form of measurement. The interview is a common method of collecting research data in sport management. It is a traditional form of collecting information in market research. It is used less often in academic research in sport business, but could be used more often.

The interview can be conducted using the in-person, digital video, or conducted by telephone. The in-person interview is used more frequently because it has a number of advantages over the telephone method. In a personal interview, the interviewer can gain a better understanding of the interviewee because the interviewer can watch for body language, voice inflection, changes in eye contact, and other observations that can only be made with an in-person interview.

Conducting an interview might seem an easy task. How hard can it be? Just ask some questions and record the responses, right? Wrong! Interviewing for research purposes is not so simple. The purpose of the interview, the questions posed, the discussion encouraged, and even the interaction of the interviewer with the interviewee can affect the quality of information collected (Berstell & Nitterhouse, 2001). It is also important that interviewers be trained to recognize their own biases so that recording responses can be done without bias.

Additionally, it must be understood that the interview is also a form of survey research (Kahan, 1990), which means it can be used as a quantitative research tool, or used alongside the survey questions for a mixed-method approach. A set of questions about a particular topic is developed in order to collect information from subjects. A survey is developed in a specific manner, and so is the questionnaire that will guide the interview. Detailed discussion on personal interviews as a survey technique is provided in the chapter on survey research.

Objectives of the Interview

The primary objective of the interview is to get unrestricted comments and opinions of the subject's thoughts or why they behave a certain way. This information will help the sport management researcher get an in-depth understanding of these opinions and behaviors and the underlying reasons for them. For instance, Wagstaff, Hanton, and Cardiff (2007) interviewed 10 athletes for the sake of determining how individuals actually cope with organizational strains or stressors and their responses to them. The results of the interviews allowed the researchers to build a model of the athletes' stress story using a diagnostic approach.

It is very important to compile information and identify common themes among the opinions and behaviors. These themes will help the researcher understand consumer thought and behavior that can be used in decision-making about products, the company, pricing, and more.

Exhibit 5.1. The Continuum of Qualitative-Quantitative Interview Method

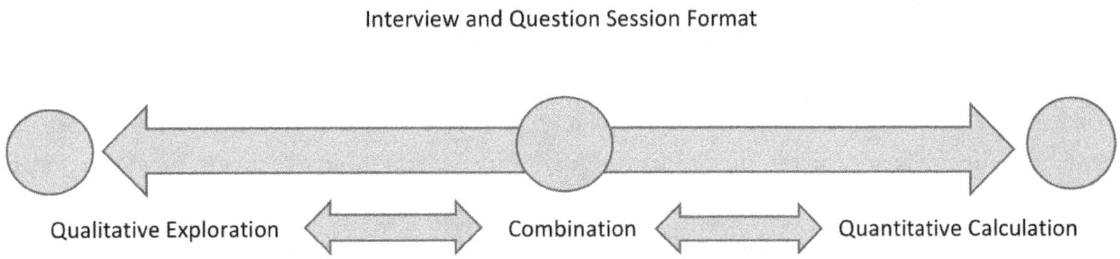

Factors Affecting the Use of Interviews

The sport management researcher must decide when it is best to use the interview method as compared to using any other method. The interview is a versatile method, but it requires careful planning, preparation, and execution. The interview method provides an opportunity for the researcher and the subject to interact and to carry on a conversation, whereas survey or other research methods are primarily one-way communication from the subject to the sport management researcher. Using the interview method is best when in-depth discussion and information are needed. The interview is most appropriate in research when certain situations are to be considered. According to Hague, Hague, and Morgan (2004), these situations include when

- the research issues are not clear,
- issues, opinions, attitudes, and motivations need in-depth exploration,
- processes and other data need to be described in detail,
- detailed explanation is needed,
- detailed understanding is needed,
- product testing is needed,
- personal reaction is needed,
- respondents are scattered and not available for group research,
- a subject's story needs to be followed from beginning to end,
- extensive comment and discussion is needed from the subjects, and/or
- the topic is sensitive.

Interviewing is usually a question-and-answer session involving a planned set of questions designed to elicit information from the respondent. Another factor to consider is what kind of information is desired—qualitative or quantitative. If the research project is planned to gather opinion, insights, attitude, perception, values, and other such exploratory information, the interview questions and the session can be loosely structured and flexible enough to provide freedom to explore the topics without limitation. However, if the sport management researcher wants to measure and quantify the to-be-collected information, the interview should be structured with a formal and restrictive set of questions designed to gather quantitative data.

Exhibit 5.1 provides an illustration of the continuum of different types of interviews. According to the continuum, if the research project need is to explore a topic through discussion and discourse, then the interview questionnaire is prepared with the flexibility for conversation and discussion of that topic. On the other hand, if the research project need is for information to measure something, then the interview questionnaire is prepared with specific questions to elicit measurable responses.

Types and Formats of Interviews

There are three primary types of interviews: in-person, telephone, and digital video call. Additionally, there are different formats of interviews that can be structured and conducted. These include open-ended, fill-in-the-blank, binary choice, scaled-response, and unscaled-response interview format. Some of the formats are mainly for collecting qualitative information; others are for a quantitative purpose. The sport management researcher must determine what type and what format will be used. This decision will affect the type of information the respondent will provide as well as the possible alternative responses. Depending on the type of information desired, the interview format will generally have a combination of all of these formats.

An *open-ended question* is one in which the respondent is asked to make a response with no limitations or restrictions on the response. This allows for a spontaneous, unstructured response. However, the question should be tightly focused to elicit the kind of information that the sport management researcher needs. After the researcher has asked the question, that individual should be ready to record the response as accurately as possible. Sometimes, an audio recording of the interview is used so that the researcher can focus on the question and answer instead of having to be burdened with writing every response. The audio recording allows the researcher to record responses at a later date. Exhibit 5.2 provides a sample open-ended question interview questionnaire.

A *fill-in-the-blank question* offers a sentence with one or more blank spaces that the respondents need to fill in their answers. This question format offers a relatively restrictive method for question and answer and is good for obtaining information needs of short-answer or simple-answer responses. For instance, the purpose of the question might be to get the respondents' information about attending sports events at a facility, and the fill-in-the-blank sentence would be "I have attended {how many} sports events at this arena, and they included {list of events}." In the middle or at the end of each question, if it appears as though interviewees may be having trouble remembering what they have done, the interviewer may use probes or prompts—ways of getting respondents to recall experiences and other information. In the previous fill-in-the-blank question, if the interviewee begins to answer with "Well, now, let me see if I can remember that. I think I remember that I went to a basketball game, but I'm not sure if that was this year or last year." Then the interviewer may prompt with triggers such as, "Okay. Do you remember what teams were playing? Was it a pro game or a college game? Was it a tournament? Or was it just a regular season game? Was it men or women?" When the interviewee begins to recall with these prompts and offers more information, the interviewer can continue to prompt until it appears as though the respondent has finally opened the box of memories of those events and can offer more detailed information.

Binary question format means that there is only one of two possible answers. For example, the question would be obtaining a response of either yes or no, or true or false. This type of response provides clear-cut information of a brief answer nature. A sport organization might be collecting information about its experiences with staff. It could be looking for specific information related to staff training. For example, the Atlanta Arena has provided a full-day workshop for staff who work on event nights at the arena. The facility has trained the staff in specific behaviors desired so as to provide a welcoming atmosphere for every person who attends the event. Part of the training included requiring staff to make eye contact and greet each person who enters the arena for an event with a "Hello" and "How are you this evening?" In addition, the facility wants the staff to then ask

Exhibit 5.2. Sample of Personal Interview Questionnaire

Interview Questionnaire for Atlanta Arena

"Hello, my name is ________________. I am conducting a survey for the new Atlanta Arena on how the Arena can better serve your interests as a sports fan. As you know, the Atlanta Arena is being built right now; it will be brand new and is a privately owned facility that can stage a variety of sports events. We want to know what kinds of facility amenities you would like so that we can incorporate them into the facility. And we want to know what sports you like and would like to see in the Arena.
So, let me ask you some questions. Feel free to take as long as you want to answer them.

Question 1.
To start, did you know that the Atlanta Arena is being built? Yes____ No____
Do you know where it is being built? Yes____ No____ Where?
Do you like the area where it's being built? Yes___ No___ Why/why not?

____

____

How did you find out about it?____

____

Do you like this idea? Yes____ No____ Why? Or why not?

____

____

Question 2.
What do you know about the plans (what facilities are being included) for the Arena?

____

____

____

Do you think these are okay? Yes____ No____ Why/why not?

____

____

____

Is there something not in the plans that you would like to see? Yes____ No____
Explain. ____

____

____

individuals if they need any help with directions or in finding their seat in the arena. The facility also wants to know that the training of the concessions staff has worked. The concessions staff was trained to smile at, make eye contact with, and say thank you to every person at the counter.

To measure the training, the interview questionnaire would include questions such as the following, all of which are yes-no questions.

- Did Atlanta Arena staff greet you at the door with a hello? Yes or no.
- Did Atlanta Arena staff smile and ask if you needed help? Yes or no.
- Did Atlanta Arena staff answer any questions you had? Yes or no.
- Did you find the information to be helpful? Yes or no.

- Did you find the information to be correct? Yes or no.
- Did concessions staff greet you with a smile and a hello? Yes or no.
- Did they say thank you after you completed your order? Yes or no.

Although at first glance this might seem simple, it is designed to obtain information to answer specific research questions that the facility has. With the information from the interview, the company can determine if its training is working, or if it needs to reconsider what to do.

Scaled-response question format consists of a question for which a list of alternative responses increase or decrease in intensity in an ordered format. Often, these types of questions are called ranking questions because the question asks the respondent to rank something. Exhibit 5.3 shows how the scaled-response question responses are reported. This response contained a scale of 1 to 5. In this example, the interviewer would ask: "On a scale of 1 to 5, with 1 being 'least influence' and 5 being 'highest influence,' how would you respond to the following features about a graduate program in athletic training? The first item is 'identified points of distinctiveness' of the program. Does this have little influence on high influence for your decision to attend this program? Remember to answer with a number from 1 to 5" (Seegmiller, 2006).

An often-measured area of research for companies in the business of sports entertainment, such as professional sports, is consumer behavior. Specifically, these companies are in the business of selling sports as a spectator product—hence, spectator sports. Exhibit 5.4 provides an example of the questionnaire in scaled-response format often used to evaluate the factors that affect an individual's attendance at a sports event. The market researcher can then use this information to provide feedback to the company. The company will then use the information to make any needed changes to the product—in this case, the facility, concessions, or event.

In *unscaled-response question* format, the respondent is asked to choose one or more options from a list. For example, the interviewer will ask: "Which of the following would you be willing to pay for a ticket to the Paralympic Games Opening Ceremonies: $50.00; $100.00; $150.00; $200.00; or other?" In this format, the interviewer needs to include a possible answer of "other" so that the respondent doesn't feel forced to select one of the provided answers. If the interviewee answers "other," the interviewer should then ask what "other" would be. In this situation, the respondent can then feel comfortable to respond with the following example: "Well, I really think about $75.00 would be about right." With this, the interviewer can prompt with the follow-up, open-ended question, "Why do you think that amount is about right for this event?" This prompt will encourage the interviewee to explain his or her perspective about that amount being about the price for this particular event. This follow-up question will provide a look inside the thinking of the respondent. In this particular example, the interviewee now has the opportunity to explain why this amount is how the ticket ought to be priced.

Interview Settings

The interview may be conducted in a number of settings. The most common settings are a residence, the street, the mall, a venue, and at an office designated for the interview research.

Residence. Using the interviewee's residence as a place to conduct the interview has advantages and disadvantages. The interviewee will most likely be more comfortable because of the familiar surroundings. This setting encourages the interviewee to open up for discussion and to share thoughts and ideas that might not be as forthcoming in an unfamiliar setting. However, typical

Exhibit 5.3. Example of Reported Interview Questions Using Scaled-Response Format

Measures of Central Tendency in Rank Order by Mean

Item #	Contributor	Mean	+ SD
11	Adequate numbers of qualified faculty	3.86	0.39
9	Program director's strong academic orientation and interest in student professional preparation	3.75	0.46
4	Goals and objectives related to enhancing students' critical thinking skills	3.72	0.51
12	Adequate numbers of qualified athletic training staff and other allied health personnel	3.70	0.51
3	Goals and objectives related to increasing students' depth and breadth of understanding	3.64	0.58
13	Faculty and staff familiarity with program goals and objectives	3.61	0.56
17	Clinical experience	3.61	0.60
21	Ongoing program evaluation	3.59	0.58
22	Analysis of evaluation results with periodic program revision	3.59	0.56
6	Adequate administrative support	3.58	0.59
5	Providing students with advanced knowledge and skills to prepare them for leadership roles	3.52	0.61
20	System of evaluating student performance	3.46	0.59
7	Day-to-day oversight of program operation, coordination, supervision, and evaluation by the program director	3.45	0.61
1	Identified points of distinctiveness	3.43	0.53
19	Facilities and equipment	3.43	0.58
2	Long-term and short-term goals and objectives	3.39	0.62
18	Admission criteria of NATABOC certification and an earned baccalaureate degree	3.37	0.70
14	Courses leading to a research experience	3.27	0.70
8	Program director's NATABOC certification and 3 years of teaching and research experience	3.26	0.77
15	Research experiences designed to expand the body of knowledge in athletic training	3.25	0.68
10	Program director's ongoing involvement in athletic training research and other scholarly work	3.20	0.73
16	Sufficient time and opportunity for research	3.16	0.75

Source. Seegmiller, J.G. (2006). Perceptions of quality for graduate athletic training education. *Journal of Athletic Training, 41* (14), 415–421.

Exhibit 5.4. An Example of the Interview Questionnaire in Scaled-Response Format

Factors That Influence Tournament Attendance – Interview Questionnaire

1. Think about the many factors that influenced your decision to attend this event and please share your thoughts with me on a scale of **1 as No Influence,** to **5 as Strong Influence,** or **NA (Not Applicable).**

   Which of the following factors influenced your decision to attend the NCAA Women's Final Four Tournament? Please tell me what you think best reflects your opinion on a scale of **1 (No Influence)** to **5 (Strong Influence),** or **NA (Not Applicable).**
   (As the interviewee answers each question, circle that response on the sheet.)

1. Entertainment value of the games	1 2 3 4 5 NA
2. A chance to attend a championship tournament	1 2 3 4 5 NA
3. Availability of ticket(s) to purchase	1 2 3 4 5 NA
4. Price of ticket(s)	1 2 3 4 5 NA
5. Promotion of the event/media advertising (TV, radio, newspaper, Internet, etc.)	1 2 3 4 5 NA
6. Support of the teams playing in the tournament	1 2 3 4 5 NA
7. Dates games are scheduled	1 2 3 4 5 NA
8. Days of the week games are scheduled	1 2 3 4 5 NA
9. Tournament location/the City of Atlanta as a destination for this event	1 2 3 4 5 NA
10. Other activities taking place in the City	1 2 3 4 5 NA
11. The Georgia Dome as a venue	1 2 3 4 5 NA

2. What do you think about the Georgia Dome as a facility for this event? As I ask you about a particular item, rate it on a scale of **1 as Poor,** to **5 as Excellent,** or **NA (Not Applicable).**
   (As the interviewee answers each question, circle that response on the sheet.)

• Facility location	1 2 3 4 5 NA
• Directional signage	1 2 3 4 5 NA
• Professionalism/courtesy of staff	1 2 3 4 5 NA
• Food quality	1 2 3 4 5 NA
• Food and beverage service	1 2 3 4 5 NA
• Seating comfort	1 2 3 4 5 NA
• Sound	1 2 3 4 5 NA
• Content and clarity of video/message display boards	1 2 3 4 5 NA
• Stadium temperature	1 2 3 4 5 NA
• Housekeeping/building services	1 2 3 4 5 NA
• Facility security	1 2 3 4 5 NA
• Parking	1 2 3 4 5 NA

residence activities can be a distraction, such as the phone ringing, kids playing, people coming in and out curious about the visitor, and a need to make the interviewer comfortable. Another disadvantage is the travel involved for the sport management researcher, who must determine the address and location of the interviewee's residence, then hope that typical travel setbacks don't happen, such as car trouble, heavy traffic, or other complications that could make the interviewer late or even miss the appointment altogether.

Street. Conducting interviews on "the street" also has a set of advantages and disadvantages. One of the advantages is that it is a good method for getting the opinion of the "person-on-the-street." This is often seen on television news or human-interest programs in which the purpose of the research is to gauge opinions or gain insight from the average person. The street interview, however, has its disadvantages. One disadvantage is that there is no control for sampling. The sport management researcher is left to get interviews from whoever is willing to stop and take the time to be interviewed.

Mall. There are times that the purpose of research and the sample needed will be people in a mall setting. For instance, the sporting goods retailer with a store location in a mall might want to measure shoppers' reactions to mall signage. The researcher will organize to intercept individuals in the mall and ask them to be interviewed. This is typically called the "mall intercept" method and is also used in survey research. The "mall intercept" principle is commonly applied in a sport economic impact study when the sport management researchers are strategically located in various locations of a sports venue to intercept spectators randomly to conduct interviews. For example, Bernthal and Regan (2004) conducted an economic impact study of a NASCAR racetrack on a rural community and its surrounding region. More than 1,000 spectators were intercepted and interviewed at a numerous locations throughout the racetrack. Additional discussion on the mall intercept interview method can also be found in Chapter 9.

Venue. In the sport industry, there are a number of different types of sports venues at which research may take place. There are sports arenas, recreation areas, fitness facilities, gyms, sports clubs, golf courses, stadiums, tracks, trails, parks, complexes, and centers, to name just a few. Often, the purpose of the research project calls for the study to be conducted on site—that is, at the site of the event or business. Often, sport management professionals need to know what consumers think about their facility, business, or event. In that case, the research will need to be conducted on site.

Office for interviews. Interviews may take place at an office especially set up and designed for the interview project. This might be temporary based on the length of the project, or, if space is available, it could be permanent. For example, to determine the effectiveness of the LG Electronics' on-site sponsorships, Choi, Stotlar, and Park (2006) conducted interviews at a temporary office set up at a sport facility used for the LG Action Sports Championships.

The Interviewer's Role

The interviewer's role is complex and multifaceted. There is an art to the science of conducting interviews. Many of the basic skills needed to conduct interviews can be taught, but some of the characteristics needed are mostly innate. An individual is either comfortable conducting interviews with complete strangers, or not, although over time the interviewer can gain confidence and eventually the comfort level desired for conducting interviews. An interview is successfully effective when the interviewer

- is knowledgeable about the topic;
- can explain the purpose of the interview;
- can structure the interview process;
- keeps the interviewee interested;
- can communicate clearly and comfortably;
- is tolerant of, sensitive to, and patient with provocative or unconventional opinions;
- can control the course of the interview;
- is unbiased in recording the responses of the interviewee;

- can retain the information from the interviewee;
- can interpret what the interviewee is attempting to say;
- can engage personally with the interviewee;
- can remain neutral;
- is adept at transitioning between topics; and
- shows appreciation for the interviewee's time and effort.

Engaging people in conversation is an art. The most successful interviewer will be the person who has the ability to put interviewees immediately at ease and to coax them into opening up and sharing their thoughts and ideas. If interviewees don't feel comfortable with the interviewer, information will not be forthcoming. The interviewer's first responsibility is to make interviewees feel comfortable at the interview setting. The interviewer must be able to communicate clearly and effectively and at a level that the respondent can understand and be able to motivate the interviewees to do a good job. Explaining the research is a good beginning, but the interviewer needs to be able to convince the interviewees to want to participate fully and be forthcoming with responses. The interviewer must be motivated enough about the purpose of the research and the interview process that this is evident to the interviewee. The interviewer must be able to clarify any questions, confusion, or concerns on the part of the interviewee. This means that the interviewer must understand the research and all of its aspects so as to be able to answer any questions with confidence. The interviewer needs to be able to conduct a good interview. Each interview will have a life of its own, because each interviewee is a different person with a different personality and set of characteristics. The successful interviewer will be able to assess this quickly and make adjustments in style and method so as to get the most out of the interviewee. The interviewer's biases will have a negative influence on research if the researcher is not able to identify them, keep them in check, and keep them from seeping into the research project. Everyone has biases. The difference is when people know how to work around them and not allow them to prejudice their work or their research. The interviewer should always be confident. Respondents will feel easier about opening up if they believe that the interviewer is competent. The interviewer must understand that difficulties will arise at some point, should be prepared to notice these and to work around them or through them, and should keep the interest level up, be vigilant about the energy level of the respondents, and make moves that will lower the chances of their losing interest (Hague, Hague, & Morgan, 2004).

The interviewer should close the session with an honest showing of thanks and appreciation for the time, work, and energy that the interviewee has shared for the project's purposes. The interviewer should be sure to say that they are aware of the amount of time the interviewee has taken for this and is truly thankful for it. The interviewer can also show appreciation by giving interviewees a surprise gift. Interviewees are expecting only the offered incentive in exchange for their time and effort, and an additional surprise gift can let them know that the interviewer appreciates the interviewee going the extra mile.

Training the Interviewer

Considering the characteristics and skills needed by the interviewer as outlined in the previous section, the interviewer can be offered training that will train the beginning interviewer and, secondarily, enhance the skills of the intermediate or advanced interviewer. The training can be in the form of workshops or seminars for those individuals who will be interviewers. Additionally, training can be offered through local organizations, and the potential interviewers can attend their training workshops. Either way, an individual who will become an interviewer will need some training. The following are some of the major topics that should be covered in training.

- *What is the research project?* The interviewer should be well versed in the details of the research project for which the interviews will be conducted. It is important that the interviewer understand the purposes of the research in order to ensure a more successful interview.
- *What is the interview method?* The interviewer needs to learn about the interview method chosen to gather the needed information.
- *What is sampling?* The process of sampling is important. Interviewers need to understand this process so that they can relate to the interviewees during the process.
- *What is interviewer bias?* Interviewers need to know the many different possible ways that they can inadvertently, or even blatantly, bias the results. They must understand why it is important that they not bias the study. They must know that it is not appropriate to slightly slant the recorded responses so as to bias the results.
- *How does the interviewer conduct an interview?* Interviewers need to know the procedures of conducting an interview, from beginning to end.
- *How does the interviewer communicate with the interviewees?* A very important skill of the interviewer is to know how to communicate effectively with the interviewees. Some basic skills specific to interviewing should be offered in the training.
- *How does the interviewer record interviewee responses during the interview?* The interviewer needs to know the different possible methods for recording responses during the interview and should be skilled in these methods. Training should include practice interviewing and practice in the use of the different methods of recording responses.
- *How does the interviewer analyze responses?* If it will be expected that the interviewers will be analyzing the responses, then they must be taught how to do this. Refer to other sections of this chapter and book on analyzing qualitative data.

Conducting the Interview

When all preparations have been made, it is time to conduct the interviews. This section gives a summary of the process for conducting an interview.

- *Arrival and preparation.* Do not be late to arrive at an interview. In fact, it is important that you arrive early and prepare the room and any materials for the arrival of the interviewee. Have everything ready to go. Have drinks and snacks on hand to offer to the interviewee. Be courteous and thoughtful. Thank the interviewee for taking the time for such important research.
- *Opening remarks.* Once settled in, begin the session with prepared remarks. This opening should include such topics as the purpose of the research, confidentiality, feeling comfortable, directions to the restrooms, instructions about timing, and an overview of what to expect in the session.
- *The interview.* Work your way through the interview questions. Use prompts and probes where needed. Allow the respondent time to answer questions completely. Ask every question—don't skip. Don't finish sentences; allow the respondent to finish. Remember to ask

the interviewee to elaborate where necessary. Record responses according to the interview protocol for that study.

- *Conclude the interview.* When you have completed the interview portion of the session, thank the respondent. Don't be hasty. Allow a comfortable amount of time for closure.
- *Post-interview notes.* Just after the interview is completed, there should be time planned for going through your notes and making more notes, filling in where you left blanks, and completing sentences where you stopped. Do this until you are satisfied that your notes are complete.

Advantages of the Interview Method

There are some advantages in conducting interviews. In a personal interview, the interviewer gains a better understanding of the subjects' responses. The well-trained interviewer can use the face-to-face time to observe the subjects' body language and can engage them in delving deeper into explanations. This gets farther into the subjects' thinking as compared to other research methods.

In a personal interview, the subject has more time to respond to the questions and can ask for clarification when needed. Time is taken to be sure respondents understand the purpose of the research, the meaning of the questions, and what they need to think about when responding.

The interview also provides a slightly greater level of legitimacy than some other research methods. The interviewee gets to see the researcher, has the opportunity to ask questions, and will often open up for discussion more easily than through a written survey method. This way, it is easier for the respondent to explain why a question was given a certain answer.

Disadvantages of the Interview Method

Increased Cost. Cost is a major disadvantage for the interview method. Personal interviews are nearly always more expensive than other methods. Whereas phone interviews can run up quite a phone bill, they are usually less costly than the in-person interview.

Increased Time. Whereas a paper survey can be completed in minutes, a personal interview can take up to three hours of time. Additionally, the amount of time needed for the researcher/interviewer is greatly increased per respondent. In a paper survey, hundreds of respondents can be surveyed in a small amount of time; an interview process will result in a very low number of respondents in a large amount of time.

Decreased Sample Size. When a research method per respondent can take between one to three hours, the number of subjects will most likely be decreased, resulting in a lower sample size.

Increased Organization Time. Organizing an interview takes time. The respondents must be recruited, they must be enticed, and the interviews must be scheduled around their calendars.

Geographic Limitations. The in-person interview has geographic limitations. That is, the researcher is limited to those respondents who are local, unless the company is willing to pay for travel of either the interviewer or the subject. If interview research is needed by the sports company's headquarters in Miami, for instance, and the customers or respondents for the interview are in Seattle, then the company will need to cover the expenses of the researcher's travel to Seattle, or for the interviewees' travel to Miami.

Summary

In this chapter, fundamentals for interview as a research method are presented. The interview is a qualitative research method that involves a planned question-and-answer discussion session with an individual to collect information. The interview can be conducted using the in-person method or can be conducted by telephone. The personal interview is used more often because it has a number of advantages over the telephone method. In a personal interview, the interviewer can gain a better understanding of the interviewee because the interviewer can watch for body language, voice inflection, changes in eye contact, and other observations that can only be made with a personal interview.

There are three primary types of interviews: in-person, telephone, and digital video call. Additionally, there are different formats of interviews that can be structured and conducted. These include open-ended, fill-in-the-blank, binary choice, scaled-response, and unscaled-response interview format.

The interview may be conducted in a number of settings. The most common settings are the residence, the street, the mall, a venue, and at an office designated for the interview research.

The interviewer's role is complex and multifaceted. There is an art to the science of conducting interviews. Many of the basic skills needed to conduct interviews can be taught, but some of the characteristics needed are mostly innate. Considering the characteristics and skills needed by the interviewer as outlined in the previous section, interviewers can be offered training based on their level of experience with the technique.

Study Questions

1. What is the purpose of the interview as a research method? How does it differ from other qualitative or quantitative research methods?
2. Why is setting objectives important for interview research?
3. What are the different formats of interviews? Describe each one.
4. What are the different places where interviews might take place?
5. What does an interviewer do in order to conduct a successful interview?
6. What are the major topics that need to be covered in an interviewer training?
7. What are the advantages and disadvantages of the interview method?

Learning Activities

Assume that you want to conduct a study to determine whether a new professional football league will be well received by the sport consumers. You have decided to use the interview as research design to gather information. Prepare a list of questions.

References

Bernthal, J. M., & Regan, T. (2004). The economic impact of a NASCAR racetrack on a rural community and region. *Sport Marketing Quarterly, 13*, 26–34.

Berstell, G., & Nitterhouse, D. (2001). Asking all the right questions. *Marketing Research, 13*(3), 14–20.

Choi, J., Stotlar, D., & Park, S (2006). Visual ethnography of onsite sport sponsorship activation: LG action sports championship. *Sport Marketing Quarterly, 15*(2), 71–79.

Djaballah, M. (2017). Giving sense to corporate social responsibility in sporting events: A case study of the Quiksilver Pro France. In James Zhang and Brenda Pitts (Editors) (2017). *Contemporary Sport Marketing: Global Perspectives.* Oxfordshire, UK: Routledge Publishers.

Hague, P., Hague, N., & Morgan, C.A. (2004). *Market research in practice: A guide to the basics.* Kogan Page Ltd: London.

Kahan, H. (1990). One-on-ones should sparkle like the gems they are. *Marketing News, 24*(18), 8–9.

Seegmiller, J. G. (2006). Perceptions of quality for graduate athletic training education. *Journal of Athletic Training, 41*(4), 415–421.

Wagstaff, C., Hanton, S., & Cardiff, D. (2007). Coping with stressors encountered in sport organizations. *Journal of Sport and Exercise Psychology* (supplement), *29,* S210–S211.

6

Focus Group Research

After completing the study of this chapter, you should be able to:

- Understand what focus group research is and what its objectives are.
- Differentiate various types of focus group research.
- Understand how to select and recruit focus group participants.
- Identify the common places where focus group research is conducted.
- Appreciate the roles of a focus group moderator.
- Follow the process to conduct a focus group.
- Appreciate the advantages and disadvantages of the focus group method.

Focus group research has been used by sport management researchers to search for solutions to research problems for many years. In the academic field of Sport Business Management, the *Journal of Sport Management* published the first article explaining the use of focus groups as a useful qualitative methodology in sport management in 1992. In the article, Inglis (1992) argued that focus groups are particularly suited to research and practice in sport management, and reviewed the features of qualitative methodology and merits of focus groups. Since then, many sport management research projects have adopted this methodology.

For sport management professionals working in the sport business industry, the use of focus group research is vital. It is a popular and reliable method. It is a "method in which a small group of individuals are brought together and guided through a discussion about a topic" (Pitts & Stotlar, 2013, p. 137). The goal is to seek opinions, ideas, attitudes, and perceptions of the focus group individuals. This information guides decisions and strategies about the business and/or product so that adjustments might be made toward enhancement for success (Pitts & Stotlar, 2017).

This chapter provides a discussion of the nature of focus group research and the objectives of this type of qualitative research methodology. The common types of focus group research, the issues in the selection and recruitment of focus group participants, the common places where focus group research is conducted, and the important roles that a focus group *moderator* plays are presented. The focal point of this chapter is the process of how to conduct a focus group. In the last section of the chapter, the advantages and disadvantages of the focus group method will be reviewed.

Focus Groups as a Qualitative Research Technique

The focus group is a research method to collect information through a structured question-and-discussion activity with a targeted group of individuals within a specific period of time. The focus group participants are recruited based on the commonality of their experience by the researcher to discuss a specific topic. It is called a "focus" group because the researcher, in the role of moderator, presents a topic for a focused discussion and keeps the participants focused on the discussion of that particular topic.

The focus group method is used when a research topic can't be explored as well as other methods allow. The focus group method is flexible as compared to some other research methods that must adhere to a specific set of logistical steps. Other methods are limited in what information can be collected because the research instrument is limited to specific questions and, usually, a given set of answers from which the study participant must choose. In a focus group, however, there will be much in-depth discussion and exploration of a topic. Therefore, the focus group method is used when the researcher needs unstructured, unlimited, and spontaneous discussion about a topic. Additionally, the focus group information is qualitative data that is "rich in words and descriptions, rather than numbers" (Hague, Hague, & Morgan, 2004, p. 49).

The researcher uses a structured method but allows for flexibility as the discussion progresses. Typically, the size of a focus group ranges from five to 10 people, although it may be larger. The group needs to be small enough to allow each person the opportunity to share thoughts, reactions, experiences, perceptions, and opinions. Larger groups can inhibit input from every individual because shyer ones will talk less, allowing the group discussion to be controlled by the more outspoken individuals in the group. The disadvantage in this is that the researcher doesn't get the interaction and diversity of opinion and experience.

An example of the use of focus groups as main research design in sport management research is the study that was conducted by Taylor, Marcy, Hoye, and Guskelly (2006). Using psychological contract theory to explore the set of expectations and obligations that community sport club volunteers regard as part of their volunteering experience, the researchers conducted 16 focus groups with 98 community sports club administrators about the methods used to manage volunteers and the organizational expectations of the volunteers. The participants consisted of board or committee members, including presidents, secretaries, and registrars, as well as one paid employee with responsibility for managing club rugby.

In another example, a team in the National Football League (NFL), the Houston Texans, used focus group research in regard to their new enormous video board installed in the stadium (Kaplan, 2013). This monstrous-sized television monitor is 53 feet high and 277 feet wide. The business wanted to learn from fans what the fans would like and not like to see on the video board. Some of what they learned was fans preferred to see "replays, in-game stats, out-of-town scores with periodic highlights, and fantasy stats" (p. 1); what they did not want to see was "too much…they do not want the board to overpower the live viewing experience" of the game being played (p. 1).

Objectives of Focus Group Research

There are three main objectives of focus group research: (1) to generate ideas, (2) to provide in-depth qualitative information about the focus group participants, and (3) to discover the participants' needs, attitudes, characteristics, and perceptions about issues of interest to a sport business.

To generate ideas means that the focus group will be used to discover and develop new ideas about the sport business, its products, its activities, and many other characteristics of the business. For example, to determine what consumers thought about a particular sports facility, the sport management researcher working with sport marketing professionals at the facility used focus group research. It was discovered that consumers liked the seating arrangements of the facility and some of the food choices. However, consumers stated that sound in the facility was terrible—considered to be too loud, and it warbled, echoed, and reverberated too much for it to be enjoyed ultimately. The consumers revealed that in today's world of digitally mastered sound, the sound in the facility doesn't measure up. Some of the individuals in the group said that the sound was distracting from the event, so much that for certain events, the consumers said they would not attend additional similar events in the future. With this new information and feedback from its consumers, the sport business marketing professionals at the facility had information that could be used to make changes to a certain element of its product.

To provide in-depth qualitative information about the consumer means that the sport management researcher will collect a deeper set of quality information about the focus group participants than would be possible with other research methods. Often, after some research has been conducted, the researcher is left asking more questions based on the data collected. However, with these methods, the data set is final. That is, the researcher must attempt to interpret the data without further input of explanation from the consumer. In focus group research, the researcher has the opportunity to question the participants directly and to ask them to explain the meaning behind their responses.

To discover the participants' needs, attitudes, characteristics, and perceptions about issues of interest to a sport business means that the researcher can collect information directly from those people who participated in the focus group about what they like or dislike, what they think works or doesn't work, how they perceive a product, or how they perceive the image or reputation of a sport business.

Types of Focus Groups

The focus group method is adopted based on the purpose of the research. A focus group session can have the purpose of discussing a television commercial, obtaining feedback on testing, determining reaction to and perception of a product idea, brainstorming about the modification of a product to meet consumer needs, or seeking suggestions from the employees of a sport organization for reform. This can lead to different types of sessions, as follows.

In the *stand-alone method*, the data and information collected from the focus groups are the singular set of data for the research. No other type of research is conducted to supplement the focus group research. Thus, the sport management researcher must be careful that this research method will provide all that is needed.

With a *survey supplement*, the data collected from the focus groups are supplementary to another research method, such as a survey. This combination of research methods provides the opportunity for further information that can enhance and contribute to the overall conclusions and recommendations.

In a *multiple-method design*, the focus group research is one of several methods of research being used to study a topic. This will add even more depth and breadth of information to contribute to the conclusions and recommendations. A good example of this approach is the study conducted by Funk, Ridinger, and Moorman (2003). They used focus groups comprising women's basketball season ticket holders to clarify the conceptual framework of the Sport Interest Inventory (SII) and to identify if any contextual motives related to basketball exist that have not been found in the literature.

Using the Internet or other media devices to conduct an *electronic focus group* is a possibility, but should be limited to those situations for which actual people cannot be assembled for a face-to-face focus group. In some instances, various blogs or chat rooms have been used to conduct the focus group study. The primary advantage of the electronic focus group method is that it seeks information from the participants who are not physically located in the same place. For example, in a study that examined the consequences of work-family conflict with 41 mothers who are Division I head coaches, Bruening and Dixon (2007) used asynchronous online focus groups in data collection. They believe that the online focus groups not only have similar properties to the traditional in-person focus groups (i.e., in-depth discussion and the ability to build off of each other's discussion), but also allow sport management researchers to gather a group of participants who could not otherwise meet.

Such methods, however, do come with a specific set of limitations. The participants are limited to their skills of expressing their opinions through writing. Additionally, the moderator is limited in observing the participants and thus is limited in the use of effective group dynamics.

Another electronic method is the *video conference,* or *streaming media, focus group.* For example, applications such as Skype and FaceTime provide video-conferencing. However, this is limited to those who have the equipment needed for video-conferencing. On the other hand, this method can be used to connect a company representative in one area of the country to a moderator and focus group meeting in another area of the country.

Selecting Focus Group Participants

The participants in a focus group are selected specifically for their similar characteristics based on what the research is exploring. Other relevant attributes may play a part in the selection process if the topic warrants. For instance, a city's sports council is considering whether to bring a WNBA franchise to the city. The type of participant needed for open discussion about the feasibility and potential popularity and support for this product should come from individuals who are sports fans. These individuals might be found as local season ticket holders of other professional or college sports in the area—especially basketball fans. The focus of the discussion would be to determine if there would be enough interest for the product to start with an initial supporting fan base, and build ticket sales and fan support from there.

The participants in the group should have similar demographics or other characteristics because they may well be strangers who are meeting for the first time and will be expected to openly share their ideas. The more they have in common, the more they will feel comfortable with each other and thus sharing their thoughts.

Recruiting Focus Group Participants

Focus group participants should be recruited based on the purpose and focus of the research. The researcher will first determine where such individuals can be found. For instance, using the example above, to find basketball fans who could give feedback on the aspects of bringing a WNBA franchise to a city, the sport management researcher could ask local professional and college basketball organizations for their season ticket holder list. Other groups who might be interested in a WNBA franchise would be high school basketball players and fans. Additionally, because there is crossover interest in watching different sports, fans of any other local professional sports teams could be considered for selection.

Incentives should be a serious consideration. The researcher will be asking an individual to give up one or two hours of their time. Thus, offering an enticement for the person to show up is common practice. Some incentives typically used include a gift, dinner, free products, gift certificate, and monetary compensation. Using an incentive doesn't ensure that the participant will show up, but it can help in sealing the deal. In fact, "no-shows" are a problem that must be included into the plan of the research. One way to overcome the problem is to over-recruit potential participants. Another method is to use a backup list of individuals who have promised to be substitutes. A confirmation call is made as a way to find out whether or not an individual can make it to the session. If something has come up, then the list of backups will be consulted and someone from that list will be called and invited to join the group.

The Number of Groups or Sessions

There are no rules about how many focus groups to get or how many sessions to hold. One focus group might give information that the sport management researcher deems reliable and useful; another similar group might turn out to be a flop. Multiple groups can be planned if different market groups are sought. To acquire information that will increase the researcher's confidence level, the best plan is to have three or four focus group sessions. However, the higher the number of groups, the more difficult it is to draw conclusions from and analyze the information.

A Place to Stage the Focus Group Session: Planning and Logistics

Staging the session is important. For the traditional, face-to-face focus group sessions, the researcher must plan to conduct the session in a setting that is clean, bright, warm, and welcoming. This will help the group members feel comfortable and thus help in getting them to open up for discussion. It is the researcher's job to create an atmosphere in which the focus group participants can feel secure, intimate, and comfortable about expressing their opinions openly in this type of setting.

Traditionally, focus group sessions can take place in a variety of settings: a hotel meeting room, a conference room, a convention area, or an exhibition room. There are no rules about what kind of setting is required. It should be a venue that is of good quality, convenient, accessible, and one in which the focus group participants are taken good care of upon arrival.

The ideal setting is a dedicated focus group facility. This can be a cluster of rooms dedicated specifically for the purpose of the focus group gatherings and sessions. It can also be comprised of only one room. Most focus group rooms are set up similarly to a conference or board room with a large table and comfortable chairs around it. Some are configured with a set of chairs facing a screen where film can be shown for feedback. In the cluster setting, there may be one room dedicated as a reception area, another room set up board meeting style, another room as a viewing room, and another room set up as a break room with kitchen and resting areas.

The Focus Group Moderator

The person who will run the focus group session is the moderator. The opening remarks by the focus group moderator are important because they will set the stage and tone for the session.

Focus group moderators need to be influential individuals, as the effectiveness of a focus group session depends to a great extent upon the ability of the moderator to keep the session on track. Accordingly, the moderator must be skilled in observation and interpersonal communication, as well as group dynamics, in order to manage a productive focus group session. The moderator must understand exactly what information is needed in this research project and the purpose and objectives of the research in order to be successful at leading discussion and phrasing questions so as to get the individuals to effectively participate. The focus group moderator's role is to

- steer the discussion through a range of topics relevant to the research question;
- be a catalyst to encourage responses from the participants;
- encourage those who are quiet to participate in the discussion, and to limit those who are outspoken to keep them from monopolizing the conversation;
- provide audio or visual materials if needed for the discussion;
- keep the discussion on track;
- manage the group dynamics; and
- steer the discovery of information by guiding the participants with questions of why, how, and what.

Exhibit 6.1 presents moderator suggestions for enhancing a positive experience for the group.

The Focus Group Process

As with any research method, there is a general process to be followed when conducting focus group research. Exhibit 6.2 offers an illustration of a process in setting up a focus group.

The first step in the focus group organization process is developing a focus group research proposal with a precise and specific purpose statement and research objectives. Based on the purpose statement and research objectives, the sport management researcher can develop the right questions to collect the needed information through the focus group and identify the right group of participants.

Develop a research time. The sport management researcher must have a clear sense of the length of executing each of the tasks involved in the focus group research. For example, how long will it take to identify the participants, develop and test the focus group questions, invite and follow up with the participants, etc. The following are timelines suggested by Simon (1999) for a number of preparatory activities for a focus group session:

Exhibit 6.1. "Tricks of the Trade" for the Focus Group Moderator

Question	Tricks of the Trade
How do you make your groups great every time?	Be prepared. Be energized. Be nice but firm. Make sure everything about the experience is comfortable.
How do you build rapport quickly?	Make meaningful eye contact during each person's introduction. Learn and remember names. Let them create their own name cards. Welcome folks as they come into the room and use small talk.
How do you bring a drifting group back into focus?	Tell them that topic is for another group or time and that they need to focus on the topic for this group and time. Make a note and tell them that they will come back to this topic if there is time. Suggest that they can talk about it on their own after the focus group is over.
How do you get them to talk about deeper things than top-of-the-mind answers?	Play naïve or dumb and ask them to help you understand by explaining. Use probes such as "Tell us more about that," or "Can you go deeper on that?" Ask for specifics such as "Tell me about the last time that you . . ." Pair them up and give them 10 minutes for each pair to come up with a solution or suggestion. Orient clients with a 10-minute overview of focus groups, research objectives, and what to expect.
What about management of the "back room" where your clients are observing?	Check with the clients during breaks, written exercises, and so on to make sure things are going well. Have an associate or colleague there to work with the clients. If you don't have an associate for the back room, ask the client to select one person to be the point person to communicate with you.

Source. Qualitative Research Consultants Association (www.qrca.org).

- *Six to eight weeks prior.* Write the purpose statement and identify the participants.
- *Four to five weeks prior.* Select a moderator, develop the questions, and write a script.
- *Three to four weeks prior.* Write and send out invitations to the participants:
- *Two weeks prior.* Follow up the invitations with phone calls.
- *Two days prior.* Place a reminder call to the participants.

As part of the proposal process, develop participant guidelines and process. The characteristics of focus group participants will be considered and clearly defined. What are the characteristics of individuals needed for these discussions? Where will these people be found? What incentives will be available to entice participants? How will the participants be approached and asked to participate?

Develop a *moderator's discussion guide*, and formulate the questions and topic areas that the moderator will need to cover during the focus group. These should be questions specifically developed to guide the discussions toward the needed information. Since a focus group session usually lasts one to two hours, the moderator may only have time to ask and probe four to five in-depth questions. The focus group questions should be open-ended to invite thorough discussion. The moderator needs to proceed from the general to the specific.

The guide begins with an introductory section that describes the nature of a focus group to the participants. Then it outlines and describes the topics to be covered. The third part of the guide

Exhibit 6.2. Focus Group Organization Process

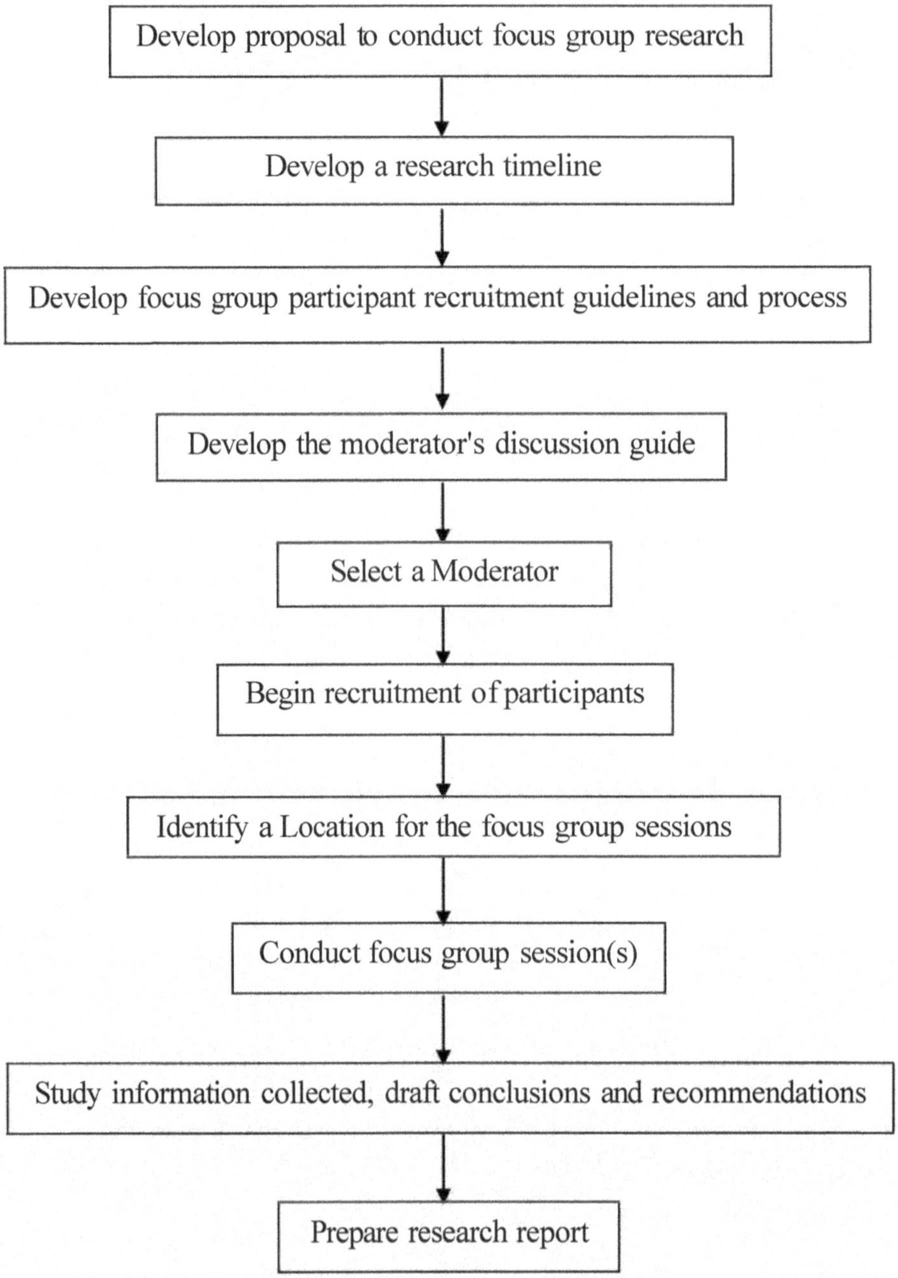

is to thank the participants, provide them with the moderator's contact information should they have additional comments or input, and explain how the information collected will be used. The reasons for using a discussion guide are several-fold. First, it helps to ensure that all sessions will be conducted in a similar manner so that the results will be reliable. Secondly, the discussion guide will help the moderator focus on the tasks (Simon, 1999).

As mentioned, the success of a focus group session depends greatly on the moderator. Thus, the selection of skilled and experienced moderator is one of the steps in focus group research whose importance should not be overlooked. The moderator could be the researcher or a professional who specializes in focus group moderation.

At the beginning of and throughout the participant recruitment process, it needs to be conducted until enough participants have agreed to participate. The researcher should be sure that enough information has been given to the participants about the activity, where to meet, what time to meet, what to bring and not bring, and any other information they will need before attending the group session. To ensure that an adequate number of people will participate in the focus group session, the researcher must follow up with the invited participants by phone calls or emails to remind them of the scheduled focus group session.

Identify a location for the meeting. As discussed previously, a focus group session could be conducted in a number of different locations. Regardless of the chosen location, the sport management researcher should keep in mind that the number one criterion in location selection is that it be a setting where participants will feel comfortable to provide their opinions. Simon (1999) suggests that the following questions be asked while choosing a location for the focus group session (pp. 5–6):

- What message does the setting send? (Is it corporate, upscale, cozy, informal, sterile, or inviting?)
- Does the setting encourage conversation?
- How will the setting affect the information gathered? Will the setting bias the information offered?
- Can it comfortably accommodate nine to 15 people? (six to 12 participants, plus the moderators), so that all can view one another?
- Is it easily accessible? (Consider access by people with disabilities, safety, transportation, parking, etc.)

After everything has been planned and coordinated, it is now time to conduct the focus group session(s). The researcher must prepare beforehand with the venue and other logistical needs for the sessions. Having a welcome area for the participants as they arrive provides them with a good first impression and helps make them more comfortable.

After the sessions have been conducted, it is time to study the data collected. As with most qualitative research, great care must be given to this process so that researcher bias can be kept to a minimum. The researcher needs to formulate conclusions based on the information and develop recommendations based on the conclusions.

Preparing a final report is the last step of the process. In this stage, the researcher can put together the full report of the focus group research.

Advantages and Disadvantages of the Focus Group Method of Research

Every method of research has advantages and disadvantages. The focus group method is quick, inexpensive, and relatively easy to conduct. It can obtain rich data in participants' own words and develop deeper insights, and the focus group session allows participants to build on one another's responses and come up with ideas they might not have thought of in one-on-one interviews. Focus group research also has disadvantages, limitations, and shortcomings similar to those of other forms of qualitative research. As a specific disadvantage, focus group research must have a competent and experienced moderator. Without a skilled moderator, the session will be ineffective and the results will be questionable. Secondly, focus group participants are chosen for their similarity of characteristics. This is a limitation because such a homogeneous group may not be representative of the whole population. Thirdly, a few dominant focus group members can skew a session.

Summary

Focus groups are usually comprised of five to 10 participants who are recruited specifically to participate in a discussion session led by a skilled moderator with a specific agenda. The moderator guides the group so that opinions, ideas, or reactions are effectively received from the participants. The findings of focus group research provide in-depth understanding of the consumer's opinions, ideas, or reactions to a specific topic.

A small number of sessions are usually held to ensure that all the topics and questions are covered. Typically, the number of sessions is three to four.

The most important element of focus group research is the moderator. This person must be skilled and competent at guiding a discussion and in using questions and other activities to prompt participants to express their thoughts openly and comfortably.

Study Questions

1. What is focus group research? What are its objectives?
2. What are the different types of focus group research?
3. How are focus group participants selected and recruited?
4. Where are the common places to conduct focus groups?
5. What are the important roles that a focus group moderator plays?
6. What is the process to conduct focus groups?
7. What are the advantages and disadvantages of the focus group method?

Learning Activities

1. Develop a hypothetical situation in which focus group research would be the ideal research method for studying this situation.
2. For activity 1 above, develop the focus group topics and questions that need answering for the focus group session.
3. Using the class as a sample, create a focus group method and conduct the session. Report the results in class.

References

Bruening, J. E., & Dixon, M. A. (2007). Work–family conflict in coaching II: Managing role conflict. *Journal of Sport Management, 21*(4), 471–496.

Funk, D. C., Ridinger, L. L., & Moorman, A. M. (2003). Understanding consumer support: Extending the *Sport* Interest Inventory (SII) to examine individual differences among women's professional sport consumers. *Sport Management Review, 6*(1), 1–31.

Hague, P., Hague, N., & Morgan, C. (2004). *Market research in practice: A guide to the basics.* London: Kogan Page Ltd.

Inglis, S. (1992). Focus groups as a useful qualitative methodology in sport management. *Journal of Sport Management, 6*(3), 173–178.

Kaplan, D. (2013, July 29–August 4). Focus group helps Texans program new video boards. *SportBusiness Journal.* Retrieved from http://www.sportsbusinessdaily.com/Journal/Issues/2013/07/29/Franchises/Texans.aspx

Pitts, B. G., & Stotlar, D. K. (2013). *Fundamentals of sport marketing.* Morgantown, WV: Fitness Information Technology.

Simon, J. S. (1999). *How to conduct a focus group.* Saint Paul, MN: Amherst H. Wilder Foundation.

Taylor, T., Marcy, S., Hoye, R., & Guskelly, G. (2006). Using psychological contract theory to explore issues in effective volunteer management. *European Sport Management Quarterly, 6*(2), 123–147.

7

Observation Research

After completing the study of this chapter, you should be able to:

- Explain under what circumstances a sport management researcher should use observation as a research design to collect data.
- Compare and contrast the six types of commonly used observation methods in sport management.
- Explain the sequential steps used to design an observational study.
- Elaborate on the advantages and disadvantages of observation as a research method.

As mentioned in Chapter 3, observation is an essential form of qualitative research designs used by sport management researchers in acquiring information. Observation "is the only scientific approach that is capable of gathering data directly from participants (athletes, coaches, trainers, etc.) in both the training and competitive contexts without eliciting a response from them" (Anguera, & Mendo, 2013, p. 136). In recent years, *participant observation* has been a common tool in many qualitative studies (Bennett & Hardin, 2002; Chalip & Scott, 2005; Fairley, 2003; Shaw, 2006). With any qualitative technique (e.g., case study, interviews, and focus groups), the use of observation as an instrument in data collection requires the researcher to have a clear understanding of certain skills. This chapter will discuss observation as a qualitative research method in five topics: the nature of observation, objectives of observation, types of observation, factors affecting the use of observation, and how to design and conduct an observation study.

The Nature of Observation

The observation design is used in solving certain research questions for which it is the most appropriate technique (Gay & Diebl, 1992). Observation, as a research design, refers to "the unobtrusive watching of behavioral patterns of people in certain situations to obtain information about the phenomenon of interest" (Johnson & Christensen, 2000, p. 147).

According to Black and Champion (1976), observation as a research data collection technique should have a number of basic characteristics as opposed to any casual, spontaneous observation in which the observer does not intend to obtain information. Those characteristics:

1. It captures the natural social context in which a person's behavior occurs.
2. It grasps the significant events and/or occurrences that affect social relations of the participants.
3. It determines what constitutes reality from the standpoint of the worldview, philosophy, or outlook of the observed.
4. It identifies regularities and recurrences in social life by comparing and contrasting data obtained in one study with that obtained in studies of other *natural settings.* (p. 330)

Clearly, those characteristics imply that to collect research-worthy data, the observer must follow a well-designed, systematic process to execute the observation technique.

Objectives of Observation

Observation, as a research design, is used to achieve three basic objectives (Black & Champion, 1976). These are (1) to capture human behavior in real time as it occurs and give the researcher the opportunity to view the demonstrated behavior in process, (2) to "provide more graphic description of social life than can be acquired in other ways" (p. 332), and (3) to explore a situation to provide the researcher some exploratory information.

Observation allows the researcher to record information on how research participants actually behave in real time. As such, observation as a type of research design may be used specifically to address the dynamic nature of behavior (Black & Champion, 1976). For example, Bechtol (2002) passively observed five selected athletic directors in their natural environment in order to capture the managerial behaviors, roles, and activities of Division I athletic directors. The study found that most of their verbal contacts took place in their offices. Traveling, conducting, and participating in scheduled meetings took up most of their out-of-office time. They spent approximately one quarter of their time handling managerial activities in after-business hours and weekend hours. The observed athletic directors used their executive assistants only in a few managerial tasks.

In managing a sport event that sells alcohol, event managers often encounter situations caused by intoxicated fans. A study that aims at developing effective measures to deal with the impaired spectators could use observation as a technique in data collection by providing visual descriptions. The data collected through the use of observation should give the event managers such information as

- What do intoxicated fans commonly do to cause trouble?
- When should they be confronted by event security staff?
- What are their reactions if confronted?

Observation can also be used as a means of exploration to understand a specific issue. The information obtained from the exploration may then provide the researcher with some sense of direction for research. For example, Fairley (2003) conducted a study using both participant observation and interviews as methods of data collection to provide exploratory information about why fans travel to follow professional sport teams in Australia. The participant observation was intended to observe reactions of those fans who traveled to follow their beloved teams. The interviews following the observation revealed the motives of why the fans did what they did.

Conditions for Using Observations

Certain conditions or elements must be presented before observation can be used as a valid means of data collection in sport management. As such, the sport management researcher must be aware of those prerequisites prior to attempting to use the observation technique. First, the wanted information or data must be accessible to observation. In other words, the behavior must be an observable act. As mentioned previously, the objective of observation is to obtain information about behavior (i.e., what people do and how they do it). If the purpose of a study is to seek information about what people think and believe and why they think and believe that way, observation is not an appropriate research technique for the objective.

Second, the behavior to be observed must be repetitive, frequent, or predictable. The condition requires that the behavior be an act that has a certain level of consistency (e.g., an intoxicated fan usually acts in a certain manner) and is demonstrated a number of times during the observation for the researcher to record. Predictability implies that the act will occur under certain circumstances. In that way, the researcher can anticipate and be ready to record when the behavior of interest may occur. Third, the behavior is demonstrated in a relatively short duration. Otherwise, it would be extremely costly to study. According to Black and Champion (1976), a number of other factors may also affect the selection of observation as an appropriate method of data collection. These factors include the problem itself, the researcher's skills and characteristics, and the characteristics of the subjects to be observed.

In some situations or social settings, observation may be a better research design option to choose. "Observation is especially suited to what might be called 'micro-social' as opposed to 'macro-social' questions" (Black & Champion, 1976, p. 336). Micro-social questions are specific and immediate to the life experiences of the subjects of interest, whereas macro-social questions are raised to address more global issues. For example, it may not be effective to study issues related to sport gambling through other forms of research design, such as surveys.

It is believed that the validity and usefulness of the information obtained from observation depend heavily on the skill levels and characteristics of the observer. It is especially true in participant observation, as the technique has a stringent requirement on the observer. That is, the observer's act of recording the information may interfere with his or her participation; the participation also may impede the results of observation.

The characteristics of the subjects are also another factor dictating whether or not observation is the appropriate method to use in data collection. "The possibility of using the techniques of observation increases as the ability of the observed to control their privacy decreases" (Black & Champion, 1976, p. 339). The ability of the subjects to control their privacy is related to several factors, such as occupation, economic and political positions, subcultural factors, and normative standards (Black & Champion, 1976).

Types of Observation

There are commonly six different ways of classifying observation methods by researchers (Black and Champion, 1976; Gay & Diebl, 1992; Johnson & Christensen, 2000): (1) *participant* and *nonparticipant observation*, (2) *obtrusive* and *unobtrusive* (or physical trace) *observation*, (3) *observation in natural* or *contrived settings*, (4) *disguised* and *nondisguised observation*, (5) *structured* and *unstructured observation*, and (6) *direct* and *indirect observation*.

Participant and Nonparticipant Observation

Depending on whether or not the researcher is part of the natural setting in which observation takes place, the observation technique can be classified as participant or nonparticipant observation. One of the most fundamental distinctions that differentiate observational strategies concerns the degree to which the observer adopts an active role and becomes an integral part of the phenomenon being studied. The task of observation can be pursued along a participation continuum. On one extreme, the observer remains detached, separated from the activities being studied...An alternative to this kind of description is to become a part of activities being investigated (Seymour, 1988, p. 58).

The participation continuum of observation is depicted in Exhibit 7.1. In the following section, a detailed description is provided to each of the observation methods. In addition, the strengths and weaknesses of each method will also be examined.

The methodology of participant observation is exceptional for studying processes, relationships among people and events, the organization of people and events, continuities over time, and patterns, ..." (Jorgensen, 1989, p. 12). A researcher who wants to study fan behavior and violence and thus becomes a spectator in a sporting event is an example of the participant observation. Participant observation requires the researcher to actively engage in the setting (e.g., to truly be part of the rowdy fans, the researcher has to demonstrate a similar level of boisterous behavior) and his or her behavior becomes an integral part of the fan behavior. The rationale for using the participant observation technique is simple. That is, in some situations, participant observation is a better technique to use to obtain quality data, as it can provide the researcher with a better perspective on the issue being studied. "Participant observers commonly gather data through casual conversations, in-depth, informal, and unstructured interviews, as well as formally structured interviews and questionnaires" (Jorgensen, 1989, p. 22).

To understand the dynamics of conflicts among youth sport clubs in a swimming league, Chalip and Scott (2005) utilized a number of qualitative research methods, including participant observation, in their study. One author served as a consultant to the league. The other helped with the establishment of a club and served on the executive board of the league. As active participants in the league operations, they were able to obtain first-hand information about the dynamics in decision making. This example demonstrates the other use of participant observation in sport management research.

One strength of this type of observation is that the researcher not only observes what occurs, but also feels as a participant involved in the situation (Mackellar, 2013). The observer's sharing

Exhibit 7.1. The Participation Continuum of Observation

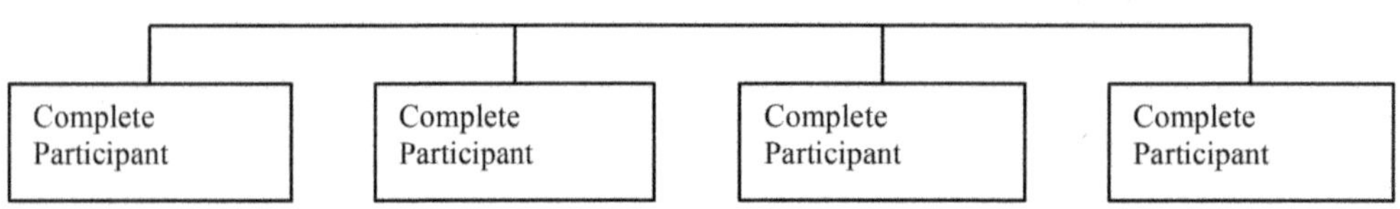

Source. Johnson, B. & Christensen, L. (2000). Educational research: Quantitative and qualitative approaches. Boston. MA: Allyn & Bacon.

of "feelings, thoughts, intentions, and meanings" with the true subjects of the study is called "intersubjectivity" (Seymour, 1988, p. 58). Seymour (1998) maintains that intersubjectivity gives the researcher the opportunity to make better inferences on the data collected through observation. Making better inferences is always a challenge to researchers using the observation technique. Being a part of the situation, the researcher is more able to make sense of the data than a bystander does. Being a part of the situation, the researcher is also more sensitive to "situational variables which may be important to understanding the situation under investigation" (Seymour, 1988, p. 59). Another strength of participant observation is that the researcher can "get fresh impressions, right as things are happening...[and] see how the experience evolves, how the impressions change, how people navigate a situation" (Morrison, 2002, p. 31). The main disadvantage of participant observation is that there may be some role confusion occurring to the researcher, subsequently affecting his or her keenness of observation and taking for granted some of the behaviors under observation.

In a nonparticipant observation, the researcher does not get involved in the situation to be observed or interact with the research participants as required in participant observation. Instead, the researcher observes the behavior of the subjects of an observational study in a natural setting as a bystander. The subjects are usually not aware of being observed. It is believed that if the setting in which observation takes place is not natural, the behavior then demonstrated by the subjects of their studies may not be the behavior that would occur in a natural setting (Gay & Diebl, 1992). So, the major advantage of nonparticipant observation, due to its unobstructive design, is that the behaviors are recorded in a natural setting; there is limited reactivity (i.e., people being observed are not aware that they are being observed, so they do not intentionally alter their behaviors" (Johnson & Christensen, 2000).

Obtrusive and Unobtrusive Observation

Based on whether or not the subjects being studied can detect the presence of the observer or the devices used in observation (e.g., hidden microphones or cameras used to observe behavior), observation can be referred to as either obtrusive observation or unobtrusive observation. In an obtrusive observation, the subjects to be observed are made aware of the presence of the observer or the observational devices stationed in the setting in which the observation is conducted. Participant observation, when the observer does not intentionally disguise himself or herself, is an example of obtrusive observation. In a study conducted to obtain an understanding of the complexity of the body and physicality in women's rugby, Chase (2006) adopted the method of obtrusive participant observation in collecting information. The researcher not only attended team meetings, but also stayed with players on the teams involved in the study. The researcher also sometimes traveled with a team to an out-of-state tournament.

Unobtrusive observation, also referred to as ethnographic observation, occurs when the researcher wants to observe people doing things in an environment that has not been intentionally altered. People perform their daily routines without noticing the existence of an observer. Nonparticipant observation is in fact a great example of this type of observation. To avoid alerting the subjects so that they will exhibit their behaviors freely and naturally, researchers usually focus their attention on physical traces that people leave behind. For example, to observe the drinking behavior and determine alcohol consumption of college football fans, the researcher can use garbage audits to accomplish the goal.

Observation in Natural or Contrived Settings

Observation can be classified as observation in natural settings or contrived settings. Observation in natural settings occurs when the behavior of interest is observed unobtrusively when and where it naturally occurs. The observer does not alter the setting in which observation will take place. The people being observed are not aware that they are under observation (McDaniel & Gates, 1991). To verify the validity and reliability of the Youth Sport Behavioral Assessment System (YSBAS), an instrument designed to assess the behaviors of parents and coaches at youth sport games, Apache (2006) and colleagues attended 284 youth sport games in various sport leagues and involving several sports to observe and record comments made by coaches and parents during regularly scheduled games (natural settings). On the other hand, observation in contrived setting refers to the kind of observation in which the situation is recreated or manipulated by the researcher to speed up the appearance of the behavior of interest. The researcher performs an observation in contrived settings like executing an experiment. The researcher first alters the environment in advance for the sake of observing the intended behavior demonstrated by the subjects so as to determine the causal relationship between the treatment (i.e., alteration to the environment) and the behavior.

For example, a sporting goods company selected a number of its chain stores to carry out an experiment to determine if the ways in which athletic shoes are displayed and arranged would affect consumers' purchasing decisions. Observation would take place to examine the actual behaviors exhibited (e.g., body language such as facial expressions) and the sales records. By contriving and manipulating the situation, the management of the company would be able to use the exhibited behaviors by consumers to make a managerial decision in a timely manner.

Disguised and Nondisguised Observation

One of the decisions that the researcher must make once observation is chosen as a research design is to decide whether or not the observer should disguise the observation. In disguised observation, the researcher intentionally keeps the purpose of the study and the role of the observer secret from the subjects so that the subjects being observed are not aware that they are being studied. In other words, the researcher in disguised observation typically pretends to be someone else (e.g., just another fan in the tailgates before the football games, to watch the alcohol consumption behaviors) as opposed to letting subjects know that the individual is a researcher.

Neal (2005) conducted a study to reveal what motivates gamblers to sustain their gambling behaviors. The disguised observation method was used to gather data. The researchers regularly visited a number of gambling places over a year. Accordingly, they were viewed by the frequent gamblers as members of their group. Such acceptance facilitated the researchers' observation and data collection as the gamblers were more willing to share their thoughts.

Structured and Unstructured Observation

Structured observation refers to the kind of observation in which guidelines or checklists are used to record the exhibited behavior of interest. It is typically used when the researcher already has prior knowledge of the behavior of interest. Then the researcher can design a detailed reporting sheet or checklist to register the data. For example, Bechtol (2002) collected data with a chronological record form and a record system for verbal contacts that were made beforehand to examine the managerial roles and activities of selected NCAA Division I athletic directors. Researchers sometimes also refer to structured observation as systematic observation.

On the other hand, in an unstructured observation, the observer does not have a preconceived objective or hypothesis in the observation. Instead, the observer simply is open-minded to observe what happens as it actually occurs. This type of observation is usually used in the exploratory stage of a research project (Seymour, 1988). Structured observation can be evaluated and reported as any other numerical/statistical data.

Direct and Indirect Observation

Depending on whether the behavior is being observed as it occurs or after the fact, observational methods are classified as direct or indirect observation. Most observational methods used by sport management researchers are direct observation, as the observers record the behavior of interest in real time. While using direct observation, the observer uses recording devices to record the behavior of interest first and analyzes the obtained information. Gee and Sullivan (2006) use a videotaped observation (VO) approach on the study of aggressive behavior in hockey. They taped a group of male varsity hockey (n = 79) players during a game and used two independent observers to code the tapes. The authors noted a number of limitations using a direct observation approach to study aggressive behavior in sport, which include the time-consuming and heavy burden on the researcher, especially used in large scale studies involving many games and seasons, and presence of social desirability (people being observed modify their behaviors).

Designing and Conducting an Observational Study

To conduct a study that employs observation as a research design, the researcher should use a research process that not only follows the essential steps required by scientific research, but also reflects the nature of observation. Specifically, the process of an observation study consists of a number of sequential steps: (1) defining the problems and variables involved in the observational study, (2) selecting an observation method, (3) training observers, (4) conducting and recording observation, and (5) interpreting the results of observation (Gay & Diehl, 1992).

Definition of Observational Problems and Variables

Like any other research design in sport management research, defining problems and variables is always an essential step, as it helps the researcher narrow down the focus of the observational study to a few variables. Subjects who are under observation usually demonstrate a variety of behaviors, and it is difficult to record every behavior they exhibit during the observation. So the researcher should base the objectives of the observational study on identifying a small number of variables.

Selecting an Observation Method

The selection of an observation method is probably the most essential step in designing an observational study, as it provides the researcher with information that will help answer several specific questions related to the methodology of an observational study. These questions:

Exhibit 7.2. Basic Decisions for Observation Studies

Natural versus Contrived Observation	To what extent will the situation to be observed be "set up" rather than occurring without intervention from the researcher?
Open versus Disguised Observation	To what extent will the respondents be aware that they are being observed?
Structured versus Unstructured Observation	To what extent will the observer be told which aspects of the overall situation to observe?
Direct versus Indirect Observation	Will the observations be made on the behavior of interest or on the results of that behavior?
Human versus Mechanical Observation	Will the observations be made by humans, by machines, or by some combination?

- *Where is the observation taking place?* To answer the question of where the observation is taking place, the researcher must first have a clear understanding of the behavior to be observed, particularly where it most likely occurs. Does it happen at a meeting, on the playing field, or at the spectator stand? The researcher also must decide how many places the behavior of interest could be observed. As mentioned in the section on the disadvantages of observation, representativeness is an issue that should be carefully dealt with in order to increase the generalizability of the study. As such, to reduce the sampling error, the researcher needs to identify a number of points of observation.
- *From whom is information being obtained?* While finalizing the methodology for the observational study, the researcher should determine who the subjects of the study would be (from whom the behavior of interest could be observed). The subject determination also affects the choice of sampling technique, ultimately impacting the validity of the study.
- *What types of data are being collected?* Basically, the question of what to observe dictates the types of information or data to be collected. If the behavior of interest is concerned with the purchasing intention of a certain type of golf clubs, then information should be collected about the number of questions asked about the product, the number of clubs purchased, etc.
- *How is the information or data being collected?* This question basically asks what specific method of observation will be used in data collection. As Exhibit 7.2 shows, the researcher has to contemplate a number of questions before selecting the correct observation method. Different methods may require different ways of data collection. One of the issues that the researcher must decide while selecting a specific kind of observation method is, to what extent will the situation to be observed be manipulated or altered so as to increase the occurrence of the behavior of interest, or is intervention from the researcher to the situation not required? If the situation calls for the researcher's intervention, then it is logical to choose the type of observation that is appropriate to use in a contrived setting.

Training Observers

To increase the reliability of an observational study, researchers should attend training workshops on observation techniques to improve their observational skills if they are not proficient in the use of observational techniques. On the other hand, if other observers are used, the researcher must make sure that the observers also participate in a training program to improve their knowledge and skills in observation. It is always a great idea to conduct a number of observation practice sessions in which the observers will have the opportunity to observe and record the behavior of interest in a setting or situation similar to the one in the actual study (Gay & Deibl, 1992).

One of the critical methodological elements in using observation as a research design is to have at least two observers involved in the same observational tasks. The results of their recorded observations can be used to assess the reliability of the study. This is another justification for having involved observers participate in the training program. One aspect of the training program would be to make sure the observers develop the same level of understanding of the specific behaviors to be observed, what constitutes the behaviors of interest, and how behaviors are coded. It is commonly agreed that in order for an observational study to have reliability, a score of 80% inter-observer agreement must be achieved (Gay & Deibl, 1992, p. 268).

Conducting and Recording Observation

The next logical step in executing the observational study is to conduct the observation to collect the data. Because the behavior the researcher wants to record usually exhibits in a short period of time, to efficiently record and collect the information of interest, the researcher needs to employ a recording or coding system that is simple and easy to use (Gay & Diebl, 1992; Seymour, 1988). "Most observation studies facilitate recording by using an agreed-upon code or set of symbols, and a recording instrument" (Gay & Diebl, 1992, p. 264). There are a variety of types of recording forms. Some can be used like a checklist, and some serve as a rating scale. For example, the study that examined the managerial roles and activities of selected NCAA Division I athletic directors conducted by Bechtol (2002) collected data with a chronological record form and a record system.

According to Anguera and Mendo (2013), the researcher should establish some guidelines for an observation study before it is implemented in order to prevent observation errors. Such guidelines:

1. Maintaining constancy between sessions. In order to ensure a maximum degree of homogeneity across the different observation sessions it is essential to draw up a list of minimum requirements that these sessions should fulfil in accordance with the established objectives. The criteria used may vary, provided they are adequate (days, place, time, activity, characteristics of the setting, absence of outside interruptions, etc.). For example, one might decide to observe all the away matches played by a given basketball team.
2. Maintaining constancy within sessions. Constancy within a single session may be affected by an unforeseen event or circumstance that leads to the observed activity being interrupted (the action stops). In the above example of a basketball match, this could be the result of a player being injured.
3. Dealing with temporal disruptions. Temporal disruptions occur when an unforeseen event or circumstance interrupts the observation session, but without the action itself stopping. This temporary lack of observability generally results from one of two things: the observed participant is momentarily outside the field of observation (e.g., a player moves out of camera shot), or there is a technical issue (e.g., part or all of the field of observation is out of camera shot, poor camera orientation at a given moment).

4. Labelling the observation session. In addition to identifiers such as date and time, it is important to include information regarding four basic aspects of all contexts: a) physical environment (surface area, lighting, implements, etc.); b) type of activity (training session, competitive match etc.); c) social data regarding the observed participant(s); and d) information of an institutional or organizational nature (start and duration of activities in their habitual context).

Interpreting the Results of Observation

After the behavior of interest has been recorded, the information obtained will be analyzed to determine if the behavior displayed by the subjects demonstrates a certain pattern. After such determination, the researcher should then attempt to make sense of the observed behavior with a theory or interpretation.

Ethnography and Observation

Sport management researchers often combine participant observation and other data collection methods, such as interviews, in a natural setting to obtain a comprehensive view of a particular situation. Such an approach in gathering information, with the use of participant observation in conjunction with other research methods, is generally referred to as ethnography (Gay & Diebl, 1992). In a study designed to gain an understanding of how political rivalries, policy coalitions, and conflicts emerge in an amateur sport league, Chalip and Scott (2005) adopted an ethnographical approach in data collection. They frequently attended the league's executive meetings to observe the group dynamic among the executives of the organization, interviewed the league's executives, and examined the league's records, including minutes and reports. It is the authors' belief that the ethnographical approach is the most appropriate in obtaining the kind of information to answer their research questions.

Several advantages of ethnography are commonly recognized by scholars. One of the advantages is that it can provide the researcher with "a much more comprehensive perspective than do other forms" of data collection (Fraenkel & Wallen, 2000, p. 546). The method allows the researcher to gain insight into a particular situation that is not possible to get with the use of other research methods is another advantage of ethnography. The major limitation of the ethnographical approach is similar to that of observation. That is, the findings obtained from ethnography usually lack generalizability.

Using Ethnography Research Techniques in Sport Management: A Case Analysis

To determine the effectiveness of the LG Electronics' on-site sponsorships, Choi, Stotlar, and Park (2006) utilized a visual ethnographic inquiry method to investigate what an average spectator at a sporting event visually records in a two-hour span in order. They purposefully selected 17 adult spectators (18 years or older) who were in attendance at the LG Action Sports Championships in Pomona, Georgia, between September 10 and 12, 2004. The participants were approached and agreed to voluntary participation. Information about their right to volunteer and the procedure of the study were clearly explained to the participants.

Each participant was provided with a digital camera phone with instructions to take pictures throughout the two-hour duration. A one-on-one interview was conducted with each participant when the camera was returned to the researchers. The participant was asked first to show the researcher all the pictures taken. During this process, the researcher asked the participant questions, such as the motives and reasons behind taking each picture. The researcher also asked questions related to LG's on-site activities, other off-site brand activities, and the overall experience. Other types of information, such as brand loyalty, cell-phone experience, and general thoughts on LG, were also obtained in the interview.

Based on repetition and relevance to the purpose of the study, the researchers broke down the collected data into several categories or themes. They compared the pictures that the spectators took and the activities that the sponsor (LG) used to activate its sponsorship at the venue to determine whether or not the sponsors had achieved their intended purpose.

Advantages and Disadvantages of Observational Research

Like any data collection method, observation also has its advantages and disadvantages. One obvious advantage is that it can be used to collect data in nondescriptive studies (Gay & Diebl, 1992). For example, the management of a sporting goods store decided to use a courtesy training program to improve the quality of customer services. Before the training program, employees were observed to determine the level of courtesy they demonstrated. After the training program, the level of courtesy exhibited by the employees was again observed. The management could then compare the results of the two observations.

It is a common belief that there is always some incongruence between attitudes and behavior (Johnson, & Christensen, 2000). In other words, the data collection from other research designs, no matter whether they are qualitative (e.g., interviews and focus groups) or quantitative in nature (e.g., survey and experiment), may reflect only a portion of the true feelings and perceptions of those people from whom the data were collected. As such, the other advantage of observation is that it gives the researcher the opportunity to record in real time the behavior of the research participants, instead of relying on the self-report of their preferences and intended behavior. In other words, observation records the actual behavior of research participants, not what they say they said or did, or not what they say they will do. The results of observation can then be used to compare the statements made by the research participants to check for the validity of their responses. This is especially important when dealing with behavior that might be subject to certain social pressure (e.g., people are less tolerant than they deem themselves to be) or conditioned responses (e.g., people say they value nutrition but will pick foods they know to be fatty or sweet), the observation technique can provide greater insights than the survey technique in this regard.

The third advantage of observation is that it does not rely on the memory or willingness of the research participants to gather data. Exhibit 7.3 is a comparison of behavioral and attitudinal research designs that may reveal the intended differences between observation and survey. As shown in Exhibit 7.3, observation is particularly appropriate for action research and can be used to obtain information concerning behavior.

Additional advantages of the observational methods are that observations are not dependent on recall and can take place in natural settings rather than research settings. They also allow for the identification of behaviors that may not be apparent to the person and may not have been

Exhibit 7.3. Comparison of Observation and Survey Research Designs

	Observation	Surveys
What is learned?	Actions	Intentions
What is the dimension of focus?	Behavioral	Attitudinal
Is there interaction with participants?	None or limited	Some
How is its reliability?	High	Uncertain

discovered through interviews and for the identification of behaviors that the person may not wish to disclose (Edexcel Limited, 2009, p. 8).

Since the observer, in most cases, is not in direct contact with the research participants while collecting data, the researcher can collect data related to social processes or problematic issues without the awareness of the participants. As such, the data collected are more reliable.

Observation is not without shortcomings. First of all, observation does not help the researcher understand what is happening within a research participant. For instance, what are their motivation and attitudes (emotions and perceptions) toward an issue? Particularly, it does not provide the researcher with any insights into what the research participant may be thinking or what might motivate a given behavior or comment. This type of information can be obtained only through a survey that asks the research participant directly or indirectly. To overcome this disadvantage, the researcher must understand that observation can obtain behavioral cues only to answer the question of what, not why. So, if the objective of the research is to seek information about what and why, a second stage needs to be added to the research after the conclusion of observation to acquire data in the attitudinal dimension, such as beliefs, feelings, and motives (Seymour, 1988). For example, a study that is designed to understand the kind of behaviors (e.g., shouting at the players of the visiting team, charging onto the field, climbing up and shaking down the goal post) that are typically demonstrated by students during a football game involving a rival institution and why they behave that way, the researcher should apply a two-stage research design that involves both observation and interviews. A study conducted by Shaw (2006) also demonstrated the importance of using the two-stage model in sport management research. However, observation and interviews were used in an intertwined matter. In that study, designed to analyze social processes within a sport management context, observation was utilized to clarify the data collected during interviews. The information collected through observation also enabled the researcher in developing the context and depth of interview questions as the study progressed.

The second disadvantage is that observation requires the researcher have well-developed observational skills in order to collect valid data. The researcher must know what to observe as the behavioral outcome could occur quickly. Also, as the data collected from observation are what the researcher observes, selective perception is a potential problem. Unskilled observers tend to observe behaviors that are related to their background and experience, and omit the important behavior cues and details that seem to be irrelevant to them, consequently presenting bias to the data. Training and the use of multiple observers are suggested approaches to limit the impact of this problem on the validity of the data acquired through the use of observation.

During observation, the researcher tends to believe that the observations are representative of the behaviors that are also demonstrated by a larger population of which the individuals or groups under investigation are a part (Seymour, 1988). As a matter of fact, there is a high tendency that the individuals or groups whom the researcher is observing may be unrepresentative. The behaviors they demonstrated may be "isolated incidents" (Seymour, 1988, p. 70). This is another disadvantage of the observation technique. In other words, the generalizability of the results obtained from an observational study could be low due to the issues involving sampling. To reduce the sampling error, the researcher should carefully examine the situation and circumstances when and where observation will take place to ensure that the subjects to be observed can reveal the general characteristics of the overall population. For example, in the case about observing student behaviors in a home football game involving a rival institution, the researcher should randomly select a number of institutions as settings of observation.

Reactivity is another disadvantage of observation (Seymour, 1988). In most of the cases involving observation as a research technique, the presence of an observer may change the behavior of the group being observed, and accordingly reduce the reliability of the observation outcome when the research participants know they are the objects of observation. As Seymour (1988) put it, "observation influenced behavior" (p. 70). Also, in a participant observation, the researcher may lose objectivity as a result of being a participant. The results of observation then may be biased because of his or her interactions with the research participants. To control this particular disadvantage, the researcher should determine if participant observation or unobtrusive measurement is required based on the objectives of the research.

Other disadvantages of observation also include the observer has little or no control over the situation or environment where observation takes place, and the observation research design has limited effectiveness for large and/or heterogeneous groups (Gay & Diehl, 1992; Ghauri, Gronhaug, & Kristainslound, 1995; McDaniel & Gates, 1991; Wrenn, Stevens, & Loudon, 2002). Costs and time are other reasons why observation is less preferred by researchers than other designs. It usually not only costs more money than survey, but also requires the researcher to devote more time conducting observation than other research approaches (Johnson & Kristensen, 2000).

Other disadvantages include the potential for the researcher to misunderstand what they are seeing, as it can be difficult to identify and record the correct type of data. The Hawthorne effect is also another possible disadvantage: if the people know they are the subject of research, they may act differently and could invalidate the whole project. The researcher must be very careful with exactly how they approach the people in observational research (Edexcel Limited, 2009, p. 8).

Summary

As one type of qualitative research design, observation has been used periodically by sport management researchers in solving their research problems. This chapter discussed several common topics related to observation, such as the nature of observation, objectives of observation, types of observation, factors affecting the use of observation, and how to design and conduct an observation study.

Observation can be the most appropriate technique for achieving a certain type of research objectives. A well-designed observational study should comprise a number of basic characteristics, such as (1) capturing the natural social context in which a person's behavior occurs; (2) grasping the significant events and/or occurrences that affect social relations of the participants; (3) determining what constitutes reality from the standpoint of the world view, philosophy, or outlook of the observed; and (4) identifying regularities and recurrences in social life by comparing and contrasting data obtained in one study with those obtained in studies of other natural settings.

Observation has its advantages and disadvantages. The advantages: it can be used to collect data in nondescriptive studies; it records in real time the behavior of the research participants, instead of relying on the self-report of their preferences and intended behavior; and it does not rely on the memory or willingness of the research participants to gather data. The shortcomings of observation are it does not indicate what is happening within a research participant; it may not produce valid data due to the use of unskillful observers; sampling errors could affect the generalizability of the study; and reactivity is a problem.

There are three basic objectives of observational studies: (1) to observe human behavior in real time, (2) to allow a more detailed description of what is going on, and (3) to provide exploratory information. Observation can be used effectively under three conditions. First, the behavior of interest must be an observable act. Second, the behavior to be observed must be repetitive, frequent, or predictable. Third, the behavior is demonstrated in a relatively short duration. In addition, a number of factors need to be considered while deciding if observation is appropriate to use. These factors include the problem itself, the researcher's skills and characteristics, and the characteristics of the subjects to be observed.

In general, observation methods can be classified into six types: participant and nonparticipant observation, obtrusive and unobtrusive (or physical trace) observation, observation in natural or contrived settings, disguised and nondisguised observation, structured and unstructured observation, and direct and indirect observation.

The process of designing and conducting an observational study usually consists of five steps: (1) define the problems and variables, (2) select an observation method, (3) train observers, (4) conduct and record observation, and (5) interpret the results of observation.

Study Questions

1. Explain under what circumstances a sport management researcher may use observation as a research design to collect data.
2. Elaborate on the advantages and disadvantages of observation as a research method.
3. Compare and contrast the six types of commonly used observation methods in sport management.
4. Explain the sequential steps used to design an observational study.

Learning Activities

1. Using observation as the method of data collection, design a study to determine the types of customers who are most likely to enter a sporting store in a shopping mall when they are walking by the store.
   a. What is your sampling method?
   b. What type of information will you collect from the customers?

2. Outline a study using observation as a research design for each of the following situations:
   a. A sports drink manufacturer wishes to determine how sports participants use its products while playing sports.
   b. A sport management researcher wants to know how many female athletes have been featured on the cover of Sports Illustrated in the past five years.
3. Watch ESPN SportsCenter for one week. Observe how much time the anchors spend on professional sports, college sports, and other amateur sports.
4. Choose a sports event for observing for approximately 1 hour (not less than ½ hour). You are required to take scratch notes to record your observation. If needed, you may participate in the scene, even ask other participants some ethnographic questions. You need to summarize the event by telling what you learned from the observation.

References

Anguera, M. T., & Mendo, A. H. (2013). Observational methodology in sport sciences. *Journal of Sport Sciences. 9*(3), 135–160.

Bechtol, D. L. (2002). Structured observation description of the managerial roles and activities of selected NCAA Division I athletic directors. *International Journal of Sport Management, 3*(1), 11–33.

Bennett, G., & Hardin, B. (2002). Management behaviors of an elite intercollegiate baseball coach. *International Journal of Sport Management, 3*(3), 199–214.

Black, J. A., & Champion, D. J. (1976). *Methods and Issues in Social Research.* New York, NY: John Wiley & Sons, Inc.

Edexcel Limited. (2009). Unit 4: Research methods for sport and exercise, BTEC's own resources. Retrieved from https://www.pearsonschoolsandfecolleges.co.uk/FEAndVocational/SportsStudies/BTEC/BTECLevel3NationalSportandExerciseSciences/Samples/StudentBook/SampleUnit4Researchmethodsforsportandexercisesciences-BTECLevel3NationalSportExerciseSciencesStudentBook.pdf

Chalip, L., & Scott, E. P. (2005). Centrifugal social forces in a youth sport league. *Sport Management Review, 8*, 43–67.

Chase, L. F. (2006). (Un)disciplined bodies: A Foucauldian analysis of women's rugby. *Sociology of Sport Journal, 23*(3), 229–247.

Choi, J. A., Stotlar, D. K., & Park, S. R. (2006). Visual ethnography of on-site sport sponsorship activation: LG action sports championship. *Sport Marketing Quarterly, 15*, 71–79.

Fairley, S. (2003). In search of relieved social experience: Group-based nostalgia sport tourism. *Journal of Sport Management, 17*(3), 284–304.

Fraenkel, J. R., & Wallen, N. E. (2000). *How to design and evaluate research in education.* (4th ed.). Boston, MA: McGraw Hill.

Gay, L. R., & Diebl, P. L. (1992). *Research methods for business and management.* New York: Macmillan Publishing Company.

Gee, C., & Sullivan, P. (2006). Using a direct observation approach to study aggressive behavior in hockey: Some preliminary findings. *Athletic Insight: the Online Journal of Sport Psychology, 8*(1).

Ghauri, P. N., Gronhaug, K., & Kristianslund, I. (1995). *Research methods in business studies: A practical guide.* New York: Prentice Hall.

Johnson, B., & Christensen, L. (2000). *Educational research: Quantitative and qualitative approaches.* Boston, MA: Allyn and Bacon.

Jorgensen D. (1989). *Participant Observation: a methodology for human studies.* Newbury Park, CA: Sage Publications.

Mackellar, J. (2013). Participant observation at events: Theory, practice and potential. *International Journal of Event and Festival Management.* Retrieved from http://hdl.handle.net/10072/53223

McDaniel, C., & Gates, R. (1991). *Contemporary marketing research.* St. Paul, MN: West Publishing Company.

Morrison, M. A. (2002). Using qualitative research in advertising: strategies, techniques, and applications. Thousand Oaks, CA: Sage Publications.

Neal, M. (2005). 'I lose, but that's not the point': Situated economic and social rationalities in horserace gambling. *Leisure Studies, 24*(3), 291–310.

Seymour, D. T. (1988). *Marketing research: Qualitative methods for the marketing professional.* Chicago, IL: Probus Publishing Company.

Shaw, S. (2006). Scratching the back of 'Mr X': Analyzing gendered social processes in sport organizations. *Journal of Sport Management*, 20, 510–534.

Wrenn, B., Stevens, R., & Loudon, D. (2002). Marketing research: Text and cases. Binghamton, NY: Best Business Books.

8

Secondary Data

After completing the study of this chapter, you should be able to:

- Understand the nature of secondary data as a quantitative technique in sport management research.
- Understand the different ways that sport management researchers can use secondary data to serve their research needs.
- Understand the common sources of secondary data for sport management research.
- Appreciate the advantages and disadvantages of secondary data.
- Follow the steps in the process of secondary data research to conduct a research project.

Data can be described as either primary or secondary. Most of the research conducted in sport management involves the collection of new data, or the primary data, to solve a particular problem. Nevertheless, sport management researchers sometimes may turn their attention to *secondary data* to help solve their research problems. This is confirmed by Li and Lawrence (2007) in their study that examines the use of secondary data in sport management. The authors reviewed the articles published in the *Journal of Sport Management* since 1992 and in *Sport Marketing Quarterly* since 2003. The results of their reviews indicate that about 10% of the 250 articles reviewed used the secondary data design. This chapter provides an overview of secondary data as a quantitative research design in sport management. The nature of this type of research design and the common ways that sport management researchers and professionals use it to facilitate their decision making will be reviewed first. Other topics, such as the sources of secondary data for sport management research and the advantages and disadvantages of secondary data, will also be discussed. The process of secondary data research will be looked at in the last part of the chapter.

The Nature of Secondary Data in Sport Management Research

Secondary data refer to the type of data that have previously been gathered and that might be relevant to the problem at hand (McDaniel & Gates, 1991). That means that, while using the secondary data design, the sport management researchers do not have to collect their own data. Instead, they can use data that already exist in the public domain to test their hypotheses or solve their problems. Secondary data may exist in many forms, including census reports, a public opinion poll, or a survey of attitudes to a particular social issue. For example, if the focus of a research project is sport participation, the sport management researcher can examine data released in the Statistical Abstract of the United States, reports from the Sporting Goods Manufacturers Association (SGMA), and the National Sporting Goods Association (NSGA), etc. Of course, the regularly gathered information on a variety of variables, such as sales, costs, and customer satisfaction by an organization, can also be used to track the trend on those variables.

Secondary data can be used to meet a number of research needs: to answer a research question, to clarify and understand a situation, to provide exploratory information, to help make a decision, to verify findings of primary research, and to build a statistical model (Blankenship, Breen, & Dutka, 1998; Wicker, Hallmann & Breuer, 2013; Wrenn, Stevens & Loudon, 2002; Zikmund, 2000).

To Answer a Research Question

One of the most common uses of secondary data in sport management research is to answer a research question or approve a hypothesis. Wrenn et al. (2002) maintain that it is essential for researchers to first explore all possible sources of secondary data prior to looking for primary data. In fact, many research questions can simply be answered with the analysis of the data collected from a secondary source. For example, to understand whether or not winning has a significant impact on alumni giving, Stinson and Howard (2004) examined the secondary data on donors making gifts of $1,000 or more between 1994 and 2002 to the Annual Giving Program at the University of Oregon. They found "in almost every year, alumni giving to athletics has increased with an associated increase in success by high profile intercollegiate athletic teams" (p. 136).

In a study conducted to approve the seven proposed hypotheses concerning personnel replacement rates in sport organizations, Theberge and Loy (1976) used secondary data included in the *Baseball Encyclopedia* published in 1969 for the 16 teams of the American and National Baseball Leagues during the period 1951–1960 to empirically test the hypotheses. Gladden and Milne (1999) also used secondary data on Major League Baseball, the National Basketball Association, and the National Hockey League to help sport management professionals understand the contribution of brand equity and winning to merchandise sales in the professional sport setting. The examples mentioned here imply that secondary data are the most appropriate data to solve a research question of this kind, which requires information in multiple years.

The results of the study conducted by Li and Lawrence (2007) examining the use of secondary data as a research design in sport management confirm that researchers often use secondary data to either answer a research question or approve a hypothesis (e.g., Allen, Bullough, Cole, Shibli, & Wilson, 2013; Ahlstrom, Si, & Kennelly, 1999; McDonald & Rascher, 2000; Orders & Chelladurai, 1994; Stinson & Howard, 2004; Zygmont & Leadley, 2005).

To Clarify and Understand a Situation

The second common use of secondary data is fact finding (Zikmund, 2000). For example, to verify the validity of such a statement, "soccer is greatly gaining popularity in the United States," the researcher could approach the research problem with the help of secondary data. Basically, the researcher can examine data obtained from a variety of secondary data sources, including the Statistical Abstract of the United States, to determine whether or not the numbers of soccer participants over the years have demonstrated an upward trend, whether or not the sales of soccer related equipment have been getting stronger, and if the numbers of soccer clubs at all levels have increased.

In a study designed to examine the economic value of professional sports franchises in North America, Lee and Chun (2008) collected secondary data from league offices and national media networks. The study revealed that while the economic significance of professional sport franchises was derived from both the traditional revenue sources, such as game receipts and sales of luxury boxes and club seats, and media revenues, the latter would increase in a greater proportion in the next decade.

To Provide Exploratory Information

Sport management researchers also utilize secondary data to provide exploratory information to a large-scale research project that requires the collection of primary data. As a matter of fact, many studies that collect primary data involve the use of secondary data to provide exploratory information to clarify the situation or help form hypotheses for the primary data research. To examine the issues of cultural diversity in women's sport, Taylor (2003) first used secondary data to generate exploratory information about the participation rates in sport and physical activities among women from minority ethnic backgrounds and Anglo-Australian women to verify that the former group of women had a lower participation rate. To further confirm the findings from the secondary data research, the researcher conducted a survey to investigate the issue with a specific sport—netball. The empirical data confirmed that the levels of female participation from culturally diverse backgrounds in netball were significantly below those of Anglo-Australian females.

This case exemplifies the substantial role that secondary data play in the exploratory phase of the research when the task at hand is to define the research problem and to generate hypotheses. The assembly and analysis of secondary data almost invariably improve the researcher's understanding of the marketing problem, the various lines of inquiry that could or should be followed, and the alternative courses of action that might be pursued.

To Help Make a Decision

A secondary data research project can be executed to give a firm that wants to make a long-term investment in a particular product line useful exploratory data to make a decision. Assume that a national chain fitness company (e.g., Gold's Gym) wants to know whether or not Lancaster, Ohio is a good location for one of its franchises. To facilitate the decision, the research team of the company carefully examined the census data of the city on population and trend of growth, income profiles, etc. The examination of the data collected from secondary sources indicates the market size of that particular location may not be large enough to successfully sustain a Gold's Gym franchise. This example illustrates the use of secondary data to help sports firms make an informed decision in terms of expansion or relocation.

To Verify and Complement the Findings of Primary Research

Secondary data research not only can help the researcher verify findings that are obtained from a completed primary research project, but also can provide context for interpretation of primary research. For example, the sport council of a city wanted to convince the owners of a professional sports league that the city was an ideal location for future expansion. To demonstrate the degree of fan support and enthusiasm, the council first conducted a survey of local residents. The results indicated a strong fan support of this initiative. The council then compiled a report using the attendance records of a minor league franchise in the same sport for the last ten years to confirm the findings of the survey research.

To Build a Statistical Model

In the previous study conducted to examine the use of secondary data in sport management research, Li and Lawrence (2007) found that 12 out of the 25 identified studies using secondary data (48%) were designed to build a statistical model that has the ability to specify and make predictions about relationships between variables or to approve a conceptual model (e.g., Agthe & Billings, 2000; Boronico & Newbert, 1999; DeSchriver & Stotlar, 1996; Maxcy & Mondello, 2006; Todd, Crook, & Barilla, 2005; Yokum, Gonzalez, & Badgett, 2006). For example, Hadley and Gustafson (1991) used secondary data to establish a single-equation model of baseball salaries. The study found that variables such as years of Major League service and eligibility for arbitration had a great impact on players' salaries in Major League Baseball.

The following is an example demonstrating the conceptual process used in building a statistical model that has the ability to specify and make predictions about relationships among variables. A researcher established an empirical model/equation for explaining National Football League (NFL) franchise operating income with the use of available secondary data. Nine variables were included in the consideration, and they were believed to have a direct impact on the operating income of an NFL franchise. The following formula shows such a statistical (multiple regression) model and the nine variables included in the model:

$$\text{NFL operating income} = f(X_1, X_2, X_3, X_4, X_5, X_6, X_7, X_8, X_9)$$

Where

X_1 = Median Family Income

X_2 = Average Cost of Living

X_3 = Average Unemployment Rate

X_4 = Total Population of the Metropolitan Area

X_5 = Usability of the Public Transportation

X_6 = % of 65 or older Population in the Metro Area

X_7 = Franchise Win/Loss % in a Given Year

X_8 = Minimum Face Price on a Single Game Ticket to A Regular Season Game

X_9 = Capacity of the Stadium/Dome

All the data required for the establishment of such a statistical model can be obtained from secondary data sources. The multiple regression analysis yields some interesting results.

$$\text{NFL operating income} = -0.59 + 4.31X_1 - 0.076X_2 - 0.22X_3 + 5.66X_4 + 6.37X_5 + 6.78X_6 + 9.88X_7 - 1.21X_8 + 0.45X_9$$

The results show that an NFL franchise's win/loss percentage in a given year is the biggest predictor of the operating income of the franchise under investigation, followed by the percentage of 65 or older population in the metropolitan area, the usability of the public transportation system, the total population of the metropolitan area, etc. The average cost of living in a particular metropolitan area is the weakest predictor and has an inverse relationship with the franchise operating income.

Sources of Secondary Data

In general, secondary data are available at two levels, the internal and the external. The *internal-level secondary data* are the information originating within a particular firm or organization. The NCAA annual expense report is an example of the secondary information compiled by a sport organization but available for the public use. The secondary information obtained from outside sources is referred to as the *external-level secondary data.*

Internal Records of and Data Released by a Public Organization

Exhibit 8.1 shows a section of the NCAA EADA report released by the NCAA that discloses financial information about member institutions. Sport management researchers who are interested in issues in intercollegiate athletic administration may find the data useful, as a number of statistical procedures can be applied to analyze the financial data. The Institute for Diversity and Ethics in Sport at the University of Central Florida periodically releases information related to gender and race in amateur, collegiate, and professional sports. The released information is a good source of secondary data for researchers who have a research focus on gender and race in the sport industry. In other words, the data can be reanalyzed to achieve a particular objective of a secondary data research project. For example, in 2003, the center released Classroom Counts, a study on the impact of Division I-A football realignment on conference graduation rates (Institute for Diversity and Ethics in Sport, 2003). Description statistics were used in data analysis. To further understand the issues surrounding conference realignment and its effect on graduation rates, more

Exhibit 8.1. Example of the NCAA EADA Report Released by the NCAA

Institution	Classification	Male	Female	Total	A4	A5	A6	A7
Alabama A & M University	NCAA Division I-AA	2187	2435	4622	1727412	1141431		2868843
University of Alabama at Birmingham	NCAA Division I-A	4441	6843	11284	2944657	1709816		4654473
University of Alabama in Huntsville	NCAA Division II (without football)	1976	1907	3883	676855	539253		1216108
Alabama State University	NCAA Division I-AA	1598	2360	3958	1003671	681966		1685637
The University of Alabama	NCAA Division I-A	7421	8410	15831	3317599	2697787		6015386
Central Alabama Community College	NJCAA Division I	435	758	1193	128700	108900		237600
Auburn University-Montgomery	NAIA Division I	448	1204	1652	440040	303254		743294
Auburn University Main Campus	NCAA Division I-A	9746	9150	18896	3674886	2615544		6290430
Birmingham Southern College	NCAA Division I-AAA	614	797	1411	1109150	1457679		2566829
Chattahoochee Valley Community College	NJCAA Division I	349	708	1057	90400	88000		178400
Concordia College	Other (USCAA)	343	498	841	0	0		0
Enterprise-Ozark Community College	NJCAA Division I	626	428	1054	190124	147037		337161
James H Faulkner State Community College	NJCAA Division I	808	1258	2066	311993	292720		604713
Faulkner University	NAIA Division I	644	866	1510	483407	304681		788088
Gadsden State Community College	Other. Some NJCCA Div. I &II	1319	1645	2964	192593	232657		425250
George C Wallace Community College-Dothan	NJCAA Division I	781	1144	1925	119024	106483		225507
George C Wallace State Community College-Hanceville	NJCAA Division I	1161	1853	3014	318425	200108		518533
George C Wallace State Community College-Selma	NJCAA Division I	340	857	1197	111026	35804		146830

Huntingdon College	NCAA Division III (with football)	371	362	733	0	0		0
Jacksonville State University	NCAA Division I-AA	2459	3255	5714	1296901	1005119		2303020
Jefferson Davis Community College	NJCAA Division I	284	733	1017	167375	126683		294058
Jefferson State Community College	NJCAA Division II	1162	1628	2790	113509	104518		218027
Lawson State Community College-Birmingham Campus	NJCAA Division I	565	1169	1734	263648	159482		423130
University of West Alabama	NCAA Division II (with football)	752	987	1739	474853	255935	58248	789036
Lurleen B Wallace Community College	NJCAA Division I	536	908	1444	178465	161398		339863
Marion Military Institute	Other	USCAA	230	26	256	92268		92268
Miles College	NCAA Division II (with football)	733	890	1623	353724	126093		479817
University of Montevallo	NCAA Division II (without football)	748	1589	2337	318004	353414		671418
Northwest Shoals Community College-Muscle Shoals	NJCCA Division I	1861	2921	4782	214882	260473		475355
University of North Alabama	UNA Box 5001 NCAA Division II (with football)	1955	2489	4444	755791	397993		1153784
Alabama Southern Community College	NJCAA Division I	338	524	862	108246	92483		200729
Bishop State Community College	NJCAA Division I	813	1566	2379	138801	136802		275603
Samford University	NCAA Division I-AA	1032	1859	2891	2033470	1427007		3460477
Shelton State Community College	NJCAA Division I	1419	1643	3062	157363	213315		370678
Sneed State Community College	NJCAA Division II	464	630	1094	134096	148556		282652
University of South Alabama	NCAA Division I-AAA	3067	4501	7568	818434	1047937		1866371

sophisticated statistical analyses could be applied to analyze the obtained data, which may include, but are not limited to, a two-way ANOVA to determine the interaction between realignment and the race of student-athletes.

To comply with the Equity in Athletics Disclosure Act (EADA), all co-educational postsecondary institutions that receive Title IV funding (i.e., those that participate in federal student aid programs) are required to submit athletics related data to the Office of Postsecondary Education (OPE) of the U.S. Department of Education. Using the data OPE created an Equity in Athletics Data Analysis Cutting Tool to allow users to not only download related data concerning equity in athletics, but also generate customized reports (USDE Office of Postsecondary Education, 2017). The data could be downloaded in either an Excel or SPSS file format.

Internal Data Released by a Private Trade Organization

Industry associations often publish data on size and characteristics of their industry. As mentioned earlier, the SGMA periodically releases information about the trends of the sporting goods industry and sport participation.

Internal Data from a Private Firm

All sport businesses record their daily operations in a number of ways, such as the orders received and delivered, the sales invoices, and the cost records of producing, storing, transporting, and marketing each of its products or services. Some of the secondary data studies that sport management professionals of those businesses can conduct using the information include the determination of the most profitable products and customers, and the tracking of sales trends in their existing customer groups.

Data from Previous Surveys

Data from previous surveys are sold as printed reports or raw data by survey research firms. A good example of such firms is Nielson Scarborough. Nielson Scarborough Research specializes in marketing research focusing on shopping patterns, lifestyles, and media habits of consumers locally, regionally, and nationally. It has data sets on 120 local market studies in the United States, which examine consumers' retail shopping behaviors, lifestyle characteristics, in-depth consumer demographics, and media usage patterns. Interested businesses can purchase a secondary data set from Nielson Scarborough so as to obtain a better understanding of consumers in a particular local market (Nielson Local, n.d.). Let's assume that the sponsorship sales staff of the Columbus Blue Jackets wants to present a sponsorship proposal to a local corporation. To prepare the proposal, the sales staff purchased a specific marketing data set for the Columbus metropolitan area. They analyze the data set and include the results of their analysis in the sponsorship proposal, which includes a section on shopping patterns, lifestyles, and media habits of consumers in that market.

Previously Published Research

The data from previously published research projects are reported in a summary form, but the original data may be available from the authors upon request. It is important that the researcher communicate with the authors of the original data to see if the latter is willing to share the information and under what conditions the data could be reused.

Publications Released by Government Agencies

The most common external sources of the secondary data are those publications released by a variety of government agencies. The following is a list of government agencies from which researchers in sport management can obtain the secondary information:

- National Center for Education Statistics (NCES)—https://nces.ed.gov/surveys/
- Social Capital Community Benchmark Survey—https://ropercenter.cornell.edu/2006-social-capital-community-benchmark-survey/
- U.S. Bureau of Economic Analysis, Survey of Current Business—http://www.bea.doc.gov
- U.S. Bureau of Labor Statistics—http://www.stats.bls.gov, good data source and excellent links
- U.S. Census Bureau—http://www.census.gov
- U.S. Economic Census—http://www.census.gov/econ/census02/
- U.S. Government Materials—http://www.fed stats.gov
- U.S. Office of Trade and Economic Development—http://www.ita.doc.gov/tradestats
- U.S. Statistical Data International and National Stat-USA—http://www.stat-usa.gov/, excellent links and information
- U.S. White House Economic Briefing Room—http://www.whitehouse.gov/fsbr/esbr.html

In addition to the secondary data released by government agencies, online databases can also be external sources for secondary data researchers. For example, sport marketing researchers may find SecondaryData.com a useful source.

Process of Secondary Data Research

To properly execute a secondary data research project, a sport management researcher should follow a four-step process. The steps involved are (1) defining the data needed, (2) locating the data, (3) evaluating the data, and (4) verifying the data. Exhibit 8.2 depicts the process of secondary research.

Defining the Data Needed

The sport management researcher who wants to use the secondary data design should always start with the definition of the problem, followed by the determination of the type of information or data that are required to solve the problem. For instance, if the problem at hand is to obtain an understanding of the trend of usage of water sports facilities by American consumers, the nature of the problem dictates the application of historical data collected by various federal and local government agencies. Obviously, the analysis of those historical or secondary data will help the researcher outline the trend.

Locating the Data

As mentioned previously, one of the issues with secondary data research is the availability of the data. So, once the type of secondary data needed for the research has been decided, the researcher

Exhibit 8.2. Process of Secondary Data Research

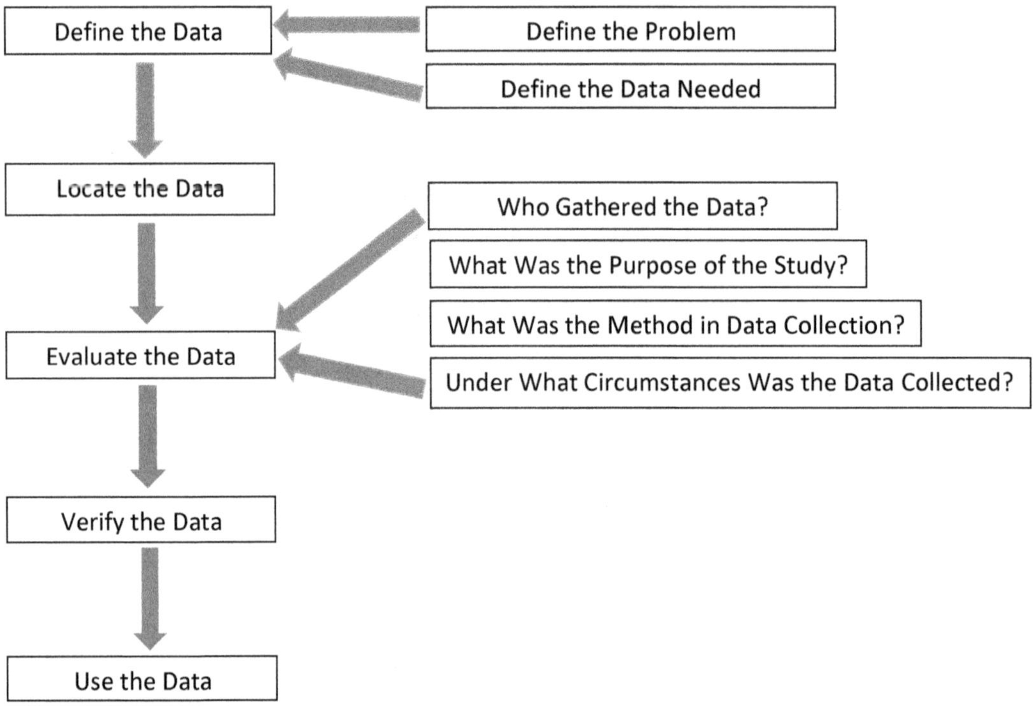

then needs to investigate where the data are located. As discussed in the prior section on the sources of secondary data, they can be found through either online databases (e.g., the U.S. Census) or printed indices, such as the American Statistics Index or the Statistical Reference Index, that are available at most libraries.

Evaluating the Data

After a secondary data source has been located, the researcher must evaluate the source to determine its validity, reliability, relevance, appropriateness, and accuracy. The evaluation also indicates whether or not the data are relevant to the research problem and situation at hand. The following are a number of questions to which the researcher should seek answers while evaluating the located secondary data.

Who gathered the data? The first legitimate question that the researcher who wants to use available data must ask: Where did the secondary data come from? In other words, who was the individual or organization responsible for collecting the data, and was the individual or organization reputable and creditable? "The source of the secondary data is the key to its accuracy" (McDaniel & Gates, 1991, p. 124).

What was the purpose of the study? It is always an issue in secondary data research that the data compiled for another research project with different objectives may not completely fit the problem at hand (Ghauri, Gronhaug, & Kristianslund, 1995). Understanding the reason the data was collected in the first place provides the researcher with clues about the quality of the data. For example, the researcher should be cautious while reviewing the secondary data compiled by a municipal sports commission on the attendance and consequent economic impact of a sports event on the local community, as the data could have been manipulated to serve some political purpose.

What was the methodology or research design used in data collection? Scrutiny should focus on such areas as the conceptual model used, the variables and their operational definitions, the measures employed, the sampling design used and the sample obtained, the data collection methods, and response rate obtained. Any flaw in the research design used in collecting the data will introduce inaccuracy to the data, consequently making the data unusable.

Under what circumstance was the information collected? The intent of this question is twofold. First, the researcher must understand that the data collected for another research project may not fit with or be relevant to the research problem at hand. It is important for the researcher to examine carefully the secondary data to determine if the unit of measurement is the same and the sampling frame is relatively similar. Second, the researcher must understand the circumstances in terms of what, when, and how the information was obtained.

After the evaluation is completed, the researcher still has to decide whether or not the obtained secondary data are sufficient to address the research problem. If not, the secondary data research must to be supplemented with primary data.

Gathering and Verifying the Data

If the result of the evaluation shows that the secondary data source appears to be valid, reliable, and accurate, the researcher should then secure a copy of the data. Before attempting to use the data set, the researcher should verify whether or not the data obtained are the right ones.

Cases in Sport Management Research Using Secondary Data

Gladden and Milne (1999) used secondary data as a research design to gain a better understanding of brand equity as a tool for guiding strategic marketing decisions in the sport industry. Specifically, the authors wanted to expand the conceptual understanding of brand equity and compare the importance of brand equity and winning in the realization of a marketplace outcome in the professional sport setting.

Secondary data were gathered from a variety of sources that provide comprehensive information on Major League Baseball, the National Basketball Association, and the National Hockey League. A number of logistic regression analyses were conducted. The results revealed that both brand equity and winning significantly contributed to the attainment of merchandise sales. However, the impact of brand equity and winning differed by league.

Advantages and Disadvantages of Secondary Data

Secondary data have several advantages. They include saving time and money, better research design, and availability. The most obvious advantage of secondary data is its readiness to use, as the name of this type of data implies—they were obtained by others but are available and accessible (Zikmund, 2000). Thus, the researcher does not have to go through the process as required in primary data collection, which usually involves the design of a questionnaire or data collection instrument, verification of its validity and reliability, the collection of data through field work, etc. The completion of those steps takes time and resources. When compared to the cost in executing a research project, the secondary data research is substantially lower than the primary data research. The cost typically associated with secondary data is to acquire the published source. There is no design cost (Wrenn, et al., 2002). The time commitment required of a researcher to complete a research project that uses secondary data starts only after the data have been extracted from a source.

As mentioned previously, the NCAA releases annually the expense report that includes financial records of most of its affiliated institutions. Researchers who are interested in a project that investigates the financial strengths of each of the three NCAA Divisions may find the report extremely useful, as it is very difficult, if not impossible, for any individual researcher to collect such primary data through a survey. The response rate is usually quite high due to the sponsorship of the NCAA in this type of research.

There is some information that is only available in the secondary data form (Wrenn, et al., 2002). For example, the internal level data originated within a particular sport firm or organization, such as personnel and financial related data, can only exist and be accessed in the secondary data format. Also, as mentioned previously, secondary data research sometimes is the only research design appropriate and available to analyze longitudinal data.

Although the great benefits that secondary data can provide sport management researchers should be recognized, their shortcomings should also be acknowledged and discussed so that the sport management researchers can try to avoid those disadvantages while designing their secondary data research and be cautious in their application of secondary data. The limitations of secondary data are found in the following four areas:

The researcher who decides to use the secondary data research design should carefully examine the definitions that have been used in generating the secondary data before attempting to use the data, as those definitions could be different from what the researcher has in mind. Assume that a researcher is interested in using secondary data to make comparisons of the sport participation rates between residents of America and residents of China. The concept of regular participation was used in the secondary data sets published respectively by the government agencies in these two nations. What is the definition of regular participation in sport? Does it mean three times a week and thirty minutes each time? Without obtaining a clear understanding of the definitions used in generating the data, it is methodologically inappropriate to pursue a study that will analyze the data comparatively.

Sometimes, even though it would be better to use the secondary data design to solve the research problem at hand, the researcher has no choice but to collect primary data to deal with the problem, due to the unavailability of secondary data (McDaniel & Gates, 1991). If ESPN wants to know how college students react to a new anchor for the Sports Center program, it must conduct a survey to obtain primary data to evaluate their reactions, as no secondary data have been generated yet.

It has always been a concern that some secondary data could suffer from a lack of accuracy. It is quite possible that inaccuracies have occurred during the original collection of the data. For example, the data might have been collected from an unreliable source. It is possible that the inaccuracies in measurement were estimated by the original researcher. Nevertheless, these inaccuracies were sometimes not made available for the researcher who wants to reuse the data. To avoid using inaccurate secondary data, a number of guidelines have been suggested for the researcher to follow before attempting to adopt data from a secondary source (McDaniel & Gates, 1991; Wrenn, et al., 2002).

The data produced could be out-of-date at the time when the researcher wants to make use of the statistics. For example, the U.S. Census Bureau conducts the Economic Census every five years. The 2002 Economic Census showed that the number of professional sports establishments (i.e., teams and franchises) was 388. The number would likely be much higher in the next 10–15 years.

Summary

Secondary data research is one of the research designs commonly employed by sport management researchers for several purposes: (a) to answer a research question, (b) to clarify and understand a situation, (c) to provide exploratory information, (d) to help make a decision, (e) to verify findings of primary research, and (f) to build a statistical model.

Sport management researchers can access secondary data from two levels: internal and external. The internal level of secondary data includes the internal records of and data released by a public organization, the data released by a private trade organization, the data from previous surveys, and the previously published research. On the other hand, the external level secondary data primarily come from those publications released by a variety of government agencies.

There are some advantages and disadvantages of secondary data. Saving time and money, better research design, and availability are recognized advantages of secondary data. The commonly mentioned disadvantages of secondary data include differences in definitions, the lack of availability, the lack of accuracy, and the age of secondary data. To conduct a secondary data research project, it is suggested that the researcher follow a four-step model: defining the data needed, locating the data, evaluating the data, and verifying the data.

Study Questions

- Give a summary description of the nature of secondary data as a quantitative technique in sport management research.
- How many different ways can sport management researchers use secondary data to serve their research needs?
- What are the common sources of secondary data for sport management research?
- What are the advantages and disadvantages of secondary data?
- What are the steps in the process of secondary data research?

Learning Activities

- The NBA wants to expand its development league, NBDL. Your consulting firm was just contracted to provide the NBA with recommendations in terms of which cities meet the expansion criteria. Based on the Census data, what city is most qualified for an expansion NBDL team?
- The national headquarters of a sporting goods retailer wants to expand its market coverage into two Ohio cities by August 1 of the current year. The company's management knows from

its past experience that a city must have a population of at least 100,000 people to support a store and that the bulk of its sales is to people between the ages of 27 and 32. Recommend two cities in Ohio, based on information available in the Sales and Marketing Management's Survey of Buying Power.

References

Agthe, D. E., & Billings, R. B. (2000). The role of football profits in meeting Title IX gender equity regulations and policy. *Journal of Sport Management, 14*(1), 28–40.

Ahlstrom, D., Si, S., & Kennelly, J. (1999). Free-agent performance in Major League Baseball: Do teams get what they expect? *Journal of Sport Management, 13*(3), 181–196.

Allen, K., Bullough, S., Cole, D., Shibli, S. & Wilson, J. (2013). The Impact of Engagement in Sport on Graduate Employability: Final Report. Sport Industry Research Center, Sheffield Hallam University. Retrieved from http://c1593.r93.cf3.rackcdn.com/BUCS_Employability_Research_Report.pdf

Blankenship, A. B., Breen, G. E., & Dutka, A. (1998). State of the art marketing research. Lincolnwood, IL: NTC Books.

Boronico, J. S., & Newbert, S. L. (1999). Play-calling strategy in American football: A game-theoretic stochastic dynamic programming approach. *Journal of Sport Management, 13*(2), 103–113.

DeSchriver, T. D., & Stotlar, D. K. (1996). An economic analysis of cartel behavior within the NCAA. *Journal of Sport Management, 10*(4), 388–400.

Ghauri, P. N., Gronhaug, K., & Kristianslund, I. (1995). Research methods in business studies: A practical guide. New York: Prentice Hall.

Gladden, J. M., & Milne, G. R. (1999). Examining the importance of brand equity in professional sport. *Sport Marketing Quarterly, 8*(1), 21–29.

Hadley, L., & Gustafson, E. (1991). Major league baseball salaries: The impacts of arbitration and free agency. *Journal of Sport Management, 5*(2), 111–127.

Institute for Diversity and Ethics in Sport. (2003). News Release: Conference graduation rates shift with realignment. Retrieved from http://www.bus.ucf.edu/sport/cgi-bin/site/sitew.cgi?page=/ides/index.htx

Lee, S., & Chun, H. (2008). Economic values of professional sport franchises in the United States. *The Sport Journal, 19.* Retrieved from http://thesportjournal.org/article/economic-values-of-professional-sport-franchises-in-the-united-states/

Li, M., & Lawrence, H. (2007, June 1). Use of secondary data in sport management research. Paper presented at the 2007 NASSM Annual Conference in Fort Lauderdale, FL.

Maxcy, J., & Mondello. (2006). The impact of free agency on competitive balance in North American professional team sports leagues. *Journal of Sport Management, 20*(3), 345–365.

McDaniel, C., & Gates, R. (1991). *Contemporary marketing research.* St. Paul, MN: West Publishing Company.

McDonald, M., & Rascher, D. (2000). Does bat day make cents? The effect of promotions on demand for Major League Baseball. *Journal of Sport Management, 14*, 8–27.

Nielson Local (n.d.). Your Participation Councts. Retrieved from http://www.scarboroughsportsmarketing.com/services/consumer-insights/sports

Orders, S.A., & Chelladurai, P. (1994). The effectiveness of Sport Canada's Athlete Assistance Program from 1980–1989. *Journal of Sport Management, 8*(2), 140–152.

Stinson, J., & Howard, D. (2004). Scoreboards vs. mortarboards: Major donor behavior and intercollegiate athletics. *Sport Marketing Quarterly, 13*, 129–140.

Taylor, T. L. (2003). Diversity management in a multicultural society: An exploratory study of cultural diversity and team sport in Australia. *Annals of Leisure Research, 6*(2), 168–188.

Theberge, N., & Loy, J. (1976). Replacement processes in sport organizations: The case of professional baseball. *International Review of Sport Sociology, 11*(2), 73–93.

Todd, S. Y., Crook, T. R., & Barilla, A. G. (2005). Hierarchical linear modeling of multilevel data. *Journal of Sport Management, 19*(4), 387–403.

USDE Office of Secondary Education. (2017). *Equity in Athletic Data Analysis.* Retrieved from https://ope.ed.gov/athletics/#/

Wicker, P., Hallmann, K., & Breuer, C. (2013). Analyzing the impact of sport infrastructure on sport participation using geo-coded data: Evidence from multi-level models. *Sport Management Review, 16*(1), 54–67.

Wrenn, B., Stevens, R. E., & Loudon, D. L. (2002). *Marketing Research: Text and Cases* (2nd Eds). New York: Haworth Press.

Yokum, J. T., Gonzalez, J. J., & Badgett, T. (2006) Forecasting the long-term viability of an enterprise: The case of a minor league baseball franchise. *Journal of Sport Management 20*(2), 248–259.

Zikmund, W. G. (2000). *Business research methods* (6th ed.). Fort Worth, TX: Harcourt College Publishers.

Zygmont, Z. X., & Leadley, J.C. (2005). When is the honeymoon over? Major league baseball attendance 1970–2000. *Journal of Sport Management, 19*, 278–299.

9

Survey Research

After completing the study of this chapter, you should be able to:

- Describe different types of survey research techniques.
- Understand the process for conducting survey research.
- Understand various types of survey research and the advantages and disadvantages of each.
- Understand the types of errors that may incur from the use of survey as a method of data collection.
- Determine the most appropriate survey research technique with given research circumstances.

Surveys are everywhere. They sit on the restaurant table requesting our feedback on service. They are in our hotel room, asking about the facilities. They reach us by mail, asking what we think about the product we just bought. They fall out of our magazines. They are thrust on us as we enter a mall or store requesting feedback on our experiences. They pop up on our computers as we surf the web, wanting to know what we think about a website. How many times have you actually completed one of these? More importantly, were you completing it to offer praises about a service? Or were you angered by the lack of service and wanted to voice your dissatisfaction? If you answered yes to the last question, you are an average customer—you typically don't think about service until you are negatively affected by it. However, surveys are an important tool in collecting information directly from consumers to inform decisions, strategies, and plans, and to enhance understanding about human attitudes, perceptions, satisfaction levels, behavior, and attachments.

Survey research is a carefully prescribed measurement procedure that involves gathering information from a sample of respondents through a questionnaire. It is one of the most frequently used methods in sport management academic research, sport marketing research, and sport business industry research (Pitts & Stotlar, 2013). It is also one of the most common types of quantitative social, consumer behavior, and business research. The survey is a questionnaire (a written instrument) designed to collect data. The broad area of survey research encompasses a number of measurement procedures that involve asking questions of respondents (Trochim, 2006). With surveys, it is possible to collect information from a small or large sample of a larger population in person, by telephone, by mail, and on the Internet. This sample is usually just a small percentage of the population being studied. Surveys are also useful as a marketing practice. If conducted correctly, the sport business can gather good information that can be used in many decision-making processes in the business.

Survey research is one of the most important types of measurement in sport business research. Surveys are used widely in sport business management. A sport business might want to know what its consumers think about its products, its customer service quality, or its promotional activities. The survey is the preferred method for collecting these types of information because it involves asking participants about their thoughts, experiences, or beliefs.

Survey research is a methodology that encompasses a measurement procedure that involves asking questions of respondents. A survey can range from a short feedback survey to an intensive one-on-one, in-depth interview. The length of a survey can range from one question to an abundance of questions in a variety of formats. The survey question can be formatted to seek a simple yes or no response or to gather an array of responses in such variable formats as multiple choice, rank, rate, and fill-in-the-blank. Further, the survey method is highly flexible, which is a major reason for its high usage rate. For example, a survey study can be conducted in a variety of facilities or locations. It can be done by phone, by mail, on the Internet, or in person, and it can be organized in a number of ways.

With any research design, there are specific steps that should be followed in order to improve the scientific quality of the study. Increasing the quality of the study design increases the confidence level of the results. Accordingly, to conduct survey research, it is important for the researcher to understand the planning process. Chapter 12 details the process of designing a questionnaire to be used in the survey research.

Surveys are used for a wide variety of purposes, depending on the problem or question needing data for use in answers. Here are some examples: A running shoe company wants to know what runners think about its new model, so the company conducts a survey. A national sporting goods retail company wants to know what time of year consumers purchase more camping gear, so it uses a survey to collect information from many stores about sales and consumer needs. The WNBA wants to know if a particular city is a good fit for a new franchise, so it conducts surveys of several kinds to collect a variety of needed data. NASCAR's sponsors want to know if consumers know that they are sponsors, if consumers recognize them, and if consumers are more willing to purchase their products because they are sponsors; companies and organizations conduct surveys to collect information to answer their questions.

Surveys provide an important source of information needed to discover new knowledge, enhance existing knowledge, and inform understanding. Surveys can reveal simple, basic information, such as income numbers, as well as complex sociological behavior, such as consumer expenditure patterns. This flexibility is the primary reason that survey methodology is used in so many different fields and industries.

Worldwide and Local Work to Improve Survey Methods

There are several organizations focused singularly on improving survey research. Many are further focused on survey research. One of these organizations is the Council of American Survey Research Organizations (CASRO), an organization that "advances the business of research through standards, guidelines, professional development, and self-regulation in the process and performance of survey research" (Council of American Survey Research Organizations, 2007). In January 2017, CASRO merged with the Marketing Research Association (MRA) and changed their name to Insights Association (Insights Association, 2017). CASRO earlier created the *CASRO Code of Standards and Ethics for Survey Research*, an important document that puts forth standards related to responsibilities to survey respondents, confidentiality, privacy, avoidance of harassment, and responsibilities for reporting results.

Other organizations with an emphasis on survey research in business and marketing include the Market Research Association (MRA), Council for Marketing and Opinion Research, European Society for Opinion and Marketing Research, Market Research Society, and the American Marketing Association.

What Is a Survey?

A survey is a questionnaire aimed at collecting a variety of information from people. The word survey is used to describe a research instrument and method for gathering information from people. The group of individuals who will be included in the survey study are the study participants within a specifically selected population. Usually, only a portion of the population will be included in the survey study. This is of course called the sample. Using scientific methodology to select the sample, it is expected that the information gathered from the sample can be inferred to the whole population.

Most often, the information gathered in a survey will include typical demographics and some psychographics, and, of course, the questions regarding the specific topics of the study. Surveys are created to gather information on a number of factors, or measures, and might include measuring knowledge, attitude, opinion, values, perception, behavior, intentions, personality, and positions on issues or topics.

In a survey study, a questionnaire is developed and a sample of the population to be surveyed is selected through a scientific process. The questionnaire is developed such that the questions gather information appropriate and pertinent to the research questions or objectives of the research. The survey administration process is standardized so that individuals in the sample are asked the same questions in the same way. The sample is chosen through a sampling methodology that ensures that each person in the population will have an equal chance of selection so that the sample is a true representation of the population. The information collected from the specific sample can then be inferred to the population with greater confidence.

Exhibit 9.1 provides an example of a survey of sports fans at a sports event. This particular survey has been used at a SuperCross event—a racing circuit of the American Motorcycle Association—in a large city, and the results will be used by the facility staff to improve the event. In the questionnaire, there are questions to gather information about the factors that affect decisions to attend the event (seen in Part I of the survey), information regarding economics and spending during the event (seen in Part II of the survey), information about the consumers' perceptions of the facility (seen in Part III of the survey), information about the consumers' perceptions of the city (seen in Part IV of the survey), and demographics (seen in part V of the survey). The results will also be shared with stakeholders, such as the hotel, transportation, and restaurant industries, and sponsors of the event.

Survey research is sometimes thought of as an easy research method when compared to other more complicated methods. However, as with any research method, survey research can be simple

Exhibit 9.1. Example of a Survey Instrument

Complete this survey and receive a coupon to use at one of our concessions (while supplies last)!!!
You must be 18 years of age or older to participate. -- Your participation is voluntary. If you have questions, please write to SportsMarketingSurveyInfo@gmail.com.
Thank you!! Georgia Dome and Georgia State University Sports Marketing Fan Research Team Thank you!! - Supercross 2017

I. Factors That Influence Your Attendance and Event Experience

Which of the following factors influenced your decision to attend this event today? Please circle the number that best reflects your opinion on a scale of **1 (No Influence)** to **5 (Strong Influence)** or **NA (Not Applicable)**.

1. Price of ticket(s)	1	2	3	4	5	NA
2. Promotion/advertising of the event	1	2	3	4	5	NA
3. A chance to see the sport itself	1	2	3	4	5	NA
4. A chance to see star riders	1	2	3	4	5	NA
5. A chance to see my favorite rider perform	1	2	3	4	5	NA
6. Current standings in the series	1	2	3	4	5	NA
7. A special promotion/giveaways	1	2	3	4	5	NA
8. To entertain business client/partner	1	2	3	4	5	NA
9. To entertain family	1	2	3	4	5	NA
10. Event location--the City of Atlanta as a destination for this event	1	2	3	4	5	NA
11. Other activities taking place in the City	1	2	3	4	5	NA
12. The Georgia Dome as a venue	1	2	3	4	5	NA

II. Economic Factors

1. What is your motivation for attending this event - check all that apply?
☐ Supercross Fan ☐ Motorcycle Fan ☐ Fun ☐ Family of a rider
☐ Friend of a rider ☐ Something to do

2. With whom did you attend this event - check all that apply?
☐ Alone ☐ Family only ☐ Friends only ☐ Partner only
☐ Family, friends, and partner ☐ School Club/Organization
☐ Business associate: Name of organization ____________
☐ Other (please specify) ____________

3. How many are in your party (including yourself)? ________

4. Did you come to Atlanta specifically for this event? ☐ Yes ☐ No

5. If your answer to #4 is NO, did you stay an extra day or more so you could attend this event? ☐ Yes ☐ No If yes, how many extra days? ________

6. How many nights total did you stay in Atlanta? ________

7. If you stayed overnight, where did you stay - check below:
☐ Hotel ☐ Friends/Relatives ☐ Camping/RV ☐ Other: ________

8. Have you attended this event previously? ☐ Yes ☐ No

9. If your answer to #8 is yes, how many times? ______

10. Do you plan to attend this event again? ☐ Yes ☐ No

11. Where did you GET INFORMATION about the event – check all you used:
☐ TV ☐ Local Newspaper ☐ Friends ☐ Family
☐ Radio ☐ Supercross Website ☐ Facebook ☐ Twitter
☐ Dome Website ☐ Other: ________

12. Do you have an account of – Facebook? ☐Yes ☐No; Twitter? ☐Yes ☐No; Instagram? ☐Yes ☐No; SnapChat ☐Yes ☐No; Periscope☐Yes ☐No
13. Do you plan to post during the event? FB?☐Yes ☐No; Twitter☐Yes ☐No; Instagram ☐Yes ☐No; SnapChat ☐Yes ☐No; Periscope ☐Yes ☐No
14. Where did you purchase your ticket(s)? ____________
15. Please estimate (in even U.S. dollars) how much you spent on each of the following items during your entire stay for this event & write it in the space:

- Food and beverages (restaurants, concessions, groceries) --> $________
- Admission fees (this event and other related events) --> $________
- Nightclubs, lounges, bars (cover charges, drinks, etc.) --> $________
- Retail shopping (clothing, gifts, etc.) --> $________
- Event souvenirs --> $________
- Lodging expenses (hotel, motel, etc.) --> $________
- Private auto expenses in the city (gas, parking fees, etc.) --> $________
- Commercial transportation (public transit, taxi, car rental) --> $________
- Other expenses (please specify) ____________ --> $________

III. The Georgia Dome

1. Have you ever attended an event at the Georgia Dome? ☐ Yes ☐ No
2. What other types of events do you attend at the Georgia Dome?
____

3. Please rate the following facility services and amenities at the Georgia Dome on a scale of **1 (Poor)** to **5 (Excellent)**. **N/A = Not Applicable.**

▪ Facility location	1	2	3	4	5	N/A
▪ Directional signage	1	2	3	4	5	N/A
▪ Professionalism/courtesy of staff	1	2	3	4	5	N/A
▪ Food quality	1	2	3	4	5	N/A
▪ Food and beverage service	1	2	3	4	5	N/A
▪ Seating comfort	1	2	3	4	5	N/A
▪ Sound	1	2	3	4	5	N/A
▪ Content/clarity of video/message display boards	1	2	3	4	5	N/A
▪ Stadium temperature	1	2	3	4	5	N/A
▪ Housekeeping/building services	1	2	3	4	5	N/A
▪ Facility security	1	2	3	4	5	N/A
▪ Parking	1	2	3	4	5	N/A
▪ Overall Rating for the Georgia Dome	1	2	3	4	5	N/A

4. What form of transportation did you use to get to the Georgia Dome (car, MARTA, taxi, other)? ____________

IV. Atlanta

1. How do you rate the following city services on a scale of **1 (Poor)** to **5 (Excellent)**? **N/A = Not Applicable.**

▪ Airport	1	2	3	4	5	N/A
▪ MARTA	1	2	3	4	5	N/A
▪ Hotels	1	2	3	4	5	N/A
▪ Taxi-Cabs	1	2	3	4	5	N/A
▪ Entertainment opportunities	1	2	3	4	5	N/A
▪ Downtown security	1	2	3	4	5	N/A
▪ Restaurant selection	1	2	3	4	5	N/A
▪ Overall Satisfaction	1	2	3	4	5	N/A

V. Demographics: Please tell us about you! It's totally confidential.

1. What is your gender? ☐ Male ☐ Female
2. What is your age? ______
3. Where do you live? City________ State______ Zip Code______

4. What is your personal relationship status – check one below.
☐ Single; ☐ Married Opposite Sex; ☐Married Same Sex; ☐ Partner
☐ Other: I use the term____________
5. What is the ***highest*** level of education you have completed: check one below
☐ Some High School ☐ 1st College degree ☐ In Doctoral Program
☐ Finished High School ☐ In graduate school ☐ Have Doc. Degree
☐ Some College ☐ Master's Degree ☐ Other: ________

6. How many children (18 and under) are in your household? ________
7. What is your income: ____________ OR check one below:
☐ Under $20,000 ☐ $20,000-$39,999 ☐ $40,000-$59,999
☐ $60,000-$79,999 ☐ $80,000-$99,999 ☐ $100,000-$119,999
☐ $120,000-$139,999 ☐ $140,000-$159,999 ☐ $160,000 or more

8. What is your ethnicity?
☐ African American ☐ Caucasian/White
☐ Asian/Pacific Islander ☐ Hispanic/Latino
☐ Other (please specify) ____________

9. What is your occupation? ____________
10. Do you - ride motorcycles?☐Yes☐No 11. - enjoy this event?☐Yes ☐No
12. Would you like for it to be in Atlanta again? ☐Yes ☐No

Any additional comments? ____________

or complicated, can be developed as low or high quality, and can provide results that will be regarded as questionable or credible. It is therefore incumbent on the researcher to adhere to the highest of quality in the development of the survey instrument, especially its validity and reliability, and the implementation procedures of the survey process.

Types of Survey Research Techniques

As the above definition on survey research implies, a sport management researcher or professional administers a questionnaire to a sample of respondents from a population while conducting a survey.

The questionnaire may be a written document that can be completed by a respondent with a pen or pencil, an electronic file that is completed by the respondent online, or an instrument to be completed by the researcher during a face-to-face interview or a telephone interview. Survey research can be conducted in a number of different ways depending upon the research circumstances.

There are two broad categories of survey techniques: the *interview* and the *self-administered survey* (Blackenship, Breen, & Dutka, 1998; Wrenn, Stevens, & Loudon, 2002). In each type, questions are prepared and used. The difference is in how the questions are formatted and how the survey is conducted. Based on the degree of interaction between the researcher and the respondent, a survey technique can also be classified either as a *direct* or *indirect survey* technique. In the direct technique, the survey is conducted in person with the survey respondent. Some direct interviews include the door-to-door survey, the mall approach, and surveys administered to a group. These are conducted according to the sampling methodology selected for the particular study. In the indirect format, the researcher or the individual who collects data does not have direct contact with the respondent. The survey is conducted through such means as the *telephone interview*, *mail survey*, *respondent self-selected survey*, *group-administered surveys*, and *online survey*. Chapter 5 is focused on the interview research methodology; the other survey methods are included in this chapter.

Self-Administered Survey Technique

The self-administered survey technique is one in which a survey instrument—that is, a questionnaire—is administered by the respondent instead of the interviewer. In a self-administered survey, the information is obtained with the traditional paper-and-pencil format. This format can be conducted by getting the survey to the study participants in such a way that the participants can complete the survey on their own. This can be done in person, through the mail, through electronic mail, or as an online survey. The study participant completes the survey and returns it. This technique has four variations: mail surveys, respondent self-selected surveys, group-administered surveys, and online surveys.

Mail Surveys

In the mail survey technique, the researcher mails a questionnaire via the postal service to the intended respondent. Often, a return postage paid envelope accompanies the survey instrument to the respondent. The respondent puts the completed questionnaire in the envelope and mails it back to the researcher. The mail survey is considered "a relatively simple and straightforward approach to the study of attitude, values, beliefs, and motives" (Robson, 2002, p. 233).

There are a number of advantages of the mail survey. It is relatively inexpensive as compared to other types of survey methods. Because there is no one-on-one contact between the researcher and the respondent in the mail survey, the respondent tends to have a greater sense of anonymity and be more willing to provide sensitive information about certain topics. Also, the technique allows the respondent to work on the survey at their leisure and give thoughtful response. In addition, the researcher can utilize this type of survey technique to distribute the questionnaire to a relatively large population.

There are disadvantages to the mail survey, including its low response rate (Blankenship, Breen, & Dutka, 1998) and relatively high nonresponse error (Wrenn, Stevens, & Loudon, 2002). The researcher also has minimal control of the sample (i.e., its size and who will or will not respond to the survey). The literacy level and language ability of the respondents, to a certain extent, affect

Exhibit 9.2. A Sample Follow-up Letter

[Date]

Dear [name]:

On [date], we started a study designed to examine institutional factors affecting the outsourcing decision of athletic administrators regarding marketing operations in NCAA Division I athletic programs. Two hundred thirty-two athletic directors were identified and contacted accordingly. Two subsequent rounds of mailing yielded a return rate of 37%. Due to the limited size of the sample, we would like to have a higher percentage of returns.

As of today, we have not received your response. Your participation is extremely important to the success of our study. Your thoughts and expertise will help us gain a better understanding of the issue. We sincerely hope that you could participate in the study by completing the survey and returning it by [date]. If you have already returned the questionnaire sent to you before, please ignore this letter and the survey instrument attached.

As a principal investigator of the research, I would be most happy to answer any questions you might have. Please call me at (xxx) xxx-xxxx or send me an email to this address: [username@domain.edu.]

Thank you very much for your assistance.

Sincerely yours,

[Name and title]

[Program]

---

the accuracy of the information corrected. Also, because there is no personal contact between the researcher and the respondent in the mail survey, no probing can occur.

Many studies have been done that focus mainly on how to improve the response rate of the mail survey. A number of techniques were found to be effective in encouraging the respondents to complete the survey instrument (Zikmund, 2000).

A *follow-up letter* serving as a reminder is a useful tool in increasing the response rate (see Exhibit 9.2 for a sample follow-up letter). There are a number of possibilities why a questionnaire is not returned to the researcher. Perhaps the respondent misplaced the questionnaire or the return envelope, or the respondent put the questionnaire aside and forgot to complete it. Follow-up mailings help overcome these aforementioned problems. Typically, the researcher sends out a postcard one week after the initial mailing of the questionnaire to either thank the respondents who have returned the survey instrument or remind those who have not completed and returned the questionnaire. In about three weeks, if the response rate is still not satisfactory, the researcher should mail out a new questionnaire with a *cover letter* that not only contains the same content as the initial cover letter, but also informs the respondent that no questionnaire has been received from him or her. No further follow ups are necessary after this point.

Prior to mailing out the initial survey instrument, the researcher may consider sending the respondent an email or a postcard to alert them that they will be receiving a questionnaire from the researcher. This *preliminary notification* should briefly mention the objective of the survey. However, it should not contain too much information to become burdensome to the prospective respondents. Kent and Turner (2002) conducted a study to determine the effectiveness of a number of pre-notification methods. They found that, in general, "pre-notification of the survey recipients significantly increased response rates, with the group receiving email pre-notification having the highest response rate among three groups [i.e., the group receiving email, the group receiving formal letter, and the control group]" (p. 230).

By offering a tangible reward (e.g., a poster with players' autographs), the researcher can induce more respondents to respond. "There is nothing inappropriate about paying respondents for providing information" (Fraenkel & Wallen, 2000, p. 445). Respondents often ask, "what do I get for completing your questionnaire?" To motivate the respondents to participate in the survey, relying solely on altruism is not enough. If it is an onsite survey, providing the participants with some incentive has been proven to be effective. If the information the researcher wants to obtain is of interest to the respondents, providing them with a summary report is also an excellent motivator. For example, a study conducted by Li and Burden (2002) on the current status of outsourcing of marketing operations among NCAA Division I institutions offered the participants (directors of athletics in those institutions) the opportunity to receive a copy of the research summary. The response rate of the study was about 65%.

If a professional organization endorses or sponsors a study, such endorsement will help increase the response rate of a survey conducted of the membership of the organization, as the respondents may feel the completion of the questionnaire is a way of fulfilling their obligation to the organization. A study that is supported and endorsed by a famous sports figure also increases response.

A well-written cover letter gives respondents a good impression. It provides the researcher with an opportunity to persuade respondents to participate in the survey by completing the survey. The cover letter should include information about the researcher, reasons or purpose for the survey, importance or significance of the study in relation to the respondents, a request of or appeal to them to complete the questionnaire, assurance of confidentiality, information about when and how to return the questionnaire, and name and telephone number of the researcher the respondents can call if they have any questions. Exhibit 9.3 is an example of a cover letter that includes all of these elements.

Respondent Self-Selected Surveys

Respondent self-selected surveys refer to the type of self-administered survey in which the questionnaire is intentionally left by the researcher in convenient places for potential respondent to pick up, complete, and drop off in a provided "depository or by a pre-paid-postage return envelope" (Blankenship, Breen, & Dutka, 1998, p. 130).

Although respondent self-selected surveys can provide quick and useful information about a certain issue, researchers must be mindful that it has a number of disadvantages. The most obvious one is that the results may lack generalization due to the high possibility of incurring sampling error in the data collection process. Because respondents are those individuals who self-select to provide feedback, the probability is high that only people who have a strong opinion about the issue, positively or negatively, will reply.

Group-Administered Surveys

The method is favored by a researcher when the researcher has the opportunity to administer the survey to the members of a particular target group in the same place (Fraenkel & Wallen, 2000).

Exhibit 9.3. A Sample Cover Letter/Informed Consent Letter

[Date]

Dear [name]:

The outsourcing of marketing operations and rights has lately become a common practice in American college athletics. It is estimated that more than half of the NCAA Division I-A athletic programs have outsourced some or all of their marketing operations and rights to nationally prominent marketing companies, such as Host Communication, International Sports Properties, and Learfield Communications. Nevertheless, a large number of Division I athletic programs are still keeping their marketing operations in-house.

This study was designed specifically to examine such polarized development in collegiate marketing. Attention was particularly directed to these two issues: why have some athletic programs decided to outsource their marketing operations, while others have not? and what are the perceived benefits of outsourcing sport marketing operations? It is hoped that the study will help those athletic programs that are currently contemplating whether or not they should outsource their marketing operations make a more rational decision. Your participation will be vital to the success of the study. Your input will help us gain a better understanding of athletic administrators' perceptions toward the issue of outsourcing. *It would be greatly appreciated if you could complete the questionnaire and mail it back to us in the enclosed, postage-provided envelop by [Date].*

You may be assured of complete confidentiality. No institution or participant will be individually named. The questionnaire has an identification number for mailing purposes only. With the help of the number, we can check your name off of the mailing list once your questionnaire is returned. The results of this study will be made available to any participant who requests a summary of results by writing "copy of results requested" on the back of the returning envelope. I would be most happy to answer any questions you might have. My phone number and post and email addresses are as follows:

[Name and title]

[Institution name]

[Mailing address]

[City and state, zip code]

[Phone]

[E-mail address]

Thank you for your assistance.

Sincerely yours,

[Name and title]

[Program]

Many sport management studies whose subjects were students used this type of survey technique in the data collection process. A significant advantage of this technique is that the researcher can distribute the survey instrument to all the potential respondents, answer any questions they may have, and collect the completed questionnaires during a specific period of time and at a specific place. This process ensures a high response rate.

A popular sampling method with group-administered surveying is mall-approach. For example, numerous surveyors will be at a particular sports event and will walk around certain areas of the facility or event. While carrying surveys for potential participants, they approach attendees casually. The technique ensures random sampling.

Online Surveys

The online survey technique has become the most popular survey technique in recent years thanks to the development and availability of online survey software and tools. The researcher can create a web-based questionnaire and then guide the respondents via email to complete the survey instrument online. Bennett, Drane, and Henson (2003) used the online survey to collect data in a study designed to examine students' perceptions about service learning. The students were asked to navigate to a questionnaire posted on the Internet using the hyperlink provided.

The online survey technique is the least expensive survey technique of all. Online survey software not only allows researchers to deliver questionnaires to potential respondents in any part of the world in seconds (Bartlett, 2005), but also substantially reduces the time and effort required to input the survey data because the software allows researchers to download the data directly from the web. Online survey software provides researchers with the opportunity to present survey instruments to respondents in an enriched manner with the use of skip patterns, colors, pop-up instructions, and drop-down boxes.

While employing the online survey technique, the sport management researcher must be aware of a number of its disadvantages. The biggest methodological concern is its limitation in sample demographics. In other words, only those people who have computer and online access will have the opportunity to complete the survey. Also, it is possible that the people who respond to the survey may not be the intended target population if the survey is completely open to the public. The aforementioned two disadvantages may invite sampling error.

The response rate is another issue. To improve response rate while using a mail survey, the researcher can usually employ a coding technique and directly send reminders to those identified respondents who have not responded to encourage and solicit their participation. In the online survey, it is impossible to track those who have completed the survey. In addition, the technical aspect of constructing an online survey could be intimidating if the sport management researcher does not have the needed computer expertise. Additional discussion on how to design an online survey instrument can be found in Chapter 12.

Factors Affecting the Choice of Particular Survey Methods

As mentioned previously, there are a number of survey methods that could be used by sport management researchers to derive answers to their research questions. What are the circumstances under which a particular type of survey technique will be chosen? In the following section, five factors that affect the choice of a particular survey method will be reviewed. These factors include population issues, sampling issues, content issues, bias issues, and administrative issues.

Population Issues

Population issues are concerned about the accessibility, the level of literacy, and the geographical location of the population of interest. A sampling frame, from which a sample will be drawn, is required for both interviews and mail surveys. However, a sampling frame is difficult or impossible to compile for some populations. For example, the researcher who is interested in determining how users of professional sports teams' social networking sites react to the design features of those sites does not have a sampling frame of the user population of those sites. As such, the self-administered online survey technique may be appropriate for obtaining the needed information. All survey techniques involve the use of a questionnaire. Completing the questionnaire requires the respondent to have a certain level of literacy in order to read and understand the questions. Even if a respondent can read to a certain degree, the questionnaire may still possibly contain some difficult or technical terms that will cause response error.

If the population of interest is dispersed over too broad a geographic range, it is not feasible to employ the door-to-door interview technique. On the other hand, phone interviews, mail surveys, and online surveys are good options to reach a nationwide sample.

Sampling Issues

Does the researcher have the basic demographic information about the sample, such as home addresses, phone numbers, or email addresses? Is the sampling frame of the population of interest up to date? How accessible is the sample? Who is the right respondent? How critical is a high response rate to the research project? These are some of the sampling issues that are important for a researcher to consider while choosing the survey techniques in data collection. Let us use the last two questions to illustrate the importance of sampling (McDaniel & Gates, 1991).

For example, a researcher is interested in finding out the number of hours American households spend watching sports programming on TV. The most pressing issue that the researcher faces is whether or not the researcher wants to acquire information from a specific person in the family, as a household is not a respondent. If the head of the household is whom the researcher desires to survey, how is the head of the household defined? Is the researcher willing to let any member of the household complete the survey instrument? Similar problems arise when a group, agency, or organization is chosen as a sample unit. Should any member of the group, agency, or organization be allowed to complete a study?

If a high response rate is crucial for the researcher to draw a valid conclusion, neither a mail nor an online survey is recommended, as low response rates are among the most difficult of problems in those two types of survey research. In this case, the door-to-door or mall-intercept interviews may become appealing to the researcher.

Content Issues

If the content of a question asked requires the respondent to consult records before giving the answer, the mail or online survey technique is more advantageous than the face-to-face or phone interview, as the former two types of survey research techniques give the respondents ample time to consult with their records. For example, to respond to a question asking the amount of funds spent on facility renovation in the last five years, an athletic administrator may have to review the budgetary records in order to get an accurate answer.

Bias Issues

As mentioned previously, the use of survey research will inevitably incur some biases. However, the biases may be less problematic with certain types of survey techniques. For instance, on-site, mail, or online surveys are better techniques to ask questions that involve social desirability. It is believed that people always want to look good in the eyes of others. As such, the respondent does not want to say anything that would be embarrassing. For example, a study is designed to investigate the use of iPhones to view NBA games among college students. In it, a question asks respondents the frequency with which they download and watch NBA games on their iPhones. If a student who does not have an iPhone is asked that question in a face-to-face interview, the student may not tell the interviewer the truth but instead offer a fabricated answer.

If the value of a particular research project depends heavily on the expertise of a respondent, a face-to-face interview is the preferable option, as the researcher has no idea of who actually completed the questionnaire in the mail surveys. For example, a study is designed to ask the commissioners of all the NCAA Division III conferences for their opinions and predictions in terms of the best way to reform the Division. It is likely that the surveyed commissioners may actually ask their subordinates to respond to the survey instead of responding themselves.

Administrative Issues

The administrative issues deal with a number of factors that the researcher must consider while implementing the survey research. These factors include costs, facilities, and time. It is believed that cost is the major factor that many researchers have to contemplate when choosing which type of survey method for their studies. Face-to-face interviews are probably the most expensive survey research technique. On the other hand, online surveys may cost very little, but are subject to other limitations. To conduct phone interviews requires well-equipped phone surveying facilities. If time is of essence, the researcher must decide which survey method can provide responses in a relatively short period of time. Under that circumstance, mail surveys are not recommended, as it takes time to send out the questionnaires and follow-ups, and researchers must wait for the respondents to return the survey instruments.

In short, a number of issues should be reviewed and considered by researchers before they choose a particular type of survey technique. It must be pointed out that making such a decision is not an easy task, as there is no one type of survey method that is clearly the best of all. Each survey technique has its advantages and disadvantages. It therefore requires the researcher to carefully review the research objectives so as to identify a method that is the most appropriate to help accomplish them.

Types of Errors in Survey Research

While using survey research as a tool of collecting needed data, the sport management researcher must be aware that three types of errors could be derived from such usage. These errors typically come from different aspects of the research process. The error that arises when the researcher is drawing a sample is called sampling error. The other two types of errors, nonresponse and response error, occur mainly in the data collection process. In the following sections, each of these types of errors will be reviewed in detail.

Sampling Errors

As mentioned previously, sampling errors occur in the process of sampling. When a sample does not well represent the population from which the sample was taken, a certain amount of error may come with the sample. To try to ensure that the sample is representative, the researcher often employs a particular type of random sampling procedure in the selection process. (Chapter 15 is devoted specifically to discussing various types of sampling procedures.) However, no matter how random the sampling procedure is, the result will inevitably bring some error with it. This type of error is commonly referred to as a random sampling error (McDaniel & Gates, 1991, p. 209), as it is a byproduct of the random procedure. Researchers can reduce this type of sampling error by increasing the sample size.

The other type of sampling error is systematic sampling error and is associated with either the sampling design or the execution of the sampling process (McDaniel & Gates, 1991). The most common type of systematic sampling error comes from the use of an inaccurate sampling frame, list of population elements, or database that contains the population elements from which a sample will be drawn. For example, a researcher wants to investigate the perceived importance of collaboration between sport management academicians and practitioners in the United States and Canada among sport management educators in these two countries. The researcher sent a questionnaire to each professional member included in the membership list of the North American Society for Sport Management (NASSM). The results could be biased in that there are sport management educators who are not affiliated with NAASM and are left out of the study.

No matter how accurate the sampling frame is, the results may still be biased if the execution of the sampling process is flawed. A typical mistake committed by those involved in the sampling process is to select the population elements subjectively. Here is an example to illustrate how such subjectivity occurs in survey research: A college sport management professor wants to determine the reactions and favorability of viewers toward the advertisements shown at the recent Super Bowl using a mall intercept interview/survey design to gather data. A number of graduate students are recruited to conduct the interviews in a local shopping mall. Several survey booths are set up, in the major entrances of the mall. Each booth is operated by one student. According to the sampling design, the interviewers are to approach every 20th person walking by the survey booth. The interviewer might subjectively decide to not intercept certain types of shoppers by making assumptions that those shoppers do not look like the traditional football enthusiasts.

Nonresponse Error

Nonresponse error refers to the "error that results from a systematic difference between those who do and do not respond to the measurement instrument" (McDaniel & Gates, 1991, p. 212). In conducting a research project using a survey instrument to gather data, the biggest challenge that the sport management researcher faces is how to ensure that the study will have a high response rate. It is very common to see the response rate in sport management research ranging between 30% and

50%. If the response rate of a study is about 40%, a fundamental question that the researcher must answer is systematically, how different were the responses from those people who completed the questionnaire and those 60% of people who chose not to fill out the questionnaire? In other words, would the results be altered if the response rate increased to 50%? In reality, it is very difficult, if not impossible, to receive a response rate of 100% in the studies employing the self-administered survey method. Nevertheless, the researcher should try to use all the techniques discussed previously to increase the response rate.

Response Error

This type of error occurs when respondents provide inaccurate information in their responses to the survey instrument, consciously and unconsciously. It is very possible for people to fabricate their answers when being asked to answer sensitive questions concerning their household income, their knowledge about certain things, or the level of their performance and/or ability. For example, when asked about their golf handicaps, most amateur golfers will either say they do not have one (basically do not know) or deliberately exaggerate their level to make themselves feel good or avoid embarrassment. One way to reduce this type of error is to draft the survey questions in such a way that they will not make respondents feel intimidated and threatened.

Response error also occurs when the respondent unintentionally gives an inaccurate answer to a question due to misunderstanding the question (the question could be too difficult for the respondent to understand). Let us use an example to illustrate when and how this type of error happens. When responding to the question, "What was your total family income from all sources last year?" a respondent may give the researcher an incorrect total figure because of carelessly failing to include certain amounts of income in that figure (Bradburn, Sudman & Wansink, 2004). Telescoping, "the respondent pushes events backward into a time period previous to the one being asked about" (Bradburn, Sudman & Wansink, 2004, p. 29), also creates response error.

Summary

Survey research is one of the research methods that is used to gather information from respondents through the use of a questionnaire. There are three types of survey techniques: the telephone interviews/survey, personal interviews/survey, and self-administered survey. Each of these survey techniques has its own advantages and disadvantages.

The personal interviews/survey method involves face-to-face contact. Such contact happens when the interviewer asks the interviewee(s) questions and then records the response on the survey instrument. Based on the location where the interviews take place, the personal interviews/survey method has two versions: the door-to-door interviews/survey and mall intercept interviews/survey.

Unlike the interview methods in which the interviewer records the information obtained from the interviewee in the survey instrument, the self-administered survey requires the respondents to fill out the questionnaire without the presence of the interviewer, except the survey administered to a group technique. Mail survey was once the most popular survey method, but its popularity has been eclipsed by that of the online survey. Follow-ups, preliminary notification, survey sponsorship, incentives, and cover letters are five techniques that can be used to increase response rates.

The sport management researcher should be aware of the types of errors that may occur from the application of survey research. In general, three major types of errors are found in sport management research: the sampling error, the nonresponse error, and the response error.

Each survey technique has its own uniqueness in application. As such, the sport management researcher should carefully examine the factors that affect the selection of a particular survey method. These factors include population issues, sampling issues, survey question-related issues, content issues, bias issues, and administrative issues.

Study Questions

1. Describe the various types of survey research techniques.
2. What are the advantages and disadvantages of each type of survey research?
3. What are the techniques used to enhance the response rate of mail surveys?
4. What are the types of errors that may occur from the use of survey as a method of data collection?
5. What are the criteria that can be used to select the most appropriate survey research technique?

Learning Activities

1. What type of survey method is most appropriate for each of the following scenarios? Give your justification.
   a. To understand the preferences of athletic directors of NCAA Division III institutions toward a number of proposals on how to restructure their Division.
   b. To obtain the opinions of the student population of a particular university toward the proposal calling for increasing student fees to support the athletic department.
   c. To determine the user satisfaction of an ice ring.
   d. To find out the level of consumer awareness to a new type of running shoes just introduced to the market by Nike.
   e. To examine the attitudes of local residents about raising taxes to pay for the construction of a football stadium for the professional football franchise currently residing in the community.
2. Provide your critique to each of the following survey designs:
   a. To determine what features users believe are important in a social networking site, such as Facebook, the researchers invite the users of Facebook to complete a questionnaire that they posted on SurveyMonkey.com.
   b. A sporting good store wants to find out if the service provided to customers is satisfactory. The general manager places packets containing a questionnaire, cover letter, and stamped return envelope at a number of locations in the store. Customers can pick up a packet if they wish.
3. Design a study that involves the use of each of the following survey techniques:
   a. The mail survey
   b. The mall intercept interviews/survey
   c. The online survey

References

Bartlett, K. R. (2005). Survey research in organizations. In R. A. Swanson & E. F. Holton III (Eds.), *Research in organizations*. San Francisco, CA: Berrett-Koehler Publishers, Inc.

Bennett, G., Drane, D., & Henson, R. (2003). Student experiences with service-learning in *sport management. Journal of Experiential Education, 26(2), 61–69.*

Blackenship, A. B., Breen, G. E., & Dutka, A. (1998). *State of the art marketing research* (2nd ed.). Lincolnwood, IL: NTC Business Books.

Bradburn, N.M., Sudman, S., & Wansink, B. (2004). *Asking questions: The definitive guide to questionnaire design – for market research, political polls, and social and health questionnaires (2nd. ed.).* San Francisco, CA: Jossy-Bass.

Council of American Survey Research Organizations. (2007). Mission statement. Retrieved from www.casro.org

Fraenkel, J. R., & Wallen, N. E. (2000). *How to design and evaluate research in education* (4th ed.). Boston, MA: McGraw Hill.

Insights Association. (2017). About us. Retrieved from http://www.insightsassociation.org

Kent, A., & Turner, B. (2002). Increasing response rates among coaches: The role of pre-notification methods. *Journal of Sport Management, 16*(3), 230–238.

Li, M., & Burden, W. (2002). Outsourcing Sport Marketing Operations by NCAA Division I Athletic Programs: An Exploratory Study. *Sport Marketing Quarterly, 11*(4), 227–232.

McDaniel, C., & Gates, R. (1991). *Contemporary marketing research.* St. Paul, MN: West Publishing Company.

Pitts, B. G., & Stotlar, D. K. (2013). *Fundamentals of sport marketing.* WV: Fitness Information Technology publishers.

Robson, C. (2002). *Real world research: A resource for social scientists and practitioners -researchers* (2nd ed.). Oxford, UK: Blackwell Publisher.

Trochim, W. M. K. (2006). Survey research. *Research methods knowledge base.* Retrieved from http://www.socialresearchmethods.net/kb/survey.php

Wrenn, B., Steven, R., & Loudon, D. (2002). *Marketing research: Text and cases.* New York, NY: Haworth Press.

Zikmund, W. G. (2000). *Business research methods* (6th ed.). Fort Worth, TX: Harcourt College Publishers.

10

Experimental Research

After completing the study of this chapter, you should be able to:

- Discuss the issues that a researcher must address while designing an experimental study.
- Explain how the independent variable is manipulated in an experiment.
- Define internal validity and explain each of the factors that may threaten the internal validity of an experiment.
- Define external validity and explain each of the factors that may threaten the external validity of an experiment.
- Elaborate on the differences between experiment and other research designs, such as survey.

Experimental research is one of the popular research designs frequently adopted by researchers in business research, as it helps the researcher establish the cause-and-effect relationship between two variables. Researchers in general believed that experimental research is the most scientific method among all business research methods, including surveys, secondary data, observation, and all qualitative research techniques, as it allows the variable causing a particular effect to be isolated (Crawford, 1997) and "establishing stronger and more isolated relationships among research variables" (Ko & Lee, 2017). Nevertheless, the research method has not been given adequate attention by sport management researchers; only a small number of studies found in the sport management literature have employed this technique. The purpose of this chapter is to provide a thorough discussion on *experiment* as a useful research technique. Specifically, the chapter will first review the nature of experimental research, followed by the discussion on the two basic types of validity in experimental research (*internal validity* and *external validity*) and the factors affecting them. The three types of *true experimental designs* will also be examined.

Nature of Experimental Research

In an experiment, the sport management researcher tries to arrange the experimental conditions or procedures so that a hypothesis about a cause-and-effect relationship between two events or variables can be tested. For example, the owner of a golf pro-shop wanted to determine, in an experimental research project, whether the packaging style of used golf balls affected their sales. First, the store sold the balls in an open bin for a month. After recording the sales of balls this way, the store packed and sold the balls in a slate of three. The change yielded better sales figures. The owner then wondered if the packaging caused this sales increase.

In the example illustrated, experimental research is intended to find out what caused a particular thing to happen (i.e., increase in sales) through manipulating the causes (i.e., packaging). Specifically, it helps the researcher obtain information on whether or not the packaging (*independent variable*) affects the sales of used golf balls (*dependent variable*). From this case, we know that the independent variable is the variable that makes the effect happen (the cause) and the dependent variable is the effect that depends on the independent variable. The independent variable is also referred to as "the experimental variable, the cause, or the treatment" (Gay & Diebl, 1992, p. 382). The dependent variable, on the other hand, is also called "criterion variable, effect, or posttest" (Gay & Diebl, 1992, p. 382). Experimental research is the only type of research design that allows researchers to prove a cause-effect relationship, that is, to draw conclusions about causation.

Experimentation: Basic Issues

Exhibit 10.1 illustrates the conceptual model of how an experiment is used to uncover the cause-effect relationship between the independent variable or cause (the packaging of golf balls) and the dependent variable or effect (sales of the used golf balls). Before implementing experimental research, it is imperative to develop a plan or design that will address the following issues:

- selecting subjects randomly from population,
- assigning subjects randomly to groups,
- assigning specific treatments or conditions of the experiment randomly to one or more groups, and
- measuring the dependent variable. (McDaniel & Gates, 1991)

To truly unveil the causal relationship, researchers commonly use several procedures. Such procedures include selecting subjects randomly from the population and assigning subjects randomly to the *experimental group* and the *control group* (Gay & Diebl, 1992). The experimental group is the one in an experiment that receives the specified treatment that is under investigation. The control group, on the other hand, refers to the group that is included in the experiment but not exposed to the experimental treatment. The use of the control group is to establish a baseline against which to measure the effect of the full treatment on the experimental group. In other words, the purpose of using a control group is to create equivalency among the groups involved in the experiment so that they are statistically equal to each other and no systematic difference exists among them, ultimately controlling the extraneous variables. Extraneous variables refers to all the potential variables that are external to the experiment but may affect the results of the treatment on the subject, consequently interfering in the relationship between the independent variable and the dependent variable. Exhibit 10.1 also shows the influence of the extraneous variables on the cause-and-effect relationship between the packaging of golf balls and the sales of used golf balls.

Exhibit 10.1. Conceptual Relationship Between the Independent Variable, the Dependent Variable, and the Extraneous Variable

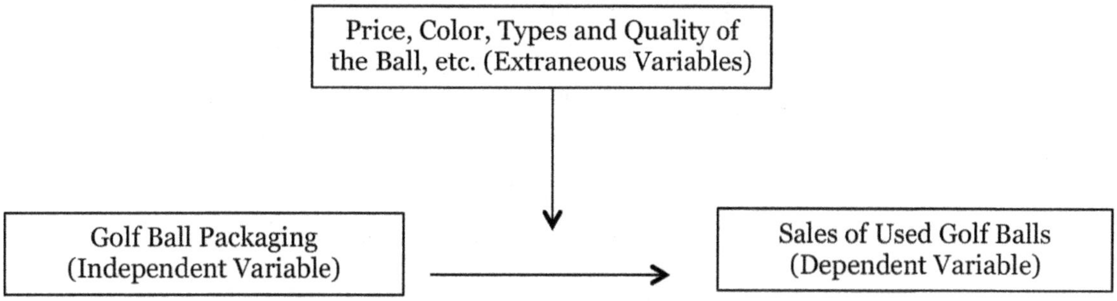

As Exhibit 10.1 shows, the cause-and-effect relationship between the packaging and sales may be compromised by other factors or the so-called extraneous variables, such as price, color, type, and quality of the balls. Specifically, under some circumstances, such as if the pro shop sells the packaged balls in a lower price per unit than that of those balls sold in the open bin, or the pro shop selects the balls with good quality and brands to be packaged (intentionally or not), the causal relationship between packaging and sales could not be clearly established. To truly discover the existence of the relationship between the independent variable and the dependent variable, the extraneous variables must be considered and controlled. To do that, the researcher must ensure that both groups come from the same population, and the subjects and the experimental treatment are randomly assigned to the groups. Random assignment of subjects to groups ensures that the differences between subjects in unsystematic (random) ways are distributed evenly so that there is no tendency to give an edge to any group. In that way, researchers can feel more secure about the results of their studies. Random assignment of experimental treatment refers to the random process to decide which group will receive the experimental treatment.

The main objective is for the randomization process to produce groups that are comparable in unknown as well as known factors that are likely to influence the outcome, apart from the actual treatment under study (Ghauri, Gronhaug, & Kristianslund, 1995). The existence of any systematic difference may weaken the outcome of the experiment and reduce the internal validity of the experiment.

The experimental treatment is the manipulation of the independent variable being investigated (Zikmund, 2000). The process to apply specific treatments or conditions of the experiment to the experimental group is called manipulation. For example, a minor league baseball team wants to know whether promotional activities (independent variable) will increase game attendance. To test if such a relationship exists, the team first identifies 10 home games, the opponents of which all have similar win and loss records. Then, promotional activities are randomly implemented in five of those games. The implementation of promotional activities only to five games is the treatment applied in the experiment.

Experimental Validity

Experimental validity is a term used by researchers to describe the accuracy and generalizability of an experiment. It can be determined in two aspects: internal validity and external validity.

Internal Validity Defined

The inside logic of an experiment is referred to as internal validity. Internal validity refers specifically to whether an experimental treatment/condition makes a difference or not, and whether there is sufficient evidence to support the claim. It basically asks the question: Does it seem reasonable to assume that the treatment has really produced the measured effect? In other words, the outcome of the experiment is the result of the experimental treatment, not something irrelevant to the experiment; otherwise, the experiment is not internally valid (Gay & Diebl, 1992).

Factors That Jeopardize Internal Validity

Campbell and Stanley (1972) identify eight basic factors that threaten the internal validity of an experiment. Their work has been widely cited. These factors are history, maturation, testing, instrumentation, statistical regression, selection of subjects, experimental morality, and interaction with the selection.

History refers to the specific external event (i.e., not part of the experiment) that occurs between the first and second measurement of the dependent variable that may affect the interpretation of the change to the variable. XYZ is a company that manufactures Athrink, a type of sport drinks that claims to have an exceptional effect in helping youths recover the lost minerals after an athletic activity, such as cross country running. The company has decided to introduce a new flavor. Before the introduction, it wants to determine whether or not it should use its traditional packaging design or a new one that is more contemporary. The company chooses Columbus, Ohio, to test the product. In the first two months, it supplies all the stores with the new drink with traditional packaging. Then it replaces all the drinks with the new packaging design in the subsequent two months. Pricing and distribution will be identical in these stores during the experiment. The traditional packaging will act as the control in this experiment. The sales figures will be analyzed to ascertain if the new design led to an increase in sales after the four months of the experiment. In the third month of the experiment, a sponsorship agreement is entered into by the Ohio High School Athletic Association and XYZ. The agreement designates Athrink as the official soft drink of the association's annual state cross country meet, which happens coincidentally during the third month of XYZ's test market. The sponsorship agreement has overshadowed the effect of the new packaging design on sales.

Maturation is the physical and mental changes that may occur within the subjects over a period of time or the processes within subjects that occur as a function of the passage of time. The changes subsequently affect the subject's performance on the measure of the dependent variable. In the case discussed above about the test market on new packaging, if customers prefer Athrink over other brands of athletic drinks as a result of learning more about the new drink through the publicity surrounding the sponsorship agreement, the increased sales is more an effect of maturation than of the new packaging.

Testing is the improvement in the second test as a result of having taken the first test. The improvement in performance of the latter test can be attributed to the familiarity or memory of the test contents by the subjects. This particular threat happens only to experiments that require the subjects to take a pretest and a posttest that are identical. In a sport executive training program, for instance, the participants are given a short questionnaire that determines their levels of awareness and knowledge about basic facts of trends and shifts in consumer needs due to the changes in demographics in the United States. After a week of training, they are asked to complete the same questionnaire again. The sponsor of the training program uses the improvement in the executives' performance on the test to justify the effectiveness of the training program. The improvement contains a certain degree of familiarity by the executives to the questions in the questionnaire after already taking the pretest (Gay & Diebl, 1992).

Instrumentation refers to the lack of consistency or reliability in the measure of the dependent variable as a result of changes occurring in the instrument, observers, or scorers. If the researcher decides to use two separate instruments (claimed to be substitutable to each other) to measure if improvement occurs in a pretest and posttest design to avoid the testing effect as discussed above, the researcher may encounter the instrumentation effect, another type of threat to the internal validity of the experiment (Zikmund, 2000). The instrumentation effect results from the non-equivalency or difference in the two instruments used. Such non-equivalency prevents the researcher from showing that the experimental treatment actually has the true effect on the dependent variable as indicated in the performance of the posttest. The instrumentation effect also presents when observers are used to evaluate the behavior of subjects in an experiment using the pretest and posttest design. The observers "may unconsciously tend to see and record what they know the researcher is hypothesizing" (Gay & Diebl, 1992, p. 389), may become less interested, or may improve their observation skills as time goes by. Consequently, the reliability of data collected by the observers from the pretest and posttest suffers.

Statistical regression is also known as regression to the mean. It occurs when the subjects are selected on the basis of their extreme scores. It is a common phenomenon that the subjects who perform the worst in the pretest would most likely improve their performance after the treatment applied to them. However, such improvement is not the outcome of the treatment, but the statistical regression instead. For example, after a test consisting of multiple-choice questions, a professor in a sport management class identifies all the students whose test scores are below the 15th percentile of the entire class. The professor immediately administers the test again to the identified students. Without any intervention (such as review sessions), the students' performance may be improved. These students are chosen because they are expected to perform better (Gay & Diebl, 1992).

Selection of subjects refers to the threat caused by the selection of comparison groups (i.e., the experimental group and the control group) that are different before the experiment even begins. The existing difference between the groups interferes with the interpretation of the difference in performance of the comparison groups in the posttest that could truly be caused by the experimental treatment. Randomization of group membership and randomization of assignment of treatment are methods that can reduce this threat.

Experimental mortality is the threat caused by the loss of subjects, that is, the subjects voluntarily or involuntarily drop out of the experiment. The reduction in the number of subjects in the experimental group may cause the characteristics of subjects in the group to be different from those in the control group. The reduction in the number of subjects in the experimental group also may cause it to be less or non-representative of the population. In both cases, systematic errors will increase that blur the causal relationship between the independent variable (the experimental treatment) and the dependent variable. For example, an Internet-based introduction to sport management class started with 161 students, and only 95 of them completed the entire class. Those who stayed in the class to the end may have been more motivated to learn and thus achieved higher performance.

Interaction with selection refers to the threat originating from the selection of comparison groups and interactions with other factors, such as history, maturation, testing, etc. The interaction may then lead to confounding outcomes and the false interpretation that the treatment caused the effect.

External Validity Defined

External validity, on the other hand, refers to the proposed interpretation of the results of the study. It asks the question: With what other groups could we reasonably expect to get the same results if we used the same treatment? External validity implies the generalizability of the findings of an experiment or the degree to which the conclusions in an experiment could be repeated by other researchers in different settings at different times.

Factors That Jeopardize External Validity

There are a number of factors that threaten the external validity of an experiment or cause the researcher to make the wrong generalization. Because an experiment in sport management research often involves people (fans) in a specific place (facility) and at a specific time (during the game or event), the generalizability of the results of the experiment to another context (for instance, another place, with slightly different people, at a slightly later time) could be affected by the three elements: people, place, and time. Different groups of fans in a different venue at a different time may alter the condition and thus alter the outcomes of the original experiment. The factors that may jeopardize external validity of experimental research in sport management include surrogate situations, selection-treatment interaction, reactive efforts, and multiple treatment interference.

Surrogate situations refers to the threat that exists when the experimental settings, the treatment applied, and the measurement units are different from those that would be found in the real setting (McDaniel & Gates, 1992). The use of college students as experimental subjects may potentially jeopardize the external validity of any research. For example, a classroom is not a real sport marketplace when students are used by a faculty member in an experiment to determine which brand of sports drink the sport fans prefer. This threat to external validity is caused by the lack of population validity. The population validity is referred to as "the ability to generate to and across subpopulations in the target population" (Johnson & Christensen, 2000, p. 216).

Selection-treatment interaction refers to the threat caused by the nonrandomization of experimental subjects. It will severely limit the generalizability of an experiment.

Reactive effects refers to the threat caused by the subjects in the experiment who intentionally alter their performance as a result of being aware of an experiment. As such, it is difficult to generalize to other experimental settings.

Multiple treatment interference refers to the threat that occurs when multiple treatments are given to the same subjects. As a result, it is difficult to control for the effects of prior treatments. Both the reactive effects and multiple treatment interference are threats to the external validity, the ability to generalize the results across settings, of the experiment (Johnson & Christensen, 2000, p. 216).

There are several approaches to improve the external validity of an experiment. First, the researcher must ensure that the sampling used in the experiment is drawn randomly from a population, so that it reflects the main characteristics of the population. Second, it is also important to keep the dropout rate low after the sample has been selected. Third, the researcher should conduct the experiment in a variety of settings with different people at different times. In that way, the ability to replicate the experiment is enhanced.

Selected Experimental Designs

To make it easy to describe various types of experimental designs, symbols are often used to represent various procedures or acts taking place in an experiment. For example,

X—Treatment,
O—Observation or measurement, and
R—Random assignment.

Specifically, *X* symbolizes the application of a treatment to the experimental group. *O* represents the observation or measurement given before and after the experiment. *R* signifies the randomization process applied to the assignment of treatment to the groups involved in the experiment.

Three types of designs are commonly seen in experimental research: *pre-experimental designs*, *true experimental designs*, and *quasi-experimental designs*. The following section provides detailed descriptions for these three types of designs.

Pre-experimental Designs

Three types of pre-experimental designs are commonly used to illustrate threats to validity of an experiment in literature related to experimental research. The pre-experimental designs do not select subjects and assign them randomly to groups. Two of the three common pre-experimental designs do not have a control group. Because of these two features, the impact from extraneous variables on the results of the experiment is tremendous. Consequently, it is extremely difficult to establish a logical relationship between the independent variable and dependent variable.

The *one-shot case study design* involves only a single group. The subjects in the group are introduced to a treatment or condition and then observed for changes that are attributed to the treatment. For example, a golf equipment manufacturer that produces a new driver has the players on the golf team of a university hit balls with the driver and then measure the distances they hit. This particular type of pre-experimental design can be expressed as

$$X \ O$$

where the symbol *X* represents exposure of the experimental group to the treatment of interest or the opportunity to hit balls with the new driver, and *O* refers to the measurement of the dependent variable or the distance hit with the new driver. In this type, treatment comes before measurement.

The most obvious problems with this design include the lack of control. Also, it is of very little scientific value in terms of securing evidence to make comparisons, and recording differences or contrasts. There is also a tendency to have the error of misplaced precision, in which the researcher engages in tedious collection of specific details, careful observation, testing, and so on, and misinterprets this as obtaining good research. A detailed data collection procedure does not necessarily equal a good design. History, maturation, selection, mortality, and interaction of selection and the experimental variable are all threats to the internal validity of this design.

In *one-group pretest and posttest design*, the subjects are given a pretest and then exposed to a treatment. After that, they are given a posttest. The difference between O_1 and O_2 is compared to determine whether or not the treatment has brought some change to the subjects. The symbolic expression of this design is as follows:

$$O_1 \, X \, O_2$$

Let us use an example to illustrate this type of pre-experimental design. A researcher wants to assess the effects of a promotional campaign on consumer purchasing intention toward a sports drink. The researcher first uses a questionnaire to measure consumers' attitude toward and the consequent purchasing intention of the product. Then, the researcher launches a two-month long campaign to promote the drink. After the promotional campaign, the researcher uses the same questionnaire to measure any changes in consumers' attitudes and purchasing intention.

This type of design is better than the one-shot case study. Nevertheless, there still exist threats to the validity of this design:

- Many events may have occurred apart from the treatment X to produce the differences in outcomes between O_1 and O_2. In other words, the change observed in O_2 may be caused by events outside the experiment rather than by the treatment. The longer the time lapse between O_1 and O_2, the more likely history becomes a threat. For example, the XYZ sporting goods company required its entire sales force to participate in a sales course that introduced several new sales techniques using the Internet. The sales performance in the six months prior to the course was recorded. After the course, the sales performance in the next six months was measured. The economic condition after the initial sales figures were calculated may have changed, accordingly affecting consumers' demand for the products of the company. That will make it difficult for the company to attribute to the course any change in sales in the second measure, taken after the course.
- Between O_1 and O_2, the subjects may have grown older or internal states may have changed (e.g., getting more informed or more experienced) and therefore the differences obtained would be attributable to these changes as opposed to the treatment X. The sales representatives may be exposed to online technology through other channels after the initial observation on their sales performance has been taken. Consequently, the change observed in their performance in the second measure may be partially or completely attributed to their learning outside the course.
- The effect of giving the pretest itself may influence the outcomes of the posttest. In the case of the sales course implemented by the XYZ sporting goods company, it is possible that the sales representatives know their performance after the course will be monitored so that they may work hard to increase their performance.
- Experimental mortality could affect the interpretation of the results obtained in the posttest. The dropout of sales representatives between the first and second measures of sales performance creates disparity between the sales force before and after the sales course.

The *static group comparison* is a two-group design in which the experimental group is exposed to the experimental treatment and the results are then tested; the control group is not exposed to the treatment but similarly tested in order to compare the effects of treatment. The difference between O_1 and O_2 represents the treatment effect. This design is expressed symbolically in the following way:

$$\begin{array}{l} X\,O_1 \\ \;\;\;O_2 \end{array}$$

The major issue of this experimental design is that no pretest is used to establish a baseline to compare the performance of subjects in the experimental group before and after the treatment application. Because of that, this design inevitably inherits flaws. The selection effect is one of the threats to the internal validity of this design—the subjects who participate in both the experimental and control groups may in fact have already been different prior to the treatment. So, the disparity among the two groups between the measures O_1 and O_2 may not actually be caused by the experimental treatment.

The differences between O_1 and O_2 may also be attributed to the dropout rate of subjects from one of the groups in the experiment, which would cause the groups to be unequal. This is the so-called experimental mortality effect. In the case discussed previously, the dropout of sales representatives between the two measures of sales performance creates disparity between the sales force before and after the sales course, potentially affecting the interpretation of the results obtained in the posttest. The interaction of selection and maturation and interaction of selection and the experimental variable are also threats to the internal validity of this type of experimental design.

True Experimental Designs

To reduce the threats to both the internal and external validity of the experimental designs as discussed in the previous section, sport management researchers can seek help from the so-called true experimental designs. In a true experimental design, they not only manipulate the independent variable as they do in the pre-experimental designs, but also randomly assign subjects to either the experimental group or the control group. The random assignment of subjects to groups is the main characteristic of the true experimental designs. In general, there are three types of true experimental designs.

The *pretest-posttest control group design* involves at least two groups of subjects. They are first randomly selected and then randomly assigned to either the experimental group or the control group. So, they are considered equivalent, if not identical, in the composition of subjects. Both groups are administered a pretest. Only the experimental group, however, receives the experimental treatment. In that way, subjects in both groups may be exposed to the same extraneous variables. The experimental treatment is applied only to the experimental group. After the application of treatment, a posttest is administered to both groups again. The difference between measures O_3 and O_4 can be interpreted as the contribution of extraneous variables. The effect of the experimental treatment X, therefore, can be obtained through the determination of the difference between $(O_1–O_2)$ and $(O_3–O_4)$. Symbolically, this design can be expressed as:

Experimental Group	R	$O_1 X O_2$
Control Group	R	$O_3 \;\; O_4$

This design controls for most of the threats to validity, such as history, maturation, instrumentation, as described previously. Let us use an example to illustrate how this design controls for some of the threats. A sporting goods manufacturer wants to measure the sales effect of a point of purchase display of a particular type of baseball glove. It randomly selects 20 retail stores to participate in the experiment, 10 of which are assigned the experimental group and 10 to the control group. Before the experiment, the participating stores' sales performances on the baseball glove are recorded. The manufacturer then places the point of purchase display in those stores selected in the experimental group for three months. After the point of purchase display has been in use in the experimental stores for three months, the manufacturer measures the sales of all stores selected in the experiment. Because of the use of randomization in assigning groups and treatment, the stores in the two groups involved in the experiment are equivalent. The maturation and testing are controlled because they are manifested equally in both experimental and control groups, as they are equivalent. The regression effect is controlled because both the experimental and control groups are randomly selected from the same extreme pool and randomly assigned to groups. If this occurs, both groups will regress similarly, regardless of treatment. The randomization procedure controls the selection effect.

There are, nevertheless, two threats that are still have some effect to the validity of this design: mortality and local history. During the three months the point of purchase display was used, one store in the experimental group was forced to close due to a tax problem and another store caught fire. The dropping out of the two stores may create disparity among the stores in the two groups, as the groups are composed of different subjects after the pretest. The lack of equivalency leads to selection bias, consequently affecting the interpretation of the results of the experiment. Unless the mortality rate is equal in both the experimental and control groups, it is impossible to indicate with certainty that mortality did not contribute to the experiment results.

Local history also affects the validity. During the experiment, a baseball tournament was held in the communities of three of the ten stores. The tournament inevitably leads to an increase in sales of baseball gloves. Because of the tournament, the change in sales between O_1 and O_2 couldn't absolutely be attributed to the application of the experimental treatment X, the point of purchase display.

The threats to external validity could not be completely eliminated either in this type of design. The effect of reactive arrangements (the point of purchase display and the subject's knowledge that the store is participating in an experiment) could have an impact on the experiment results.

The *posttest-only control group design* is similar to the pretest-posttest control group design except there is no pretest. This type of design is used when a pretest is not possible (Zikmund, 2000). Again, subjects are selected and assigned randomly to the two groups in the experiment. In that way, both the experimental and control groups are equivalent prior to applying the treatment to the experimental group. The equivalency not only forms a basis of comparison but also ensures that all extraneous variables have effects on both groups. After the experimental group is exposed to the experimental treatment, a test is administered to both the experimental and control group. By determining the difference between (O_1–O_2), the effect of the experimental treatment X can be extracted. The symbolic expression of this design is as follows:

Experimental Group	R	$X\,O_1$
Control Group	R	O_2

Similar to the pretest-posttest control group design, most of the threats to validity, such as testing and instrumentation, are controlled in this true experiment design. In addition, as a pretest is not administered to both groups, it can be assumed that both groups have experienced the same current events and same developmental processes, thus minimizing both the history and maturation effects. Because the subjects are selected and assigned into either the experimental or control group, they are considered approximately equivalent. Also, the selection effect is not a problem in this type of design.

The *t*-test is the simplest form of statistical tests for this design. However, other forms of statistical analysis (e.g., covariance analysis) can also be used as they increase the power of the significance test similarly to what is provided by a pretest.

To eliminate the threats to validity of the previously mentioned two true experimental designs, the researcher can use the so-called *Soloman four-group design*. This design is in fact the combination of both the pretest-posttest control group design and the posttest-only control group design. To implement this type of experimental design, the researcher randomly selects and assigns subjects to four different groups: the experimental group with both the pretest and posttest, the experimental group without the pretest, the control group with both the pretest and posttest, and the control group without the pretest. By using experimental and control groups with and without a pretest, both the main effects of testing and the interaction of testing and the treatment are controlled. In addition, the impact of extraneous variables is also eliminated. The effect of X is replicated and isolated through several ways and comparisons, that is, $(O_2–O_1)$—$(O_4–O_3)$, $(O_5–O_5)$, and $(O_2–O_4)$. Even though the Soloman design is considered the gold standard of experimental design, due to its impractical nature (the time, effort, and financial resource required), it is rarely used in business and sport management research. The symbolic expression of this design is shown as follows:

Experimental Group 1:	R	$O_1\ X\ O_2$
Control Group 1:	R	$O_3\quad O_4$
Experimental Group 2:	R	$X\ O_5$
Control Group 2:	R	O_6

A good way to test the results is to rule out the pretest as a treatment and treat the posttest scores with a 2*X*2 analysis of variance design.

Quasi-Experimental Designs

Because of the application of procedures commonly observed in both the pre-experimental and true experimental designs to control and eliminate threats to the internal validity, such as the use of two groups in the experiment and the randomization in subject selection and assignment of treatment, the generalizability of experiments to other populations and settings is severely affected and compromised. To deal with the problem, quasi-experimental designs, also referred to as compromise experimental designs, are introduced and preferred by sport management researchers, as they are usually more feasible in real-world settings (Cuskelly & Hoye, 2013; Romsa, 2013; Vojinovic, Maksimovic, Kokovic, Raic & Doder, 2015).

In a quasi-experimental design, the randomization of selection of subjects or assignment of subjects to groups is not observed due to practical constraints. For example, in a study that involves the employees of an athletic department, it is not possible for the researcher to assign subjects into groups randomly. Instead, the researcher must use existing units within the department. Under this kind of special research circumstance, the researcher then must use a number of the so-called quasi-experimental designs that still have a certain level of control over the threats to the validity of an experiment. To ensure that an experiment has adequate validity, however, it is strongly recommended that the researcher use the true experimental design, if feasible (Gay & Diebl, 1992).

A number of quasi-experimental designs have been commonly discussed in literature in business research (Gay & Diebl, 1992; McDaniel & Gates, 1991; Zikmund, 2000). They include the *time-series design*, *multiple time series design*, and *nonequivalent control group design*.

The *time-series design* is in fact the expansion of the one-group pretest and posttest design, one of the pre-experimental designs. In this design, the subjects are repeatedly pretested and then exposed to a treatment. After that, they are again repeatedly post-tested. The difference between O_1 and O_2 is compared to determine whether or not the treatment brings some change to the subjects. This design is especially appropriate for an experiment that requires repeated measurement that is "unobtrusive" and participants that are "not reactive" (McDaniel & Gates, 1991). The symbolic expression of this design is as follows:

$$O_1\ O_2\ O_3\ X\ O_4\ O_5\ O_6$$

The time series design provides greater control over extraneous variables than the one-group pretest and posttest design, as the participants are tested repeatedly before and after the treatment. As discussed previously, history is a threat to the internal validity of an experiment using the one-group pretest and posttest design. This is because events occurring between O_1 and O_2 could affect the interpretation of treatment X on the dependent variable, as the change observed in O_1 may result from the events rather than the treatment. To illustrate how the time series design can help minimize the history effect but not completely eliminate it, let us use the example in which the XYZ sporting goods company requires its entire sales force to participate in a sales course that introduces several new Internet sales techniques. Before the course, the company records the performance of its sales force in the following manner: six months prior to the course, four months prior to the course, and two months prior to the course. It then measures the performance of its sales force with the same two-month interval after the course: two months after the course, four months after the course, and six months after the course. The repeated measures can detect any unexpected surge in sales performance. If the performance of the sales force is essentially the same in a number of measures prior to the course, and significantly improved in a consistent manner after the course as reflected in a number of measures, the company can be confident that the course is effective. The events that occur after the last pretest and before the first posttest potentially are still a problem that may confuse the company's understanding of the true effect of the course.

To improve the effectiveness of the time-series design, it is suggested that another group be added as a control group to the basic design. This design is called the *multiple time-series design*. The symbolical expression is as follows:

Experimental Group:	$O_1\ O_2\ O_3\ X\ O_4\ O_5\ O_6$
Control Group:	$O_1\ O_2\ O_3\quad\ O_4\ O_5\ O_6$

This variation of the time-series design helps control history and other threats to internal validity, such as instrumentation. The key issue for this design is to select a control group that is closely equivalent to the experimental group in the characteristics of interest.

The third quasi-experimental design is the *nonequivalent control group design*, in which there are two groups of subjects involved in the experiment (i.e., the experimental group and the control group). Both groups receive a pretest before the treatment is applied to the experimental group. After that, a posttest is administered to both groups. This design is similar to one of the true experimental designs, the pretest-posttest control group design, except subjects are not randomly assigned to the two groups. So, the researcher cannot be sure that the two groups are equivalent in terms of their composition. As shown in the following symbolical expression, it can be noticed that the *R* is missing.

Experimental Group:	$O_1\ X\ O_2$
Control Group:	$O_3\quad O_4$

The lack of randomization in group assignment presents problems to this design. Nevertheless, as mentioned previously, the quasi-experimental designs are options in research situations in which it is impossible to randomly assign subjects to groups. The use of a control group, even though it may be equivalent to the experimental group, still can help the researcher interpret the effect of treatment on the experimental results (O_2–O_4). The challenge to the researcher is to find a way to achieve equivalency of the two groups. The more similar they are, the more likely the threats to validity can be controlled effectively.

In a study aiming at examining the efficacy of organizational support strategies and its effects on the retention of early career sports officials, Cuskelly and Hoye (2013) used a nonequivalent control group experimental design to achieve its research goal. The sports officials in the experimental group were provided with a mentor during the season while the referees in the control group received no mentorship in the experiment. No randomization procedures were employed to assign participants into groups. A pre-survey prior to the season and a post-survey after the season were administered to all participants. Measures in the survey included sport commitment, stressors of officiating and POS as independent variables, and ITC refereeing as dependent variable.

A Case of Experimental Research in Sport Management

An experiment was conducted (McCarville, Flood, & Froats, 1998) to determine the effectiveness of several promotion activities on spectators' assessments of a sponsor. Specifically, they wanted to seek answers to the following research questions:

- Will different message types offered in a sponsorship context alter spectators' impression of the sponsor being promoted?
- Will different message types offered in a sponsorship context alter spectators' impression of the product being promoted?
- Specifically, will different message types offered in a sponsorship context alter spectators' impression of the quality of the product being promoted?
- Will different message types offered in a sponsorship context alter spectators' likelihood of future purchase of that product?

The experimental design involves 163 undergraduate students who are randomly assigned into one of five groups, four of which are the experimental groups and the fifth is the control group. The four experimental groups are exposed to different sponsorship messages (treatments). The specific arrangement of each experimental group and the treatment it receives are as follows:

Experimental group I: Exposed to the sponsor's logo
Experimental group II: Received the sponsor's information
Experimental group III: Received the sponsor's discount coupons
Experimental group IV: Tried the product

The control group receives no relevant sponsor-related information or treatment. Using symbolic expression, the design of this study (i.e., the posttest-only control group design), is shown as follows:

Experimental Group I:	$R\,X_1\,O_1$
Experimental Group II	$R\,X_2\,O_2$
Experimental Group III:	$R\,X_3\,O_3$
Experimental Group IV:	$R\,X_4\,O_4$
Control Group:	$R\quad O_5$

During the experiment, the subjects were given the sponsor-related information while they were watching a slide show of a volleyball match in a large lecture-theater-setting with the presence of many other students. After the slide show, the subjects in all five groups were asked to complete a questionnaire that contained questions related to the efficiency, reliability, responsibility, worth of the business, and product quality.

This is a posttest-only control group design with more than one experimental group. No survey was administered to the students before the treatments and the slide show. The experimental design successfully eliminated most of the threats to its internal validity.

- Both the history and maturation effects are controlled as the posttest (i.e., the survey administered to the subjects) was given immediately after the subjects finished viewing the slide show. No possible history events could occur that may have jeopardized the internal validity of the results. It is equally true that the time was too short to have any maturation effect on the outcome of the experiment.
- The selection bias is also effectively controlled. It is accomplished through the use of the randomization procedure in subject assignments into groups. Because of randomization, it can be assumed that the five groups are equivalently similar in demographic characteristics. The difference observed in the dependent variable can then be attributed to the treatments.
- The effects of testing and instrumentation are effectively eliminated because only a posttest was used.

The only threat to the validity of the experiment is that it was not conducted during a live sport competition. Even though the subjects watched the slide show at a large lecture theater setting with the presence of many other students to simulate a true competition venue, their experience may be quite different while watching a live event at a sport facility. This might be the drawback of the research.

Why Are Experiments Not Used More Often in Sport Management Research?

As mentioned in the beginning of this chapter, experimental research is one type of research design that can reveal a cause-and-effect relationship between two events or variables. As such, it should be a very powerful means for sport management research. However, a brief review of all the research published in the several leading sport management journals indicates that far less than 1% of all sport management research published in those journals can be described as experimental research. Why?

Three reasons experimental research is limited in business research were presented by McDaniel and Gates (1991): high costs of experiments, security issues, and implementation problems (pp. 300–301). The same reasons are also applicable to sport management research.

It is often the case that experimental research is more costly and time consuming than survey design. Experiments are also used as a primary means of test market. To implement the experiment, certain elements of the marketing plan may have to be disclosed. Consequently, competitors may learn about the business actions of the company that wants to test the market, and accordingly come up with reactive strategies. Because of being afraid of losing their competitive edge, companies are very careful in using experimental research.

Researchers may encounter some problems in implementing an experiment, which include, but are not limited to, difficulty in gaining cooperation, subject contamination, differences between test markets and the total population, and lack of a group of people or geographic area as a control group (McDaniel & Gates, 1991, p. 300).

Ethical Issues in Experimental Research

When the design of an experiment involves double-blind conditions in which subjects are not informed concerning things to which they will be exposed, a certain level of deception is required, and informed consent can be difficult to establish. This deception can cause harm to subjects and violate their rights. The research code of conduct requires that the subjects be adequately informed of the experiment. In certain experiments, to avoid the interaction between selection and treatment, researchers often hide the purpose of the experiment from the subjects.

Debriefing can be used to mitigate this tendency. Debriefing is a process in which subjects are given the information about the experiment, such as its nature and purpose, after the experiment has been completed (Zikmund, 2000).

Summary

Experimental research is one of the quantitative research designs in sport management. In an experiment, conditions or procedures are arranged so that the researcher can test a hypothesis about or determine if a cause-and-effect relationship exists between two events or variables. Well-executed experimental research often involves elements such as random selection of subjects from population, random assignment of subjects to groups, random assignment of experimental treatment or treatments to groups involved in the experiment, and measurement of the dependent variable.

Two types of validity are often observed in experimental research: internal validity and external validity. The internal validity refers specifically to whether an experimental treatment or condition makes a difference or not, and whether there is sufficient evidence to support the claim. It is about the inside logic of an experiment. On the other hand, external validity is concerned about the generalizability of the findings of an experiment or the degree to which the conclusions in an experiment can be repeated by other researchers in different settings at different times.

To improve the internal validity of an experiment, it is important for the research to control threats that may jeopardize the interpretation of the causal relationship between the independent variable and dependent variable. These common threats include history, testing, instrumentation, maturation, mortality, selection, statistical regression, and interaction with selection. Four common threats to external validity in experimental research are surrogate situations, selection and treatment interaction, reactive effects, and multiple treatment interference. Three types of pre-experimental designs are commonly used to illustrate the threats to the validity of an experiment. There are the one-shot case study, the one group pretest and posttest design, and the static group comparison. Because of the lack of control over extraneous variables, it is very difficult to make causal inference between the independent variable and dependent variable.

The researcher can employ three true experimental designs to reduce and prevent the threats of extraneous variables from contaminating the internal validity of their experiment. The three true types of experiments include the pretest-posttest control group design, the posttest-only control group design, and the Soloman four-group design. True experimental designs can effectively control most of the threats that jeopardize the validity of an experiment.

If the random assignment of subjects to groups is impermissible in some research environments, the researcher can turn to the quasi-experimental designs (they are also called compromised experimental designs) for help. The three commonly used quasi-experimental designs include the time-series design, the multiple-series design, and the nonequivalent control group design. To improve the validity of the type of experimental designs, the researcher should try to create similarity between the two groups involved in the experiment.

Study Questions

1. What are the issues that a researcher must address while designing an experimental study?
2. How is the independent variable manipulated in an experiment?
3. What is internal validity? Explain each of the factors that may threaten the internal validity of an experiment.
4. What is external validity? Explain each of the factors that may threaten the external validity of an experiment.
5. What are the differences between experiment and other research designs, such as survey?
6. Compare and contrast the pre-experimental designs, true experimental designs, and quasi-experimental designs.

Learning Activities

1. Design an experiment that uses one of the three good experimental designs to determine the effect of a sales workshop on the performance of the participants.
2. What type of experimental design will you recommend in each of the following situations? Be sure to explain the details of your design.
   a. A professional baseball team decides to use variable ticketing in formulating its ticket sales strategies. The management wants to determine the impact of the combination of price and type of game on ticket sales.
   b. A sporting goods store wants to know whether or not the presence of a professional player will attract shoppers to the store and increase its sales.

References

Campbell, D. T., & Stanley, J. C. (1972). *Experimental and quasi-experimental designs for research.* Chicago, IL: Rand McNally.

Crawford, I. M. (1997). *Marketing research and information systems* (Marketing and Agribusiness Texts—4). FAO, Rome: Agricultural Support Systems Division.

Cuskelly, G., & Hoye, R. (2013). Sports officials' intention to continue. *Sport Management Review,* 16, 451–464.

Gay, L. R., & Diebl, P. L. (1992). *Research methods for business and management.* New York: Macmillan Publishing Company.

Ghauri, P. N., Gronhaug, K., & Kristianslund, I. (1995). *Research methods in business studies: A practical guide.* New York: Prentice Hall.

Johnson, B., & Christensen, L. (2000). *Educational research: Quantitative and qualitative approaches.* Boston: Allyn and Bacon.

Ko, Y. J., & Lee, J. S. (2017). SMQ Call for Papers: Experimental Research in Sport Consumer Behavior. Retrieved from http://fitpublishing.com/journals/SMQ/smq-call-papers-experiment

McCarville, R. E., Flood, C. M., & Froats, T.A. (1998). The effectiveness of selected promotions on spectators' assessments of a nonprofit sporting event sponsor. *Journal of Sport Management, 12*(1), 51–62.

McDaniel, C., & Gates, R. (1991). *Contemporary marketing research.* St. Paul, MN: West Publishing Company.

Romsa, B. (2013). *Undergraduate Sport Management Students' Perceptions of Leadership Behaviors through Service Learning: A Quantitative, Quasi-Experimental Study.* (Doctoral Dissertation.). Retrieved from Proquest Dissertations and Theses database. (UMI Number: 3564134).

Vojinovic, P., Maksimovic, N., Kokovic, D., Raic, A., & Doder, D. (2015). Predicting the future of sports organizations. *Motriz, Rio Claro, 21*(2), 107–115.

Zikmund, W. G. (2000). *Business research methods* (6th ed.). Fort Worth, TX: Harcourt College Publisher.

Section III

The Research Process in Sport Management Research

11

The Research Process in Sport Business Management

After completing the study of this chapter, you should be able to:

- Understand the decision-making continuum and the variables affecting the confidence of a decision.
- Identify the sport business management research process and its steps.
- Explain the details of each of the steps in the sport business management research process.
- Compare and contrast qualitative and quantitative research methods in sport business management.
- Identify the common types of research designs or methods used in sport business management research.

Sport business management professionals in nearly all sport businesses typically make decisions based on information gained from research. To provide accurate information, it is important for the sport business management professional to follow the scientific model of research to derive the needed information. No matter how simple or complicated the issue or problem that needs to be studied, the scientific model of research is always involved.

The purpose of this chapter is to introduce the research process. To accomplish this goal, this chapter presents and explains the decision-making continuum and the variables affecting a sport business management professional's confidence in making a decision. A detailed discussion of each of the steps in the sport business management research process and some common types of *research designs* or methods used in sport business management research will follow.

A Primary Purpose of Research: Decision Making

A primary purpose of research is to inform decisions. Decision-making is the thought process of making a determination involving what to do about something or what to do in a situation.

Exhibit 11.1. Decision Maker's Continuum of Confidence

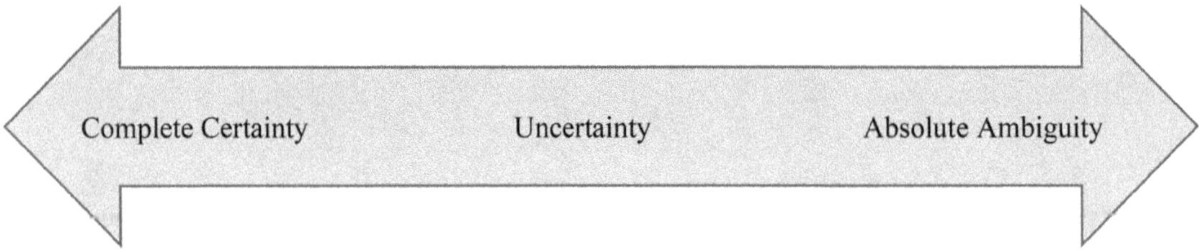

BusinessDictionary.com offers this definition: "The thought process of selecting a logical choice from the available options," and further states, "when trying to make a good decision, a person must weigh the positives and negatives of each option and consider all the alternatives" (Decision Making, 2017). Additionally, the decision maker must understand the nature and elements of the problem or issue, determine how much information is readily available to help make a decision, determine what additional information is needed, and decide on the best method for collecting any needed information.

Every sport business problem or decision-making situation can be categorized along a continuum of confidence. Exhibit 11.1 illustrates the continuum. This continuum offers a look at the effects of the decision (Daft, 1990; Schermehrhorn, Hunt, & Osborn, 1991). In other words, how confident can the decision maker be with any decision? For example, is it possible for an individual to make a statement and be completely confident that the statement is correct? The answer is yes. If the individual is able to substantiate the statement with evidence that the statement is true and correct, then that individual can say that they are 100% confident, or that the statement is true with 100% certainty.

Certainty versus Uncertainty

Complete certainty can only exist when the decision maker has the best and most correct information possible to validate the decision. When there is information from research that was conducted in which the findings of the study reveal specific conclusions, the decision maker can make a decision with a high level of certainty. For example, the sport marketing director at a collegiate athletic program wants to know the demographics of its women's basketball season ticket holders. The director knows exactly what information is needed and how to obtain it. The director can determine if the information already exists. If not, then research must be conducted to get the information. Based on the research results, the director now has the information and can state with certainty the demographics of the season ticket holders.

Uncertainty exists when the decision maker has an idea about the concept of the problem or opportunity, but lacks specific information about it, or lacks information about alternatives. In this situation, decisions will be made with uncertainty. In other words, the decision maker will be guessing about the outcome of the decision. This kind of decision making can lead to disaster. A decision made with uncertainty has the potential to be wrong.

Ambiguity

Ambiguity is a situation in which the problem to be solved is unclear. The cause of the problem is unknown or indistinguishable. Therefore, the problematic situation leads to vague descriptions, unclear objectives, and ambiguous solutions. All sports management professionals face ambiguous situations. For example, a sporting goods manufacturer will inevitably face a great deal of ambiguity when it decides to expand into a foreign market where the rules of doing business, the financial system, and business relation issues are unclear, and business strategies that have been proven to be successful in North America may no longer be applicable. It is under this situation that research becomes more desirable and advantageous. Typically, in these situations, the manufacturer making an expansion decision needs to take the time and money to conduct appropriate research that will move the situation from ambiguous toward certainty.

The Research Process in Sport Business Management

Sport business management research begins when a question arises, a problem occurs, or there is a need to gather information for a purpose. Some research is conducted by scholars, academics, and/or scientists for the purpose of discovering information to answers questions and/or to advance the existing set of information or knowledge about something. Some of this research is done to add to information to an existing body of knowledge, also called a body of literature. It is critical that the body of knowledge about a topic be as current as possible so that information is as accurate as possible. Professors and researchers in academic settings conduct research to create and develop a body of literature that will be used in education—students must have this current information in order to be appropriately educated. Thus, professors and researchers in academia are responsible for creating information that will be published in academic and scientific journals, as well as in textbooks.

In a business industry setting, workers are responsible for enhancing the odds that a business will continue to offer products that consumers want, need, or desire. To do this, accurate information is critical. Accurate information comes from appropriate research. Thus, in a commercial setting, workers are responsible for conducting research to discover information that will be used in making decisions and solving problems. For example, if a college sports sport marketing director is confronted with a situation in which 20% of the women's basketball season ticket holders did not renew their season tickets for next season, that director must conduct research to discover why this happened. A process involving specific stages and steps for determining how to find a solution or an answer to the question (i.e., what caused the decrease in season ticket renewal) should begin. Having the accurate information will guide decisions to be made to address the loss of sales in the future.

In all settings, research should be conducted that will provide accurate and appropriate results. This means that valid methods to conduct research should be followed. Although there is no one exact process to be followed in sport business management research for every research project, there are specific stages and/or steps that should be consulted and followed closely so that a higher level of certainty for accurate and appropriate results can be reached. With a higher level of certainty, the sport business management professionals can have a higher level of confidence that the information obtained from the research is valid.

There is no absolute adherence necessary in the research process regarding a step-by-step sequence. Some variation of the process can occur. Some steps can be started out of sequence, some can be omitted, and some can even be carried out simultaneously. This is acceptable as long as the integrity of the research is not compromised, so that results are valid. Exhibit 11.2 illustrates a typical research process. When an individual wants to conduct research to gain information to help solve a problem or to answer a question, the steps of the research process should be followed.

Exhibit 11.2. The Research Process

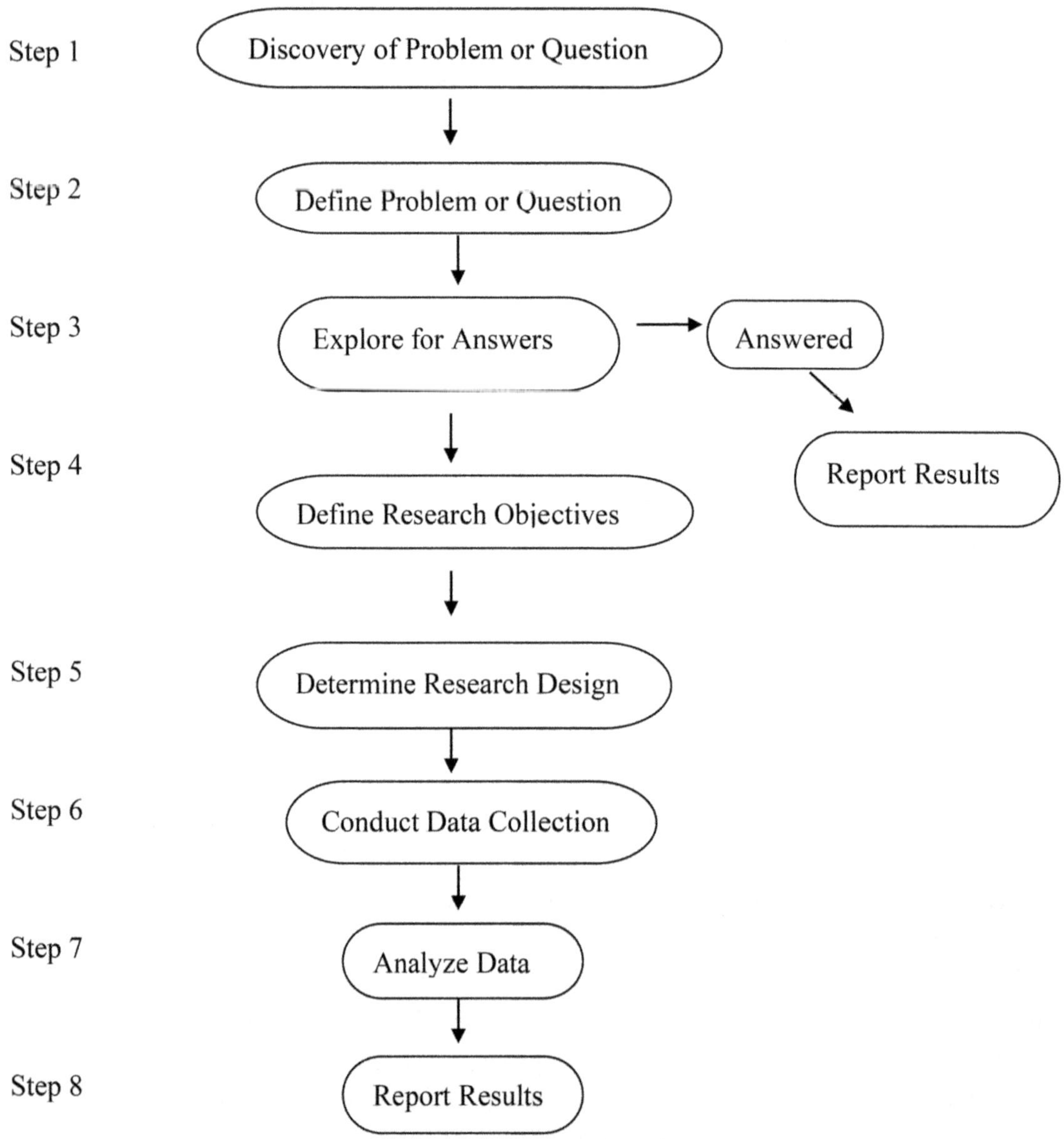

Step 1: Identify the Problem or Question

The first step is when a problem or question has been discovered. The word problem typically suggests that something or a situation is regarded as out-of-the-ordinary, not normal, wrong, unwelcome, harmful, and/or worrisome. A problem might also be a situation that needs study. A question may be a situation in which someone wants an answer to a question or inquiry. It might mean that someone in the company needs information in order to make a decision.

Here is an example to illustrate this situation. Social media has grown as a phenomenon in society in recent years. Individuals as well as companies have struggled with its purposes and uses and adopting appropriate guidelines alongside legal and ethical boundaries. As social media platforms and use have been adopted and grown, innovations have been made utilizing this technology for a myriad of purposes. One such purpose is relationship building or social networking.

Social networking has enabled Internet users to forge online relationships with individuals based on a variety of factors, such as place of residence, school attended, and occupation. Internet social networking has grown so quickly that many organizations still do not fully understand the phenomenon. Without a solid grasp of the nuances of this phenomenon, organizations are unable to leverage the technology within their respective business model. One cell phone provider wants to launch a college athletics themed social networking site to take advantage of the popularity of the social networking phenomenon. To fully understand the phenomenon, management of the cell phone provider company needs to conduct research to seek answers to a number of questions: Why do people use social networking sites? What features on social networking sites do people use the most? What features on social networking sites do people use the least? What features are missing from social networking sites that people would like to use? What social networking sites do people use the most? How can the company attract potential sports fans to their social media sites? Are sales possible with the use of the company's social media advertising? All of these questions will need proper research conducted in order to attain the information needed.

In other situations, a problem might be that the sport business decision-maker has discovered something wrong—a problem—and needs information in order to correct or determine a solution to the problem. For a sporting goods retail company, a problem might be slumping sales of a particular item. In that case the question is, why have the sales of this item decreased over the past 6 months? For a university athletics department, a problem might be that the number of spectators for the women's softball games have dropped over the past year. In that case the question is, why is the number of spectators decreasing? For a large city marathon, the number of sponsors have decreased over the past five years. In this situation, the question is, why have we lost sponsors over the past few years? These are a few common scenarios that require sport business management professionals to find out the reasons and take action.

A question might arise from simply wondering how, why, where, or what is or is not going on. An owner at a fitness center might wonder why clients are leaving clothing and keys lying around the workout areas. This could be triggered with a simple observation. When the owner asks the manager why this is happening, the manager may or may not have an answer. If the manager has no answer, then this is a question for investigation. The question formally becomes why fitness center members are putting clothing, keys, and other items on the floor and equipment while they are working out. It may seem innocuous. However, items lying around are a health and legal risk hazard. They create the potential for accidental injury, and they pose a health risk because they are laden with bacteria that can be left on the surfaces they touch. Additionally, from a purely aesthetic perspective, they make the club appear cluttered and disorganized—definitely not the image that the owner wants for the fitness center. The owner might request the manager to take some time to look into this situation and then offer solutions. At this point, the manager will need to approach the situation in a formal manner. This is the point at which the manager will determine a more formally defined problem or question.

In another example, a sports equipment company wants to find a new material with which to manufacture a new baseball and softball bat, a new basketball, or a new swimming goggle. Most of these types of companies have a research and development department, and some have laboratories dedicated to conducting scientific laboratory research full time.

In summary, the first step toward determining that research needs to be conducted is the discovery or development of a problem or a question.

Step 2: Define the Problem or Question

Simply stating that there is a problem is not enough. One must describe the problem with specificity. Otherwise, research cannot be conducted that will bring about solutions. Not all problems or questions are researchable; some may not require a formal research process. So, one must begin by determining if the problem or question is researchable.

In addition, a question might be in a form that is too simple for formal investigation. In the earlier example about the fitness center, the original question might appear to be simple enough. But it needs formulation and refinement so that the manager can look for answers that will bring about real solutions. For instance, if the manager studies the question and learns that the reason personal items are left lying about the center is because the center has no locker room facilities for personal item storage, then the search for decisions and solutions to this situation can be done with an educated purpose.

In one example, the athletic administrators of Central University realize that tickets to the men's basketball games are not selling, while tickets to the women's basketball games are always a sellout. They want men's basketball to sell all of its tickets too. Therefore, this is a problem that needs a solution.

In another example, the Acme Sporting Goods Company, a company that manufacturers golf clubs, determines that two of their six models aren't selling as well as the others. The executives of the company want all models to sell equally or they may need to terminate some models. In order to decide how to sell the models, they will have to determine why the models aren't selling as well as the others.

In the golf clubs example, the models that aren't selling as well as the others may or may not be a problem. It is up to the executives to decide if this is a problem. It could be considered a problem if the sales totals for the two models are below the cost to manufacture, thus creating a situation in which there is loss instead of profit. The executives will need to decide if this is acceptable or not. If not, then it is considered a problem. The question then is developed out of the situation: why aren't the two models selling?

In summary, this step requires defining and structuring the problem or question in such a way so that it can be researchable. Based on this, research can now be conducted to try to answer the specific question. The results of the research will then be used to create a solution to the problem.

Step 3: Explore for Answers

Once the problem or question has been defined, the researcher can begin to search for answers. There is the potential that the answers might already exist in available information. For example, if the question is what the going price for high-end tennis rackets is, this information can be found easily by conducting a web search. In that instance, the question is answered and the results can be reported. In another example, if the question is what are tennis participation rates for a specific year, it is also possible to find this information in a number of places. Most likely, a research company or a tennis organization has this information. You can access it and the question is answered. This existing information is called *secondary data*. This means that the information is second-hand information, or that it came from someone besides you who conducted the research and produced the information. If secondary data exists that will provide the information needed to solve a problem or answer a question, then no further action is needed. If, however, secondary data does not exist, research will be needed. This is called *primary data*. Primary data are information created or collected by the researcher. That is, the researcher has conducted the research personally—or oversaw the process—and the information was developed by the primary source.

In summary, this step is to determine if information exists already that will provide solutions to a problem or answers to a question. If not, then research will be needed.

Step 4: Define Research Objectives

After identifying the problem or the question, the researcher will need to develop a detailed and formal statement of the problem and the *research objectives.* The objectives will establish exactly what should be researched. This will help identify what type of information is needed and provide a framework for the research project.

Many researchers overlook this very important step. Yet, this is the critical step that delineates what is to be accomplished. Otherwise, the project has no direction and worse, it is likely that wrong or inappropriate information will be gathered. The danger of this is that using wrong information will lead to inappropriate decisions that will most likely be unsuccessful.

A question such as why aren't current university students attending the women's and men's basketball games is a typical research question. The objective then might be stated as the objective of this research is to determine why current university students do not attend the women's and men's basketball games. A research objective is a precise statement indicating the problem and what should be researched (what needs to be accomplished in further research) (Zikmund, 2000). Both the research question and objectives are exchangeable. Either the question or the objective can be used to state the problem.

The research question or objective should be one that can be measured or tested. For example, with the objective of the project to find out why university students aren't attending basketball games, this can be measured. In the simplest form, the researcher would survey the current students and ask them why they don't attend the games. This one question would provide a simple answer to the problem. Once the objective has been developed, the next step is to determine the best research design.

In summary, the purpose of this step is to develop formal and strategic statements and objectives of the research. These will help guide decisions about what type of research methodology will be best, and what research logistics will be needed to reach the objectives.

Step 5: Determine the Research Design

The research design is the master plan that identifies the research methods, procedures, and all logistics for the development of the study, collection of the data, and analysis of the data. The objectives of the study are used as the basis around which the design will be built. This will ensure that the information to be collected is appropriate.

The researcher must build into the design the sources of information, the research methodology, sampling methods, project schedule, logistics, and costs. Particularly, this step is a good place to determine what the research will cost and check the budget to see if there is enough funding to carry out the chosen design. If the money is available, then the project can be carried out as planned. If the money is not available, then the researcher must decide on a less costly design that will provide the best information possible to answer the research question without hurting the integrity of the research.

The research design begins with choosing the type, or method, of research to use. For most research conducted for academic and scholarly purposes, the researcher often begins with a search of existing studies on the topic. This is called a review of the literature. A *literature review* is a search of published studies in journals and books that have already been conducted on the same or similar topic. It is done to determine the legitimacy of the research. Specifically, the purpose of reviewing literature is two-fold: (1) to find out what other people have found concerning the topic and (2) to determine clues and suggestions on how to proceed with setting up the investigation and data gathering. More specifically, the review of literature may provide the team with a framework and clues for its questionnaire design. If previous studies can be found, then it must be determined if the studies and the findings answer the research question or provide solutions for the research problem. If not, then another approach is needed.

The researcher may also use some exploratory research to better understand the issue at hand. One of the frequently used exploratory techniques is a *pilot study*. A pilot study is a precursor to the regular study. Often, a pilot study is a much smaller trial run of the regular study. Sometimes, the findings of a pilot study offer enough information to solve the research problem, answer the research question, or meet the research objectives. In this case, no further research need be conducted at the moment. For example, to understand sport sponsorship in China, Geng, Burton and Blackmore (2002) designed a questionnaire to gather information from representatives of all major Chinese sport organizations and a large number of commercial corporations that have purchased sport sponsorship. Before putting the questionnaire in use, they conducted a pilot study to test the language and format of the instrument within the Chinese community in central Utah, using 40 university graduates working in various business corporations in the area.

Most existing research methods today have been formally and scientifically developed and designed and have been validated to assure that the methodology is measuring what it is designed to measure. Most research methods are categorized into two large areas of research: qualitative research methods and quantitative research methods.

Qualitative research methods are comprised of those methods meant to explore behavior, perceptions, attitudes, and experiences of people (Qualitative Research Consultants Association, 2017). "Qualitative research methods originated in the social and behavioral sciences" of sociology, psychology, and anthropology (p. 1). Primarily, qualitative methods involve interview or observation methodologies, some of which are case study, interview, focus group, and observation. Qualitative research is meant to make attempts to discover understanding human behavior. For example, using an interview method, a sport marketing researcher may be trying to fully understand why season ticket holders stopped purchasing season tickets, or a sports shoe manufacturer may be trying to fully understand why a specific market of consumers have stopped purchasing their shoes. These studies can lead to decisions involved in making changes to strategic marketing plans, which might mean changing the product, changing pricing strategies, or changing advertising techniques if it is discovered that these were the factors that lead to consumers changing their purchase behavior.

Quantitative research methods are comprised of those meant to discover facts derived from measurements of data collected. These types of data are analyzed through numerical calculations and comparisons and through the use of statistical analyses and inferences (McLeod, 2017). Quantitative research methods are considered to be a bit more scientifically objective because the data are subjected to statistical analysis (McLeod, 2017). Quantitative methods can include survey, questionnaire, and experiment—methods that focus on gathering numerical data for statistical analysis. For example, using laboratory methods, a golf club manufacturer can test and calculate the endurance of a material. Any club material that fails the testing will not be used for production. In another example, a sport marketing professional can determine the effects of a variety of give-away promotions. A survey of attendees' reasons for attending an event based on the item in the give-away promotion will produce results. The sport marketing professional can then use the results of the survey to eliminate the give-away item that was the least attractive to the consumer.

The complete plan for the research design should include intricate logistics. Logistics involve the details regarding the study, including such details as what day and time will the study be conducted, where will it be conducted, what equipment will be needed, what supplies will be needed, will transportation be needed, and how will data be stored as it is collected? For example, if the research design is a survey at a sports event, every detail of logistics must be planned: how many copies of the survey will be needed; how many pens and pencils will be needed; will tables and chairs be needed and if so, how many, what type, and where will they come from; will transportation be provided to the survey workers, or will parking fees be provided; how many workers will be needed, and what time; where will you be allowed to survey, and for how long; how do you want the survey workers to be dressed; what do you want the survey workers to say; will there be an incentive offered to the potential survey participant, what will it be, and how many do you need; how will the survey workers carry all the survey gear; how will you communicate with the survey workers; what is your liability for the survey workers; and, where will you put the completed surveys and how will you keep them safe and confidential. These are just a few of the logistics needed to be considered, planned, and taken care of before the study actually begins.

In summary, the purpose of this step is to develop the complete plans needed to conduct the study.

Step 6: Conduct Data Collection

In this step, the actual research study will be conducted—the procedures and logistics determined and planned in Step 5 will be carried out. In summary, the purpose of this step is to conduct the study.

Step 7: Analyze the Data

As part of Step 5, the analysis of the data will have been planned. The purpose of this step is to carry out a summarization and analysis of the data. After the data has been collected, it will need to be converted into a format that can be analyzed according to the research design selected and can be used to develop solutions to the problem, or answer the research questions. Summarization may include developing graphs, tables, charts, or other illustrations of the data and its analyses.

The analysis selected must be one that produces information that can be examined and from which conclusions can be made. Most analysis of statistical information involves making statistical inferences—deducing properties—and/or looking for benchmarks, trends, comparative data, and drawing conclusions. Several chapters of this book are devoted to the description of statistical procedures that are commonly used in sport business management research.

The analysis will include conclusions drawn based on the statistical calculations. Analysis might include answering such questions as: Is there a statistical significance? Are there any patterns in the numbers? What can you determine from the results?

Step 8: Report the Results

Once the analysis stage is concluded and recommendations and conclusions have been drawn, then the researcher must put the results together in a format that can be used for the reason the study was conducted. In sport business industry settings, most research reports include results written and illustrated in such a way so as to lead to decisions or strategies that can be made based on the results of the study. In academic research settings, most research reports are written and organized based on a traditional reporting method usually determined by a scholarly journal, or by the publisher of a book. The academic research will be written for the purpose of progressing the literature about a specific topic, and/or so as to be used in the classroom setting for the purpose of preparing students for careers in the sport business industry. Chapter 17 provides guidelines on communications research results.

In summary, the purpose of this step is to create a report of the study from start to finish, complete with conclusions and recommendations for the application of the results.

Summary

Decision making is formally defined as the process of resolving a problem or selecting from alternative opportunities. Information is a critical element of making decisions. Information derives from research.

Research begins with the recognition of a problem, or a question, that needs a solution or an answer. Findings of the research provide the information needed to develop solutions and answer questions.

Research proceeds in an eight step process, although strict adherence to the steps is not absolutely necessary and some variation can occur. Some steps can begin out of sequence, and some can be carried out simultaneously. This is acceptable as long as the integrity of the research is not compromised.

A research design is selected based on which design is best to appropriately study the problem. The most commonly used and basic types of research in sports business include exploratory, experimental, survey, interview, observation, and focus groups.

Study Questions

1. What is the so-called decision making continuum and what are the variables affecting the confidence of a decision?
2. What are the steps in the process of sport business management research?
3. Among all the steps in the research process, which step is considered the most critical one?
4. What are the differences between qualitative and quantitative research methods in sport business management?
5. What are the common types of research designs or methods used in sport business management research?

Learning Activities

1. Review carefully the following situations and identify the most appropriate research design for each.
   a. The athletic department of XYZ University wants to identify the demographic characteristics of its men's and women's basketball season ticket holders
   b. The marketing director of a major league baseball team wishes to know if in fact sales promotion increases attendance
2. Identify an issue or a topic that is significant enough in sport business management to warrant an in-depth examination of it, and then develop a research proposal that outlines detailed procedures in terms of how the in-depth study will be executed.

References

Daft, R. (1990). *Management.* Fort Worth, TX: Dryden Press.

Decision making. (2017, February 20). *BusinessDictionary.* Retrieved from http://www.businessdictionary.com/definition/decision-making.html

Geng, L., Burton, R., & Blackmore, C. (2002). Sport sponsorship in China: Transition and evolution. *Sport Marketing Quarterly, 11*(1), 20–32.

McLeod, S. (2017). Qualitative vs. quantitative. Retrieved from https://www.simplypsychology.org/qualitative-quantitative.html

Qualitative Research Consultants Association. (2017). What is qualitative research? Retrieved from http://www.qrca.org/?page=whatisqualresearch

Schermehrhorn, J. R., Hunt, J. G., & Osborn, R. N. (1991). *Managing organizational behavior.* New York, NY: John Wiley & Sons.

12

Questionnaire Design

After completing the study of this chapter, you should be able to:

- Describe the process of developing a questionnaire.
- Understand the importance of a draft plan in questionnaire development.
- Explain the advantages and disadvantages of both open-ended questions and closed format questions.
- Explain the criteria of a good question.
- Understand the factors that sport management researchers should consider when determining the order of questions and answer choices.
- Explain how to deal with the "don't know" or "not applicable" answer choices.
- Indicate the criteria that sport management researchers use to determine the usability of a questionnaire during the pretesting phase of questionnaire development.
- Explain the methods that can be used to increase response rate.

As mentioned in previous chapters, sport management researchers use questionnaires to gather data in most studies. Because of the importance of questionnaires, students should develop a clear understanding and skills for proper questionnaire design. A questionnaire is a form that contains a number of questions used to gather information from a statistically significant number of subjects in a survey (Robson, 2002). As such, it is also referred to as a survey instrument.

In this chapter, a number of important issues pertaining to questionnaire design will be discussed. Specifically, the questionnaire development process and other practical issues related to questionnaire design, such as the order of questions and answer choices, the ways to deal with "don't know" or "not applicable" answer choices, the length of a questionnaire, and ways to increase response rate, will be examined.

Exhibit 12.1. The Questionnaire Development Process

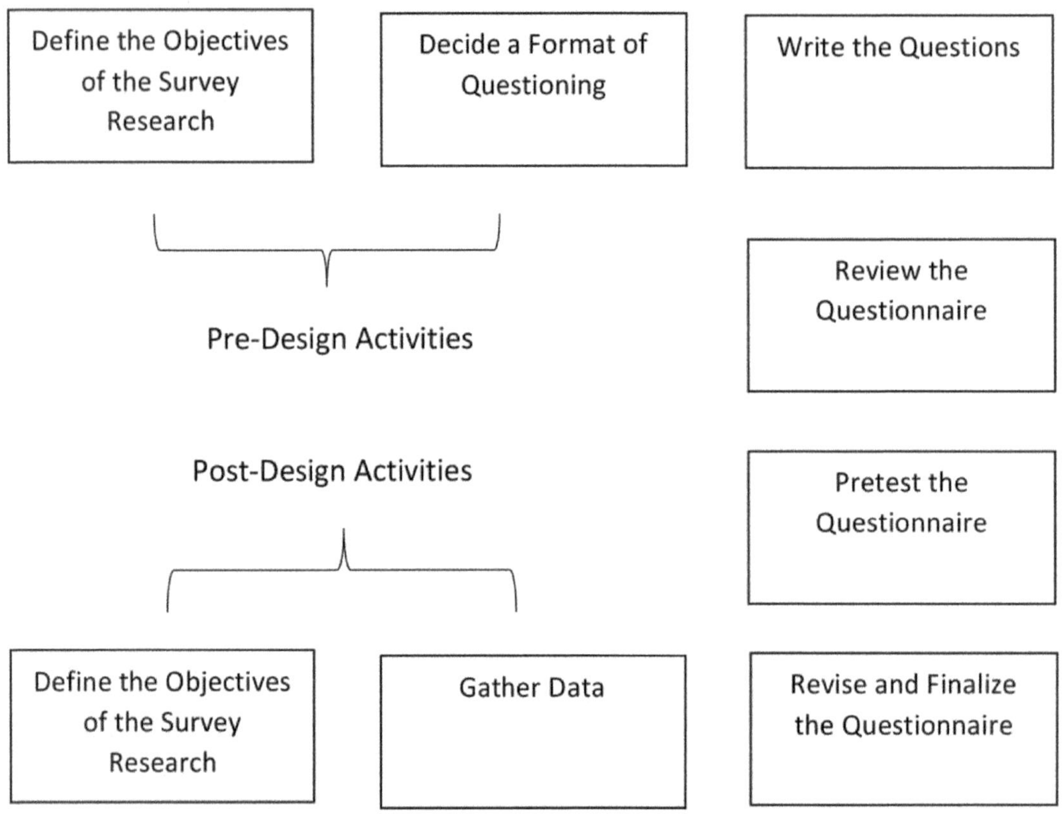

The Questionnaire Development Process

Designing an effective questionnaire is a systematic process; it requires time and careful thought (Bradburn, Sudman, & Wansink, 2004; University of Wisconsin Survey Center, 2010). Before a sport management researcher decides to design a questionnaire to solve a particular issue (e.g., to determine whether or not fans' satisfaction affects their intention to attend a sport event), the researcher should find out if questionnaires already exist that could be used or modified to gather the required information. It will save a lot of time and work if the researcher does not need to develop an entirely new questionnaire. If such a questionnaire is not available, the researcher then should follow the necessary process to design the instrument.

The development of a questionnaire is a multistage process that begins with defining the objectives of the survey research and ends with revising and finalizing the questionnaire before use (McDaniel & Gates, 1991; Patten, 2001). Every step needs to be executed carefully, as it ultimately affects the quality of the final product. Although questionnaires may be less expensive to administer than other data collection methods, they can be costly in terms of the time required to design them.

As depicted in Exhibit 12.1, the questionnaire development process includes these six steps: (1) defining the objectives of the survey research, (2) determining a format of questioning, (3) writing the questions, (4) reviewing the questionnaire, (5) pretesting the questionnaire, and (6) revising and finalizing the questionnaire (Swanson & Holton, 2005). In the following section, descriptions of the six steps and actions that should be taken in each step will be provided in detail.

Defining the Objectives of the Survey Research

The importance of well-defined objectives cannot be overemphasized. A questionnaire that is designed without clear objectives is inevitably going to overlook important issues and waste participants' time by asking useless questions. The researcher often comes to realize that the objectives of the survey research have not been specified precisely when the questionnaire is being developed. Consequentially, the questionnaire may lack a logical flow and cause the participant to lose interest, ultimately compromising the data that it is intended to collect. All questions to be included in the questionnaire must be consistent with the objectives of the research (Bradburn, Sudman, & Wansink, 2004). Thus, the biggest challenge in developing an effective questionnaire is to translate the objectives of the survey research into a well-conceptualized and methodologically sound survey instrument. To accomplish this goal, the researcher should ask questions: Why is this survey being conducted? What do I need to know? How will the information be used? How accurate and timely does the information have to be (Ghauri, Gronhaug, & Kristianslund, 1995)?

Most of the problems found in the analysis of survey results can be traced back to the design phase of the questionnaire. Well-defined objectives are the best way to assure a good questionnaire design (Bradburn, Sudman, & Wansink, 2004). When the objectives of survey research can be expressed in a few clear and concise sentences, the process in designing the questionnaire becomes considerably easier, because the questionnaire is developed to directly address the objectives of the research. Therefore, it is necessary for the researcher to put the survey objectives in writing. A good rule of thumb in this regard is that if the researcher finds it difficult to write a question, then the researcher has not spent enough time defining the objectives of the survey research (Ghauri, Gronhaug, & Kristianslund, 1995). At this point, the researcher needs to go back to review the objective again. The review will give the researcher more clues of how to phrase the question, as the question should always be derived naturally from the objective. Researchers should resist the temptation to ask a question because it would be interesting to know and should guard themselves with the principle of optimal ignorance—collect only information that is really needed.

Before designing a questionnaire, it is suggested that the researcher develop a *draft plan* that outlines the objectives of the survey research, the data requirements, and the analysis plan. This document will determine the variables to be measured, and ultimately, the survey questions and response alternatives. The analysis plan will describe the target audience and provide a list of the units to be sampled (e.g., athletes, facilities, managers). In addition, the analysis plan will also state the method of data collection to be used (e.g., face-to-face interview, telephone interview, mailed questionnaire) and explain how the questionnaire content, wording, format, and pretesting process will be developed. The methods to be used during the data processing (e.g., coding, editing) are also described in the plan. Finally, the last two important issues to be considered are the time required to complete the entire process and the budget that has been allotted to it (Ghauri, Gronhaug, & Kristianslund, 1995).

Determining the Formats of Questioning

Before determining the formats of questioning and subsequently writing specific questions, the researcher must decide the mode of information collection: whether or not the questionnaire will be self-administered or interviewer-administered, and delivered by mail, email, fax, telephone, in person, or posted on the Internet. The mode of information collection will determine how questions and response options are constructed. Also, the overall questionnaire length, question complexity, and question sensitivity must be weighed in determining the mode of information collection besides financial constraints. For example, long questionnaires may not work well on the telephone, complex questions may require an interviewer to be sure that they are understood, and sensitive questions may be best done in a self-administered format. Review the discussion in Chapter 9 about factors affecting the choice of particular survey methods.

After the mode of information collection has been decided, the researcher can proceed to determining the formats of questioning. In general, sport management researchers use two types of formats in asking questions, *open-ended* or *closed-ended question formats* (Yates, 2004). Most questionnaires contain both types of questioning formats. However, open and closed question formats are appropriate in different contexts and provide different kinds of information.

Open-Ended questions

Open-ended questions are questions that allow respondents to answer freely, however they choose, so as to receive unprompted opinions (Johnson & Christensen, 2000; Yates, 2004). There are no predetermined responses. For example, a fan attending a NBA game might be asked to respond to the following open-ended question—do you like tonight's game and why?

Open-ended questions are often used when the issue is complex. An obvious advantage of an open-ended question is its flexibility, as respondents can answer in any way they wish. Another advantage is that it is good for soliciting subjective data when the range of responses is not tightly defined. Accordingly, the variety of responses is wider and more truly reflects the opinions of the respondents (Johnson & Christensen, 2000). This increases the likelihood of receiving unexpected and insightful suggestions, for it is impossible to predict the full range of opinions. It is common for a questionnaire to end with an open-ended question asking respondents for their unabashed ideas for changes or improvements. They may be a better means of eliciting true opinions and attitudes and of identifying how strongly attitudes are held or not. The open-ended question is often used in the exploratory stage of a research project.

One main concern of open-ended questions is that they require more thought and time on the part of both the respondent and researcher (Johnson & Christensen, 2000). When respondents feel like they have been asked to answer a lot of questions, it increases the chance that they will become tired or bored. This may dramatically reduce the number of questions that the questionnaire can realistically ask. And because the researcher must read all the responses individually, open-ended questions can be costly in both time and money, and they may not be practical for lower budget or time-sensitive evaluations. Another concern about the use of open-ended questions is that it is more difficult to pool opinions across the sample, and respondents may sometimes answer in unhelpful ways (Yates, 2004). Obviously, there is no way to automatically tabulate or perform a statistical analysis on open-ended questions. The third concern of open-ended questions is that they are also open to the influence of the reader, for no two people will interpret an answer in precisely the same way. This conflict can be eliminated by using a single reader, but a large number of responses can make this impossible (Yates, 2004).

Closed-Ended questions

Closed-ended questions are questions in which all possible answers are identified and the respondent is asked to choose one of the answers provided (Yates, 2004). They are commonly used situations in which answers are known, are limited in number, and are clear-cut. In other words, closed format questions are an appropriate means of asking questions that have a finite set of answers of a clear-cut nature. As such, they usually take the form of a multiple-choice question. The following is an example of a closed-ended question:

Based on your observation, what is the level of international sport management education training that your faculty members have received?
- ❑ No or minimal training
- ❑ Moderate training
- ❑ Advanced/adequate training

There is no clear consensus on the number of answer choices that should be given in a closed-format question. Obviously, a sufficient number of answer choices need to be provided to fully cover the range of answers, but not so many as to blur the distinction between them. Usually, five to nine possible answer choices are included for each question. For questions that measure a single variable or opinion, conventional wisdom says that there should be an odd number of answer choices, as it allows for a neutral or no opinion response. Other schools of thought contend that an even number of answer choices is best because it forces the respondent to get off the fence (Bradburn, Sudman, & Wansink, 2004). It may, nevertheless, induce some inaccuracies as the respondent may actually have no opinion. Professional wisdom in questionnaire design supports the theory that it is best to use an even number of choices to prevent large numbers of neutral answers for larger questionnaires that examine opinions on a large number of items (Robson, 2002; Yates, 2004; Zikmund, 2000).

Closed-ended questions offer many advantages in terms of time and money. First, they involve minimal effort on the part of the respondent. Second, they provide the researcher with greater control over the outcomes of the survey, as the respondent is forced to answer particular questions. Third, restricting the answer set makes it is easy to calculate percentages and other hard statistical data over the whole group or over any subgroup of participants (Johnson & Christensen, 2000). Modern scanners and computers make it possible to administer, tabulate, and perform preliminary analysis in a relatively short period of time. Fourth, closed-ended questions also make it easier to track opinion over time by administering the same questionnaire to different but similar participant groups at regular intervals. Finally, closed-ended questions allow the researcher to filter out useless or extreme answers that might occur in an open-ended question (Yates, 2004).

Closed-ended questions are not problem-free. Closed-ended questions are appropriate only when the set of possible answers are known and clear-cut. If poorly designed, closed-ended questions may be misleading and frustrate respondents. Examples of inappropriately drafted closed-ended questions can be found in the following section.

Writing the Questions

The wording of questions will influence the answers that the respondent will provide. As such, the sport management researcher should keep in mind the basic principles or criteria used to write good questions in questionnaire design. The following are important criteria recommended by professional wisdom in questionnaire design to consider when phrasing questions. These criteria are used to judge whether or not a question is appropriately written (Johnson & Christensen, 2000; Schwab, 1999; Zikmund, 2000).

Be Simple

The three most important things for anyone involved in designing a questionnaire to remember are simplicity, simplicity, and simplicity (Johnson & Christensen, 2000). Ideas need to be conveyed clearly and questions should be easy to comprehend. There must be no guesswork for the respondent when it comes to understanding exactly what information is being asked. Otherwise, the reliability of the question will be questionable.

Evoke the truth in a non-threatening manner. When a respondent is concerned about the consequences of answering a question in a particular manner (i.e., feels threatened), there is a good possibility that the answer will not be truthful. For example, the following question will put the respondent in a very uncomfortable position to answer:

The use of illegal substances is rampant among players on your team.
Yes ❑ No ❑

Anonymous questionnaires that contain no identifying information are more likely to produce honest responses than those that identify the respondent. If the questionnaire does contain sensitive items, the researcher must clearly state the policy on confidentiality to ease concerns (Bradburn, Sudman, & Wansink, 2004).

Ask for an Answer on Only One Dimension

The purpose of a survey is to find out information. A question that asks for a response on more than one dimension will not provide the information the researcher wants (Bradburn, Sudman, & Wansink, 2004). For example, a researcher interested in finding out if the users of an athletic club are satisfied with the club asks, "Are you satisfied with the programming and services provided by this club?" If a respondent answers no, there is no way for the researcher to know whether the programming or service, or both, were unsatisfactory. This kind of question is often referred to as a double-barreled question. A good solution to the problem is to obtain the information with two separate questions.

Accommodate All Possible Answers

Multiple-choice items are the most popular type of survey questions because they are generally the easiest for a respondent to answer and the easiest to analyze. Asking a question that does not accommodate all possible responses not only can confuse and frustrate the respondent, but can also introduce bias (McDaniel & Gates, 1991). The following question is an example of questions that do not accommodate all possible answers:

What brand of tennis racquet do you own?
❑ Wilson
❑ Yonex

Clearly, there are many problems with this question. What if the respondent doesn't play tennis and thus does not own a tennis racquet? What if the respondent owns a different brand of tennis racquet other than those listed? What if the respondent owns tennis racquets in both brands? There are two ways to correct the problem. The first way is to make each response a separate dichotomous item on the questionnaire. For example:

Do you own a Wilson tennis racquet? (Circle: Yes or No)
Do you own a Yonex tennis racquet? (Circle: Yes or No)

Another way to correct the problem is to add the necessary response categories and allow multiple responses. This is the preferable method because it provides more information than the previous method.

What brand of tennis racquet do you own? (Check all that apply)

- ❑ Do not own a tennis racquet
- ❑ Wilson
- ❑ Yonex
- ❑ Other (specify)

The most common example for not accommodating all possible answers is to leave out such responses as neutral or don't know when, in fact, respondents may well be neutral or may actually not know. More discussion on this particular phenomenon can be found later in the chapter.

Have Mutually Exclusive Options

A good question leaves no ambiguity in the mind of the respondent. There should be only one correct or appropriate choice for the respondent to make (Bradburn, Sudman, & Wansink, 2004). Bias could be introduced if the answer choices are not mutually exclusive and only one choice may be selected. An obvious example:

Where did you grow up?

- ❑ Country
- ❑ Farm
- ❑ City

A person who grew up on a farm in the country would not know which choice to select. This question not only would not provide meaningful information, but also could frustrate the respondent. There is a great possibility that the respondent may stop completing the questionnaire while frustrated.

Produce Variability of Responses

When a question produces no variability in responses, it leaves considerable uncertainty about why this question is asked and what the researcher wants to learn from the information. If a question does not produce variability in responses, it will not be possible to perform any statistical analyses on the item (McDaniel & Gates, 1991). For example:

What do you think about this game?
- ❑ It is the worst game I have watched
- ❑ It is somewhere between the worst and best
- ❑ It is the best game I have watched

Because the respondents will most likely select the second choice, the question yields very little information. So, the researcher should phrase sensitive questions to allow for differences between respondents. Again, the following is another example of a question with very little variability in responses, as most of the respondents will choose yes:

Are you against drug use by professional athletes?
Yes ❑ No ❑

Follow Comfortably from the Previous Question

Writing a questionnaire is similar to writing anything else. Transitions between questions should be smooth. Grouping questions that are similar will make the questionnaire easier to complete, and the respondent will feel more comfortable. Questionnaires that jump from one unrelated topic to another feel disjointed and are not likely to produce high response rates.

Do not assume a certain state of affairs. Among the most subtle mistakes in phrasing questions is to make an unwarranted assumption. An example of this type of mistake:

Are you satisfied with your tennis racquet?
Yes ❑ No ❑

This question will present a problem for someone who does not currently have a tennis racquet. The researcher should write the questions that will apply to everyone. This often means simply adding an additional response category.

Are you satisfied with your tennis racquet?
- ❑ Yes
- ❑ No
- ❑ Do not have a tennis racquet

One of the most common assumptions is that the respondent knows the correct answer to the question. Industry surveys often contain very specific questions that the respondent may not know the answer to. For example:

What percent of your budget do you spend on participating in sport?

Very few people would know the answer to this question without looking it up, and very few respondents will take the time and effort to look it up. If researchers ask questions similar to this, it is important that they understand that the responses are rough estimates and there is a strong likelihood of error. So, it is essential for researchers to look at each question and decide if all respondents will be able to answer it. Researchers must be careful not to assume anything. For example, the following question assumes the respondent knows what GO bond is about.

Are you in favor of the use of GO Bond in facility financing?
- ❑ Yes
- ❑ No
- ❑ Undecided

If there is any possibility that the respondent may not know the answer to the question, researchers should include a "don't know" response category.

Do Not Use Unfamiliar Words, Colloquial Expressions, or Abbreviations

The researcher must have a clear understanding of the target population of the survey research, thus writing questions particularly for them. With that in mind, the researcher should avoid using uncommon words or compound sentences and should write short sentences (Bradburn, Sudman, & Wansink, 2004). The researcher should also avoid using colloquial or ethnic expressions that might not be equally used by all participants. Abbreviations or technical terms can be used only if the researcher is absolutely certain that all respondents understand the meanings of those terms or abbreviations. Otherwise, abbreviations should be avoided. The following question, for instance, might be okay if all the respondents are golf enthusiasts, but it would not be a good question for the general public:

What is your handicap?

Do not imply a desired answer. People sometimes have a desire to give the right answer to a question whether or not it is actually true. This type of question is referred to as a leading question. Researchers should strive for objectivity in the survey and must be careful not to lead the respondent into giving the answer they would like to receive. Leading questions are usually easily spotted because they use negative phraseology. For example:

The city has spent too much taxpayers' money in building sports facilities.
- ❑ Strongly Agree
- ❑ Agree
- ❑ Neutral
- ❑ Disagree
- ❑ Strongly Disagree

The researcher should consider asking the question in a less obvious way:

Should the city spend money in building sports facilities?
Yes ❑ No ❑

Here is another example. The spectators attending a professional football game were asked to express their overall impression of the game:

Is this the best game you have ever watched?
Yes ❑ No ❑

In this case, even if the spectators enjoyed the game, but had a favorite game that was preferred, they would be forced to answer no. Clearly, the negative response covers too wide a range of opinions. A better way would be to ask the same question but supply the following choices:

This is the best game you have ever watched.
- ❑ Strongly Agree
- ❑ Agree
- ❑ Neither Agree or Disagree
- ❑ Disagree
- ❑ Strongly Disagree

Also, the word best has strong overtones that deny the spectators an objective environment to consider the game. The signal sent to the respondent is that the researcher thinks it is the best game, and so should everyone else. Though this may seem like an extreme example, this kind of superlative question is common practice in sport management research.

Do Not Use Emotionally Loaded or Vaguely Defined Words

This is one of the areas often overlooked by researchers. Quantifying adjectives (e.g., most, least, majority) are frequently used in questions. It is important to understand that these adjectives mean different things to different people. Accordingly, the researcher should try to be as precise as possible in wring the questions to ensure that respondents understand exactly what is being asked. In other words, it is best to phrase the question empirically if possible and to avoid the use of unnecessary adjectives. For example, a question asks about frequency, but supplies choices that are open to interpretation, such as:

How frequently do you exercise?
- ❑ Very Often
- ❑ Often
- ❑ Sometimes
- ❑ Rarely
- ❑ Never

This question is fairly vague. How many times or days a person exercises is considered very often? A better way to ask the question is to provide the respondents with quantifiable choices, such as:

- ❑ Every Day or More
- ❑ 2–6 Times a Week
- ❑ About Once a Week
- ❑ About Once a Month
- ❑ Never

This example illustrates a dilemma: when questions are made more precise, they often become more limited in scope. Do not ask hypothetical questions, such as "if you were David Stern, what would you do to stop players from fighting in the NBA?" This type of question forces respondents to give thought to something they may never have considered. It does not produce clear and consistent data representing real opinion.

Try to Prevent Prestige Bias

Prestige bias is the tendency of respondents to answer a question in a way that makes them feel better. People may not lie directly but may try to put a better light on themselves. For example, a survey on alumni contributions includes this question: "Did you donate any money to XYZ University during its annual campaign last year?" To deal with prestige bias, the researcher should try to make the questionnaire as private as possible and assure the respondents of absolute confidentiality.

Do not make the questions dependent on responses to previous questions. Branching in written questionnaires should be avoided. Although branching can be used as an effective probing technique in telephone and face-to-face interviews, it should not be used in written questionnaires because it sometimes confuses respondents. An example of branching:

1. Do you play golf? Yes ❑ No ❑
   If no, go to question 3
2. How much did you spend on playing golf?
3. Why don't you play golf?

If the purpose of the survey is to find out how much golfers spent on average, these questions could easily be rewritten as one question that applies to everyone:

How much did you spend last year on playing golf? __________ (Write 0 if none)

Do Not Ask the Respondent to Order or Rank a Series of More Than Five Items

The researcher should avoid asking respondents to rank items by importance. This becomes increasingly difficult as the number of items increases, and the answers become less reliable. This becomes especially problematic when asking respondents to assign a percentage to a series of items. In order to successfully complete this task, the respondents must mentally continue to readjust their answers until they total 100 percent. Limiting the number of items to no more than five will make it easier for the respondent to answer.

In attitude assessment, how many alternatives should be given has been an issue debated among scholars. Five alternatives are the most common set, typically ranging from strongly agree to strongly disagree, unless one wishes to deny the respondent the lazy choice of a mid-point, which sometimes has no meaning. Careful thought is needed to synthesize the attitude measures into a meaningful indicator or profile. Often informants ought to participate directly in deciding the importance of profile elements.

Determination of the Question and Answer Choice Order

After writing all of the questions, the researcher's next task is to determine the question and answer choice order. The order of the questions and the order of the answer choices not only encourage people to complete the survey, but also affect the results of the survey (Schwab, 1999).

The order of questions affects the outcome of the survey. An idea mentioned early in one question can make the respondents think of it while they answer a later question. Habituation is the other way the order of questions can affect the outcome of a survey. Habituation is what has occurred when respondents give the same answer to all questions without really considering them, after having been asked a series of similar questions that have the same answer choices.

There are a few principles that experts in survey research generally have a consensus on in terms of the question and answer choice order (Bradburn, Sudman, & Wansink, 2004):

- The questions in the early sections of the survey should be easy to answer. These kinds of questions encourage respondents to continue the survey.
- Place the most important items in the first half of the questionnaire. Respondents often send back partially completed questionnaires. By putting the most important items near the beginning, the partially completed questionnaires will still contain important information.
- Grouping together questions on the same topic makes the questionnaire easier to answer.
- Difficult or sensitive questions should be asked in the end of the questionnaire. The advantage of such an arrangement is that the respondents would have already answered most of the questions if they decided to quit at that point.
- One way to reduce habitation is to ask only a short series of similar questions at a particular point in the questionnaire. Then ask one or more different kinds of questions, and then another short series if needed.
- To reduce habitation in survey instruments that contain questions asking respondents to indicate their levels of agreement, researchers can phrase some statements in such a way so that a high level of agreement means satisfaction (e.g., "My supervisor gives me positive feedback") and some others so that a high level of agreement means dissatisfaction (e.g., "My supervisor usually ignores my suggestions"). This technique forces the respondents to think more about each question.

Answer choice order can make individual questions easier or more difficult to answer. Whenever there is a logical or natural order to answer choices, researchers should use it. They should also always present agree-disagree choices in that order. Presenting them in disagree-agree order will seem odd. For the same reason, positive to negative and excellent to poor scales should be presented in those orders. When using numeric rating scales, higher numbers should mean a more positive or more agreeing answer.

The order in which answer choices are presented can also affect the answers given. People tend to pick the choices nearest the start of a list when they read the list themselves on paper or a computer screen. People tend to pick the most recent answer when they hear a list of choices read to them.

As mentioned previously, sometimes answer choices have a natural order (e.g., Yes followed by No; or Excellent—Good—Fair—Poor). If so, researchers should use that order. At other times, questions have answers that are obvious to the person who is answering them. In these cases, the order in which the answer choices are presented is not likely to affect the answers given. However, there are kinds of questions, particularly questions about preference or recall or questions with relatively long answer choices, that express an idea or opinion, in which the answer choice order is more likely to affect which choice is picked. Researchers who are using telephone, computer direct, or Internet interviewing should have their software present these kinds of answer choices in a random order.

There are different schools of thought in terms of whether or not demographic data should be collected at the beginning of the questionnaire, at the end, or scattered throughout. The proponents for having this type of question collected at the beginning of the questionnaire argue that normally background questions are easier to answer and can ease the respondent into the questionnaire. On the other hand, it is also believed that demographic questions, such as age, gender, income, and education, should be asked at the end of the questionnaire. By then the researcher may have collected enough information, even if the respondent refuses to complete the demographic questions. Some exceptions exist. If a demographic question qualifies someone to be included in the survey, it should be placed near the beginning as a filter question. For example, sport economic impact studies often exclude local residents from their studies.

It is important to ask only those background questions that are necessary. Researchers should not ask the income of the respondent unless there is at least some rationale for suspecting a variance across income levels. There is often only a fine line between background and personal information. Researchers do not want to cross over into the personal realm unless absolutely necessary. If they need to solicit personal information, they should phrase the questions as unobtrusively as possible to avoid ruffling the participants and causing them to answer less than truthfully (Bradburn, Sudman, & Wansink, 2004).

Dealing with "Don't Know" or "Not Applicable" Answer Choices

It is recommended that a "don't know" or "not applicable" answer choice be available for questions that the researchers are not sure the respondents will be able to answer (Patten, 2001). In most cases, these are wasted answers as far as the researcher is concerned, but they are necessary alternatives to avoid frustrated respondents. Sometimes "don't know" or "not applicable" will really represent some respondents' most honest answers to some questions. Respondents who feel they are being coerced into giving an answer they do not want to give often do not complete the questionnaire. For example, many people will abandon a questionnaire that asks them to specify their income but does not offer a decline to state choice. When the researcher decides to include "don't know" or "not applicable" as an answer choice, it should be treated as a legitimate response in data analysis and reporting (Huisman & van der Zouwen, 1998). For the same reason, the researcher should include "other" or "none" as a logically possible answer if the circumstance warrants.

Reviewing the Questionnaire

After the initial draft is completed, the researcher should carefully examine the questionnaire to determine that each of the questions included in the instrument meets certain standards (Fraenkel & Wallen, 2000). Extra time and resources are often wasted asking for information that will never be put to use. Reviewing the questionnaire is a fundamental step in developing a questionnaire. The following is a checklist of some key points for researchers to consider when reviewing questionnaires:

- Is the introduction informative?
- Does it stimulate respondent interest?
- Are the words simple, direct, and familiar to all respondents?
- Do the questions read well?

- Does the overall questionnaire flow?
- Are the questions clear and as specific as possible?
- Does the questionnaire begin with easy and interesting questions?
- Does the question specify a time reference?
- If the questions are closed-ended, are the response categories mutually exclusive and exhaustive?
- Are the questions applicable to all respondents?
- Are the questions included in the questionnaire clearly linked to the objectives of the survey research?
- Do the questions directly address the issues under the investigation?

While reviewing the questionnaire, the researcher should pay close attention to the last two questions, as they are the most critical ones to ask. To address them, the researcher should first review the objectives of the study, and then write down each question number next to the objective that the particular question will accomplish. After that has been done, the researcher can determine the need to either add more questions or drop some from the questionnaire.

As Exhibit 12.2 shows, the fifth question does not address any of the four research objectives. As such, it should be dropped from the questionnaire. On the other hand, none of the

Exhibit 12.2. Survey Questions in Reference to Research Objectives

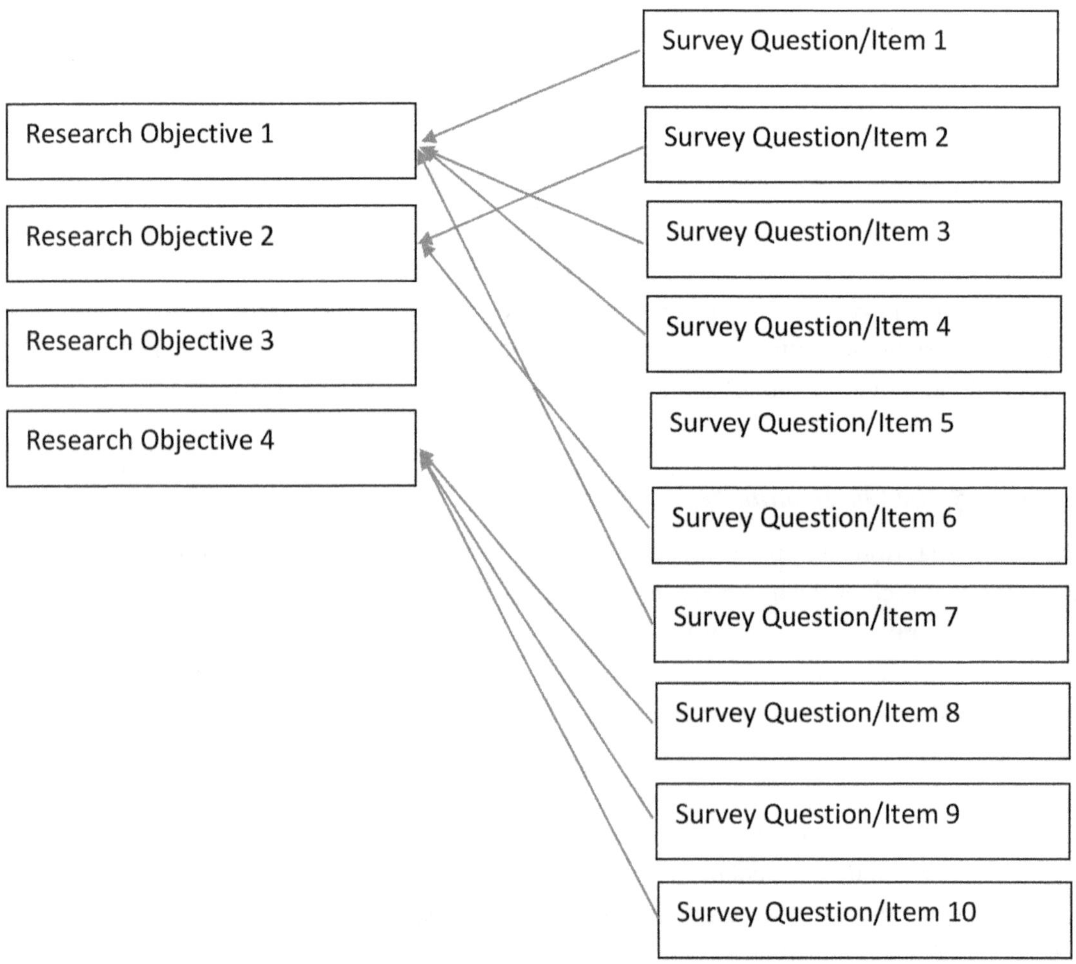

questions included in the questionnaire are written to specifically accomplish research objective 3. The researcher accordingly must add questions that are related to this particular research objective.

Pretesting the Questionnaire

Although the questionnaire has gone through a few steps in its design process, it is not ready to be sent out yet. It needs to be pretested. Pretesting is considered an essential step in questionnaire design (Fraenkel & Wallen, 2000; Johnson & Christensen, 2000). Even though the researcher has tried to ensure that all questions are clear and concise in writing the questions, invariably some items are ambiguous and some bias and error still exist in the questionnaire. Thus, it is important to conduct two pretests to see how it works. Specifically, pretesting helps discover poor wording or ordering of questions, identify errors in the questionnaire layout and instructions, determine problems caused by the respondent's inability or unwillingness to answer the questions, suggest additional response categories that can be pre-coded on the questionnaire, and provide a preliminary indication of the length of the interview and any refusal problems.

Testing can include the entire questionnaire or only a portion of it. A questionnaire will at some point in time have to be fully tested. The pretesting can be done by a small number of people (number from 25 to 75), preferably including some of the people for whom the questionnaire was designed. Prior to the first pretest, the questionnaire usually has more questions than the final version, as a key purpose of the first pretest is to identify weaker or inappropriate questions and drop them from the questionnaire. The second pretest is used simply for polishing, trimming, rearranging, and other refinements, but not for adding new questions or making major substantive changes to the questionnaire. It is possible that this step may need to be repeated more than twice depending on resources and the need for accuracy. The main purpose to re-test the questionnaire is to determine its face validity. (See Chapter 13 for detailed information about face validity.)

There are a few evaluative criteria that can be used during the pretesting to determine the *usability* of the questionnaire:

- The researcher needs to obtain information about whether or not the questionnaire is easy to administer, score, and interpret. In other words, what are the practicalities of implementing and using a questionnaire?
- Is the questionnaire capable of producing reliable results or data? It is crucial for the researcher to obtain an answer to this question during the pretesting process.
- A questionnaire should actually measure what it is designed to measure. It depends greatly on the quality of the questioning.

It should be kept in mind that no questionnaire is perfect. Therefore, a little extra time and care in the pretesting step of a questionnaire can avoid many problems and greatly improve the quality of information sought.

Finalizing the Questionnaire

After the pretest, the researcher should review the questionnaire with the respondents in terms of how they felt about the questionnaire. Review questions include:

- Does the questionnaire miss some key information?
- Should multiple-choice questions include other possible answers?
- Can respondents follow the questionnaire or do they get confused or frustrated?
- Are questions repeated unnecessarily?
- Are questions understood and answered properly, or are there confusing words or phrases?
- Is the questionnaire too long?
- Based on the feedback and comments obtained from the respondents, the researcher can revise the questionnaire before putting it into use.

Other Practical Issues Related to Questionnaire Design

Length of the Questionnaire

It has been proven in sport management research that long questionnaires in general get less response than short questionnaires regardless of whether or not it is a paper-pencil survey or online survey. So, the researchers should keep their questionnaires short, particularly those to be sent to sport management practitioners to complete. Response rate is the single most important indicator of the level of confidence of the results. A low response rate can be destructive to the quality of a study; therefore, researchers should try to do whatever they can to maximize the response rate. One of the most effective methods of maximizing response is to shorten the questionnaire.

If a survey instrument is more than a few pages, the researcher should shorten it by eliminating questions. To determine which questions to eliminate, the researcher might read each question and then ask, "Is this question critical to answer the research question?" If the information will be important to solve the research problem or to use in a decision-making process, then the question should be kept. If not, it could be omitted.

Even though long questionnaires often have fewer responses than short questionnaires, some studies have shown that the length of a questionnaire does not necessarily affect response (StatPac Inc., 2007). More important than length is question content. Subjects are more likely to respond if they are involved and interested in the research topic.

The Order of the Questions

No definitive empirical evidence suggests that the order of questions affects response in survey research; however, it is believed that the order effect may exist in interviews but not in written questionnaires (Bradburn, Sudman, & Wansink, 2004).

Regardless, the sport management researcher should attempt to order the questions in a logical manner. Grouping questions that are similar will facilitate the respondent's completion of the questionnaire as the order makes the respondent feel more comfortable. Questions that use the same response formats, or those that cover a specific topic, should appear together. Also, the order of questions should be designed so that the questions appearing earlier hold the interest of the respondents and accordingly motivate them to complete the questions in the latter part of the questionnaire (Bradburn, Sudman, & Wansink, 2004). Also, the transitions between questions should be smooth. A questionnaire with questions that jump from one topic to another may frustrate respondents and discourage them from completing the questionnaire, thus leading to low response rates.

The Layout of a Questionnaire

The layout of a questionnaire is also an important part of the questionnaire design. The general principles for the layout of a questionnaire are that the layout must be attractive, easy to understand, and easy to complete. Also, it should be easy for data entry (University of Wisconsin Survey Center, 2010).

The researcher should try to keep the answer spaces in a straight line, either horizontally or vertically. A single answer choice on each line is best. It is easier for a respondent to follow a logical flow across or down a page.

As shown in the second of the following examples, researchers like to use question-and-answer choice grids in the layout of their questionnaires. The grids not only look attractive, but also save paper or computer screen space. They also can avoid a long series of very repetitive question-and-answer choice lists. The researchers, however, must realize that such layouts may be more difficult than the repeated lists for some people to understand. As always, the researchers need to consider their subjects when creating the questionnaire. Look at the following layouts and decide which you would prefer to use:

Do you agree, disagree, or have no opinion that the benefits of outsourcing for a sport organization:

- It helps the sport organization build sustainable competitive advantage—agree/not sure/disagree.
- It brings in greater financial return than keeping them in house—agree/not sure /disagree.
- It helps the sport organization enhance the quality of its marketing operations— agree/not sure/disagree.
- It allows the sport organization to handle all aspects of your marketing operations in a more timely fashion—agree/not sure/disagree.

An alternative layout:

Do you agree, disagree, or have no opinion that the benefits of outsourcing for a sport organization include

It helps the sport organization build sustainable competitive advantage.

Agree	Not Sure	Disagree
3	2	1

It brings in greater financial return than keeping them in house.

Agree	Not Sure	Disagree
3	2	1

It helps the sport organization enhance the quality of its marketing operations.

Agree	Not Sure	Disagree
3	2	1

It allows the sport organization to handle all aspects of its marketing operations in a more timely fashion.

Agree	Not Sure	Disagree
3	2	1

The second example shows the answer choices in neat columns and has more space between the lines. It is easier to read. The numbers in the second example will also speed data entry, if a paper questionnaire is used.

Designing Web-Based Questionnaires

Because of the advantages of online surveys in the areas of research costs, access to subjects, the scope of the research, and the nature of behavior under study, they are becoming an increasingly popular choice of researchers in collecting data with a questionnaire, over paper-pencil surveys (Andrews, Nonnecke, & Preece, 2003). Although the principles used in designing paper-pencil surveys can be applicable to the construction of online questionnaires, it is important to remember that the latter requires some special attention because of the distinctive differences that exist between email and online surveys. Exhibit 12.3 is an example of an online questionnaire used in sport management.

Online Questionnaire Design

There are several important rules for researchers to keep in mind while drafting online questionnaires (Andrews, Nonnecke, & Preece, 2003; QuestionPro, n.d.; Lumsden, 2005, pp. 6–7).

The first rule is precision and concision. The respondent may lose interest in completing the questionnaire if the questions are complicated and long. This rule implies that the questionnaire should not be so long as to displease the respondent. It is suggested that the total number of questions included in an online questionnaire should be less than 60 (Lumsden, 2005).

The question and answering process should be attractive—that is, uncluttered and easy to complete. A question should never be divorced from its response set—that is, a question and all elements of its response should appear on the same page or screen. Questions relating to a given topic should be presented together and clearly sectioned from questions related to other topics. Section headings and sub-headings that are meaningful and well-designed should be used to clearly differentiate sections.

It is also important to:

- avoid using too many sections in any one questionnaire, because this is likely to become confusing to respondents; it also reflects poor questionnaire design and lack of focus.
- allows respondents to interrupt and re-enter responses and move backward and forward throughout the entire questionnaire, unless for very specific reasons.
- avoid excessive scrolling, list only a few questions per page and provide clear and easy links to the preceding and succeeding pages of a questionnaire. On the other hand, simply having one question per page is not a good idea either as it will increase the time to complete the survey, which may increase the possibility for the respondent to quit the survey.
- avoid excessive numbers of pages with complex navigational aids. As a general rule, the decision to scroll or to segment a questionnaire into a series of pages should be determined on the basis of the content and size of a questionnaire.

Formatting is also an important aspect of online questionnaire design. Text, color, graphics, tables, and frames are several aspects of formatting that deserve attention from sport management researchers. The basic guidelines, as suggested by Lumsden (2005, pp. 9–11), for online questionnaire design are as follows.

Exhibit 12.3. An Example of an Online Questionnaire Used in Sport Management

XYZ Wildcat Club Young Alumni Program

The Athletic Department Office at XYZ University is in the process of creating a young alumni program for the students of XYZ University. The program's goal is to keep the recent alumni connected to the Athletic Department at XYZ University through the Wildcat Club. All information gathered in this survey will be kept anonymous and used in the development of the young alumni program.

1. After graduation, do you plan on moving away from the XYZ area?
   - ❍ Yes
   - ❍ No
2. Following graduation, do you plan on returning to campus for any of the following next year? (Please check all that apply)
   - ❍ Homecoming (football)
   - ❍ Football game (non-homecoming)
   - ❍ Men's basketball game
   - ❍ Any other athletic event
   - ❍ Halloween
   - ❍ Graduation
3. Do you plan on checking your XYZ University email account upon graduation?
   - ❍ Yes
   - ❍ No
4. What is your preferred method of contact?
   - ❍ Mail
   - ❍ Phone
   - ❍ Text Message
   - ❍ Email
5. Do you know what the XYZ Wildcat Club is?
   - ❍ Yes
   - ❍ No
6. Do you plan on joining the XYZ Wildcat Club upon graduation?
   - ❍ Yes
   - ❍ No
7. Are you willing to join the XYZ Wildcat Club for $25 to achieve the $100 membership level?
   - ❍ Yes
   - ❍ No
8. Please rate the following benefits in order of importance for membership (most important on the left side and declining in importance to the right)

	Most Least
Opportunity to purchase away game tickets	❍ ❍ ❍ ❍ ❍ ❍ ❍ ❍ ❍ ❍
Lapel Pin	❍ ❍ ❍ ❍ ❍ ❍ ❍ ❍ ❍ ❍
Access to website content	❍ ❍ ❍ ❍ ❍ ❍ ❍ ❍ ❍ ❍
Priority for football season tickets	❍ ❍ ❍ ❍ ❍ ❍ ❍ ❍ ❍ ❍
Priority for tickets for Championships/Bowl Games	❍ ❍ ❍ ❍ ❍ ❍ ❍ ❍ ❍ ❍
Wildcat newsletter	❍ ❍ ❍ ❍ ❍ ❍ ❍ ❍ ❍ ❍
Homecoming (football) ticket	❍ ❍ ❍ ❍ ❍ ❍ ❍ ❍ ❍ ❍
Wildcat priority points	❍ ❍ ❍ ❍ ❍ ❍ ❍ ❍ ❍ ❍
Priority for Men's Basketball season tickets	❍ ❍ ❍ ❍ ❍ ❍ ❍ ❍ ❍ ❍

9. Please list any other benefits that would be of importance to you.

- In regards to text, sentences should not exceed 20 words and should be presented with no more than 75 characters per line. If elderly respondents are anticipated, then this limit should be reduced to between 50 and 65 characters per line. Paragraphs should not exceed five sentences in length. Technical instructions (instructions related to the basic technical operation of the website delivering the questionnaire) should be written in such a way that non-technical people can understand them. Questions should be easily distinguishable, in terms of formatting, from instructions and answers, and each question type should be consistent in terms of the visual appearance of all instances of that type and the associated instructions concerning how they are to be answered; in particular, the relative position of the question and answer should be kept consistent throughout the questionnaire. Where different types of questions are to be included in the same questionnaire, each question type should have a unique visual appearance.
- In regards to color, use consistent color coding throughout the questionnaire to reinforce meaning or information in an unambiguous fashion and a neutral background color that excludes patterns that can make text very hard to read. When pairing colors or using two colors in close proximity, endeavor to use colors of high contrast to ensure maximum discernibility across the target audience. This is particularly important for questionnaires targeted at audiences over 35 years of age.
- In regards to graphics, avoid cluttering the questionnaire with graphics. When using graphics, try to use small graphics that will download quickly. Individual images should not exceed 5KB in size and no single web-page should exceed 20KB of graphics in total. Wherever a graphic is used, it is essential to provide a text-only version of the web page. Minimize the number of colors that are used in any single graphic, and do not use graphics that mimic (or resemble) other items on a typical website. Do not blur pictures, gray out information, or overlap menus; use crisp, clear images to maximize accessibility for users with visual disabilities. Do not associate multimedia and audio clips with graphics if plug-ins would typically have to be downloaded by the respondents in order to access the associated media. Make it easy for users to skip, without penalty, multimedia content.

Online Survey Software

A number of companies are specialized in providing online survey services, such as, Qualtrics Insight Platform, QuestionPro, Snap Survey, SurveyGizmo, SurveyMonkey, SurveyPlanet, Zoho Survey, etc. Many of these services not only include online survey creation and programming, but also data analysis.

Increase the Survey Response Rate

In addition to those techniques discussed in Chapter 9, sport management researchers can also use the following tips relative to questionnaire design to increase the effectiveness of their survey (SmartSurvey, 2017) and the response rate of a survey:

Experts' suggestions will improve the questionnaire. Researchers should form a panel of sport management experts who have expertise in the area of relative research and ask this group to review the questions. This will ultimately validate the questionnaire.

Long questionnaires will deter respondents. So, the researcher needs to always ask what to do with the information collected from each question. Based on the examination, questions can be placed into three groups: must know, useful to know, and nice to know. The nice questions should be dropped unless the total number of questions in the previous two groups is very low.

A good questionnaire also has a coherent structure and a short, meaningful title. Where possible, researchers should collect questions under definable subject areas and develop a logical order of questions. A questionnaire with a title is generally perceived to be more credible than one without.

Researchers want the respondent to complete the questionnaire. One way to keep a questionnaire interesting is to provide variety in the type of items used. Varying the questioning format will also prevent respondents from falling into response sets. At the same time, it is important to group items into coherent categories. All items should flow smoothly from one to the next. The questionnaire should start with a few nonthreatening and interesting questions. Respondents may be discouraged to complete the survey instrument if threatening or boring questions are provided at the beginning of the questionnaire. Respondents generally make a decision of whether or not they will continue to answer the questions in the questionnaire by reviewing the first few questions.

Use the supporting text throughout the questionnaire wisely. The supporting text (1) should be used to communicate clearly to the respondents about the purpose of the survey, (2) should address the confidentiality of responses, and (3) must instruct the respondents on how to complete the questionnaire. The instruction provided in terms of how to complete each section of the questionnaire is very important.

Paper-Pencil Survey Response Rate

While using a paper-pencil survey, sport management researchers should also pay addition to the following methods in order to improve the return rate of their surveys.

Use professional production methods for the questionnaire. Researchers should be creative and consider trying different colored inks and paper. The objective is to make the questionnaire stand out from all the others that the respondent receives. Each page of a mail questionnaire should be numbered.

One criticism of questionnaires is their inability to retain the flavor of a response due to lack of adequate space for respondents to make comments. Leaving space for comments will provide valuable information not captured by the response categories. Leaving white space also makes the questionnaire look easier, and this increases response.

Use the tablet form instead of a staple to hold the questionnaire. If a questionnaire is more than a few pages and is held together by a staple, include some identifying data on each page (such as a respondent ID number). Pages often accidentally separate.

Make the envelope unique, as the first impression of the respondents is very important. Their first impression comes from the envelope containing the questionnaire. The best envelopes use commemorative postage stamps. Envelopes with bulk mail permits or gummed labels are perceived as unimportant. This perception will generally be reflected in a lower response rate.

The researcher should always include a self-addressed postage-paid envelope in the mail survey. Envelopes with postage stamps get better response than business reply envelopes. Also, print the return address on the questionnaire. It often happens that the questionnaires get separated from the reply envelopes and the cover letter. By printing the return address on the questionnaire, it gives another assurance that the respondents will send back the survey instrument.

Summary

Questionnaire design is an important process that demands careful attention. A questionnaire is a powerful evaluation tool and should not be taken lightly. Design begins with an understanding of the capabilities of a questionnaire and how they can help the research. If it is determined that a questionnaire is to be used, the greatest care should go into the planning of the objectives. Questionnaires are like any scientific experiment. Researchers do not collect data and then see if they found something interesting.

Questionnaires are versatile, allowing the collection of both subjective and objective data through the use of open or closed format questions. Modern computers have only made the task of collecting and extracting valuable material more efficient. However, a questionnaire is only as good as the questions it contains. There are many guidelines that must be met before a questionnaire can be considered a sound research tool. The majority deal with making the questionnaire understandable and free of bias. Mindful review and testing are necessary to weed out minor mistakes that can cause confusion in meaning and interpretation. When these guidelines are followed, the questionnaire becomes a powerful and economic evaluation tool.

Online survey has gradually replaced paper-pencil survey because of its quick distribution and lower cost. Although the general principles in questionnaire design can be applied to both the paper-pencil survey and the online survey, the latter has some special characteristics.

Study Questions

1. Elaborate on the process in developing a questionnaire.
2. What is a draft plan in questionnaire development?
3. Compare and contrast the open-ended questions and closed format questions in terms of their advantages and disadvantages.
4. What are the criteria of a good question?
5. What should sport management researchers consider while determining the order of questions and answer choices?
6. Explain how to deal with the "don't know or "not applicable" answer choice.
7. What are the criteria that sport management researchers use to determine the usability of a questionnaire during the pretesting phase of questionnaire development?
8. Explain the methods that can be used to increase response rate.

Learning Activities

1. The following are questions taken from several survey instruments. Based on the criteria of asking good questions, provide your critique to each of the questions. Be sure to indicate the problem first and then revise it. If you believe that a question is well stated, just write "Okay" next to the question.
   a. Questions developed by the XYZ sport marketing firm for the solicitation of public opinions about the quality of services provided at a sports facility:

(3) The services staff is courteous and knowledgeable about the facility.

Strongly Agree	Agree	Neutral	Disagree	Strongly Disagree
❒	❒	❒	❒	❒

(2) The food served by ARMARK, a company whose products are highly recognized by the sport industry as an excellent food provider, is incredible.

Strongly Agree	Agree	Neutral	Disagree	Strongly Disagree
❒	❒	❒	❒	❒

b. How many football home games have you attended in the last five years?
❒ 1–5 ❒ 5–10 ❒ 10–15 ❒ 16–20 ❒ 21–25 ❒ Every Game

c. What or who makes you believe that our football home games suffer from a lack of excitement?
❒ The media ❒ My peers ❒ My own ❒ Other source observations (specify)

d. In a study conducted to examine consumers' preference of athletic shoes, one of the questions asked is "Do you have a pair of Nike shoes?"

e. Should the team trade John Smith, who is too old and overpaid, for a younger player?
Yes ❒ No ❒

2. The following is a questionnaire drafted by the staff in the athletic department of XYZ University. It is intended to determine local residents' levels of awareness of XYZ University athletics in general. You are asked to evaluate the survey instrument and accordingly provide feedback to them of how to revise it. Show your critique.

XYZ University Athletics Awareness Survey

Level of Athletic Involvement

1. Approximately how often have you attended the following XYZ University athletic events over the past 12 months? Please circle one for each sport.
   1 = Almost all games/matches
   2 = More than half of the games/matches
   3 = More than two but less than one-half of the games/matches
   4 = Once in a great while
   5 = Have not attended
2. The number in parenthesis indicates the number of home games in that sport.

Football (6)	1 2 3 4 5
Men's Basketball (13)	1 2 3 4 5
Women's Basketball (12)	1 2 3 4 5
Baseball (30)	1 2 3 4 5
Women's Softball (18)	1 2 3 4 5
Men's Golf (3)	1 2 3 4 5
Women's Volleyball (12)	1 2 3 4 5
Men's Tennis (5)	1 2 3 4 5
Women's Tennis (5)	1 2 3 4 5
Men's Cross Country (2)	1 2 3 4 5

Women's Cross Country (2)	1 2 3 4 5
Men's Soccer (12)	1 2 3 4 5
Women's Soccer (12)	1 2 3 4 5
Men's Swimming (4)	1 2 3 4 5
Women's Swimming (4)	1 2 3 4 5
Other ____________________	1 2 3 4 5

Please Check One for Each Question

1. How long have you attended XYZ University athletic events?
   ____ Less than a year ____ 1–2 years ____ 2–3 years ____ 3–5 years ____ More than 5 years
2. Will you spend any nights away from home attending an XYZ University athletic event?
   ____ Yes ____ No If yes, how many? ____ 1 night ____ 2 nights ____ 3 or more nights
3. Why do you attend XYZ University athletic events?
   ____ Interest in that particular sport
   ____ Social/family activity
   ____ Business
   ____ Support the school
   ____ Other (please specify) ____________________
4. When was the last time you attended an XYZ University athletic event?
   ____ Week or less ____ 1–4 weeks ____ 1–6 months ____ 6 months–year ____ Over a year
5. How often do you plan your schedule around XYZ University athletic events?
   ____ All the time ____ Often ____ Sometimes ____ Rarely ____ Never
6. How far are you willing to travel to attend an XYZ University athletic event?
   ____ 50–99 miles ____ 100–250 miles ____ 251–500 miles ____ 500+ miles
   ____ Would not travel

Info-Bits

1. Can you name the new Athletic Director at XYZ University?
   ____ Yes ____ No If yes, who? ____________________
2. Can you name two head coaches at XYZ University, besides Joe Jones?
   ____ Yes ____ No If yes, who? ____________________
   a. ____________________
   b. ____________________
3. Can you name the Marketing/Promotions Director at XYZ University?
   ____ Yes ____ No If yes, who? ____________________
4. Can you name the new Sports Information Director at XYZ University?
   ____ Yes ____ No If yes, who? ____________________
5. Do you know the schedule for XYZ University athletic events?
   ____ Yes ____ No
   If yes, from what media?
   ____ Newspaper ____ Radio ____ TV ____ Mail ____ Word of mouth ____ Pocket schedules
6. Does XYZ University have a coach's show on television
   for football? ____ Yes ____ No
   for basketball? ____ Yes ____ No

7. How often do you listen to XYZ University athletic events on the radio?
   ____ All the time ____ Often ____ Sometimes ____ Rarely ____ Never
8. How many Great Sky Conference Championships did XYZ University win in 2002–2003? Please circle one.
   0 1 2 3 4 or more
9. Can you identify an XYZ University athletic marketing promotion or advertising program to which you have been exposed in the past year?
   ____ Yes ____ No If yes, what? ________________________________

Level of Athletic Support

1. When did you last purchase XYZ University merchandise? (hats, shirts, etc.)
   ____ Less than a month ago ____ 1–6 months ____ 6–12 months ____ over 12 months
   ____ Never
2. If Yes to #1, how much did you spend in the last year?
   ____ $1–$20 ____ $21–$50 ____ $51–$75 ____ $76–$100 ____ $100+
3. If Yes to #1, where did you make your purchase?
   ____ University store
   ____ Discount store (Wal-Mart, K-Mart, etc.)
   ____ At the event
   ____ Sports store (Athletic Attic, etc.)
   ____ Other (specify) ________________________________
4. Do you donate money to XYZ University? ____ Yes ____ No
   If yes, in what area?
   ____ General funds
   ____ Endowments
   ____ Athletics
   If Yes to #4, please answer #5 and #6, if No to #4, please move on to #7.
5. How often do you donate money to XYZ University?
   ____ Once a month
   ____ Once every 3–6 months
   ____ Once every 6–12 months
   ____ Once a year
   ____ Once every 1–5 years
6. Approximately how much money have you donated in the last year?
   ____ $1–$100 ____ $101–$500 ____ $501–$1,000 ____ $1,001–$5,000 ____ $5,000+
   ____ Zero last year
7. Have you volunteered in any capacity in the last 12 months for XYZ University athletics?
   ____ Yes ____ No If yes, how much time?
   ____ 1 hour ____ 1–4 hours ____ 4–8 hours ____ More than a day ____
   Other ________________________________
8. What is the likelihood that you can donate any items such as labor, equipment (maintenance, landscape), materials (uniforms, athletic equipment), or other resources?
   ____ Very likely ____ Somewhat likely ____ Likely (50-50) ____ Somewhat unlikely
   ____ Very unlikely

Level of Satisfaction

1. Please rate the following XYZ University attributes on a 5 point scale, with 1 representing totally unsatisfactory and 5 representing totally satisfactory. Please circle one for each attribute.

All athletics win-loss record	1 2 3 4 5
Schedule strength (all sports)	1 2 3 4 5
Women's athletics	1 2 3 4 5
Ticket prices	1 2 3 4 5
Ticket availability	1 2 3 4 5
Availability of good seats	1 2 3 4 5
Concession prices	1 2 3 4 5
Concession quality	1 2 3 4 5
Parking situation/facilities (all sports)	1 2 3 4 5
Overall service	1 2 3 4 5
Academic reputation	1 2 3 4 5
Community relations	1 2 3 4 5
Pregame publicity (advertising)	1 2 3 4 5
Promotions/Activities at games	1 2 3 4 5
Game times/days	1 2 3 4 5
Corporate sponsorships	1 2 3 4 5
Location of venues	1 2 3 4 5

2. If you could change one thing about XYZ University athletics, what would it be? Explain.
   ____ Male ____ Female please check one.
3. Age? ____
4. Level of education: ____ High school ____ Some college experience ____ College graduate ____ Master's/PhD., etc.

Thank you for your help and participation!

References

Andrews, D., Nonnecke, B., Preece, J. (2003). Electronic survey methodology: A case study in reaching hard to involve Internet users. *International Journal of Human-Computer Interaction*, *16*(2), 185–210.

QuestionPro. (n.d.). Preparing An Online Questionnaire: How To Effectively Conduct An Online Survey by Preparing Online Questionnaires. Retrieved from https://www.questionpro.com/article/preparing-online-questionnaire.html

Bradburn, N., Sudman, S., & Wansink, B. (2004). *Asking questions: The definitive guide to questionnaire design—for market research, political polls, and social and health questionnaires.* San Francisco CA: Jossey-Bass.

Fraenkel, J. R., & Wallen, N. E. (2000*). How to design and evaluate research in education* (4th ed.). Boston, MA: McGraw Hill.

Ghauri, P. N., Gronhaug, K., & Kristianslund, I. (1995). *Research methods in business studies: A practical guide.* New York: Prentice Hall.

Huisman, M., & van der Zouwen, J. (1998). *Item nonresponse in scale data from surveys: Types, determinants, and measures.* Technical report, University of Groningen.

Johnson, B., & Christensen, L. (2000). *Educational research: Quantitative and qualitative approaches.* Boston, MA: Allyn and Bacon.

Lumsden, J. (2005). *Guidelines for the design of online questionnaires.* Fredericton, NB: National Research Council of Canada.

McDaniel, C., & Gates, R. (1991). *Contemporary marketing research.* St. Paul, MN: West Publishing Company.

Patten, M. L. (2001). *Questionnaire research: A practical guide* (2nd ed.). Los Angeles, CA: Pyrczak Publishing.

Robson, C. (2002). *Real world research: A Resource for social scientists and practitioners-researchers* (2nd ed.). Oxford, UK: Blackwell Publisher.

Schwab, D. P. (1999). *Research methods for organizational studies.* Mahwah, NJ: Lawrence Erlbaum Associates, Publishers.

SmartSurvey. (2017). 10 tips for creating an effective online survey. Retrieved from https://www.smartsurvey.co.uk/articles/10-tips-for-designing-an-effective-online-survey

StatPac Inc. (2007). Length of a questionnaire. Retrieved from http://www.statpac.com/surveys/length.htm

Swanson, R. A., & Holton, E. F., III. (2005). *Research in organizations.* San Francisco, CA: Berrett-Koehler Publishers, Inc.

University of Wisconsin Survey Center. (2010). Survey Fundamentals: A guide to designing and implementing surveys. Retrieved from https://oqi.wisc.edu/resourcelibrary/uploads/resources/Survey_Guide.pdf.

Yates, S. J. (2004). *Doing social science research.* London, UK: Sage Publications.

Zikmund, W. G. (2000). *Business research methods* (6th ed.). Fort Worth, TX: Harcourt College Publishers.

13

Measurement

After completing the study of this chapter, you should be able to:

- Explain the four different scales of measurement.
- Recognize the levels of measurement used in articles of sport management studies.
- Define reliability.
- Compare and contrast the concepts of stability, equivalence, and internal consistency as they relate to reliability.
- Define validity.
- Compare and contrast validity based on evidence (i.e., content, internal structure, and relations to other variables).
- Interpret the coefficient of correlation based on the standards reported in major research and statistical textbooks.
- Explain why validity and reliability are critical attributes in evaluating a quantitative research report.

The purpose of this chapter is to provide students with an understanding of the concepts of *measurement*, the different scales of measurement, the concepts of *reliability*, and the concepts of *validity*. These techniques are used in basic, applied, and evaluation research and can be used in experimental or nonexperimental research (McMillan & Schumacher, 2006).

This chapter introduces the different scales for classifying levels of measurements. When using this system, researchers are instructed to follow the procedures of one or a combination of four classifications of measurement commonly known as *nominal*, *ordinal*, *interval* and *ratio levels*. The nature and scope of reliability as a feature for measuring the consistency of the scales of research instruments used in sport management studies are then discussed. Seven procedures are captured within three approaches for measuring the reliability of quantitative measuring instruments. In the later part of the chapter, the nature and scope of validity as an element for measuring the accuracy of the scales of research instruments used in sport management studies are also reviewed. Several of the most common procedures are captured within three approaches for assessing the accuracy of quantitative instruments in sport management research. In addition, coefficient correlation is illustrated to show how degrees of consistency and accuracy are calculated for reliability and validity, respectively.

Have you heard of GIGO? GIGO refers to garbage in, garbage out. It is a significant principle that without good measurement, you will not have any good results to interpret. When you conduct research, you develop your research question first, then you figure out how to answer each research question by collecting data or information. Whenever you collect data for your research, you need to measure or assess something, and if you employ poor measurement, your research will be poor.

Definition of Measurement

Measurement is the act of measuring. A number of references demonstrate a variety of definitions of measurement, such as "the process of assigning numbers to variables" (Nieswiadomy, 2002, p. 191); "assigning numbers to characteristics according to specified rules to reflect the quantity of the characteristic that the test products possess" (Shao, 2002, p. 219); "the translation of observations into numbers" (DePoy & Gitlin, 1994, p. 195); "assigning symbols or numbers to something according to a set of specific rules" (Johnson & Christensen, 2014, p. 160); and "a standardized process of assigning numbers or other symbols to certain characteristics of the objects of interest, according to some prespecified rules" (Aaker, Kumar, & Day, 1998, p. 274). By producing symbols or numerical values, we can summarize characteristics of cases under research. A measure is an instrument to enter such numerical values (Schwab, 2005). For instance, let's assume that you are investigating the level of satisfaction of college students about ESPN Gameday. In your questionnaire, if a score of 1 is assigned to represent very unsatisfied, then a measure of satisfaction should get scores of 1 for all individual college students who are very unsatisfied. College students who experience other levels of satisfaction should obtain other numerical values.

Different Levels of Measurement

When choosing a proper measurement, it is important that the researcher is knowledgeable and fully understands the different levels of measurement. There are typically four levels of measurement, known as nominal, ordinal, interval, and ratio levels. It is critical to know the level of measurement being used, as it indicates the statistical manipulations of the data and presents the type of information being communicated (Johnson & Christensen, 2014). Presented in Exhibit 13.1 is a summary of the primary characteristics and an example of each of the levels of measurement.

The Nominal Level of Measurement

The nominal level of measurement refers to data that can only be named, labeled, or categorized. This level of measurement is considered the least sophisticated degree of quantification (Triola & Franklin, 1994, p. 8). The unique aspects of nominal variables are that they allow for the researcher to classify people, objects, or events into mutually exclusive categories that are based on the qualification of having a common characteristic (Aczel, 1995, p. 8). The categories must be distinct from each other (mutually exclusive categories) and include all of the possible ways of categorizing the data (exhaustive categories). Mutually exclusive means that each item classified (e.g., male or

Exhibit 13.1. Summary of the Primary Characteristics and Examples of Each Level of Measurement

Level	Characteristic	Example
Nominal (Category)	• Data are given names or categories • Names and categories are mutually exclusive (they have no relationship to each other) • Can't be quantified • Has the lowest degree of quantification	• Gender (male or female) • Marital status • Racial differences • Employment settings (high school, college/university, clinic, industrial, professional) • Categories of sport products • Jersey numbers in basketball • A list of motives for measuring job satisfaction or organizational commitment
Ordinal (Category and Order)	• Data are given names and/or categories • Names and categories are not mutually exclusive; however, they are ordered (they have relationship to each other) • Distances between the categories are not equal and have no meaning beyond indicating a more-or-less relationship between each category.	• Years of work experience • a subject's responses on a scale (1=always, 2=frequently, 3=sometimes, 4=rarely, and 5=never) • Ranking the categories of a sport product as "good," "better," "best" • Ranking the categories of a sport product as "1st", "2nd", "3rd", and "4th" • Ranking the standards for employment performance • Managerial levels (supervisor, middle management, upper management)
Interval (Category, Order, and Spacing of Equal Intervals)	• Data are given names and/or categories • Categories are not mutually exclusive; however, ordered (they have relationship to each other) • Distances between the categories are equal and have meaning beyond indicating a more-or-less relationship between each category • No inherent starting point (no absolute zero)	• Evaluation of a sport product in relative terms on a scale of 1 to 100. (100 as best and 1 as worst) • Measure temperature in Fahrenheit
Ratio (Category, Order, Spacing of Equal Intervals, and a Zero Point)	• Data are given names and/or categories • Categories are not mutually exclusive; however, they are ordered (they have relationship to each other) • Distances between the categories are equal and have meaning beyond indicating a more-or-less relationship between each category • No inherent starting point • Has the highest degree of quantification	• Making an assessment of the numbers of study participants in specified age, height, weight, or salary ranges

female, etc.) falls into only one category. Collectively exhaustive means that there is a complete set of categories (e.g., gender) available so all items can be classified (Aczel, 1995). The NCAA football designations (i.e., Division I-FSB, Division I-FSC, Division II, and Division III) are a good example of the nominal level of measurement.

The Ordinal Level of Measurement

Like nominal variables, ordinal level variables are identified by name, classified, or categorized, but have an additional property of a logical or rank order of the variables or values. Data that can be rank-ordered, as well as placed into categories, are at the ordinal level of measurement. Although ordinal variables are arranged in order, the interval between the variables cannot be measured (Triola & Franklin, 1994; Levine, Ramsey, & Brenson, 1995; Gall, Gall, & Borg, 2003). For instance, a review committee rank-ordered a series of program proposals as first, second, third place, and so on. The members of the committee may have felt that a proposal ranked higher was better than the ones ranked lower. However, they did not give a quantifiable distinction of how much one is better than the other as the differences between intervals are not reflected in this level of measurement. So, the exact differences between the ranks cannot be specified with this type of data. The numbers that are obtained from this measurement process indicate the order rather than the exact quantity of the variables. The unique aspects of ordinal level variables are that they allow the researcher to classify as well as to rank people, objects, or events into mutually exclusive categories that are based on the qualification of having a common characteristic (See summary in Exhibit 13.1). Ordinal level variables provide only for comparison of scores or categories in terms of smaller or larger, higher or lower, or best to worst. They do not permit comparisons in terms of how much larger, lower, or better. The measurement scale that sport management professionals commonly use to assess the quality of a sports event (i.e., excellent, good, fair, and poor) is an example of the ordinal level scale.

The Interval Level of Measurement

Like nominal and ordinal scales, the interval level of measurement identifies, names, classifies, or categorizes. Similar to ordinal scales, interval scales have ordered categories or values. However, interval scales have an additional property for determining specific distances between the order or ranks of the data (Triola & Franklin, 1994). Equal distances or intervals between scores or scale values are an important property of the internal level of measurement (Aczel, 1995). For instance, when we measure temperature in Fahrenheit, the distance from 10 to 20 degrees is the same as the distance from 34 to 40 degrees. Hence, we can compute an average value for interpretation with the interval level of measurement where calculating an average value is not meaningful.

The Ratio Level of Measurement

The ratio level of measurement has all the properties of nominal, ordinal, and interval level variables plus an absolute zero point. The absolute zero point means that a scale value or score of zero represents a complete absence of the characteristic being measured (Johnson & Christensen, 2014). The ratio level of measurement is actually the interval scale modified to include the inherent zero starting point. The ratio scale is the strongest scale of measurement (Nieswiadomy, 2002).

Common examples of ratio scales are height, weight, age, salary, and net worth. Because ratio scales have equal intervals between scale values like interval scales, all the arithmetic operations that can be performed on interval data can also be used with ratio data.

Identifying a Good Measurement in Sport Management Research

When a researcher chooses measuring instruments, it is critical to confirm that those instruments will provide the best and most accurate measure of the variables in one's study. When choosing and employing a measurement instrument, a scholar must consider the issues of reliability and validity. In a general way, reliability is the extent to which a test or measuring procedure yields the same results on repeated trials. Validity is the extent to which a situation as observed reflects the true situation correctly. Reliability and validity are viewed as two different aspects for measuring the credibility of instruments in sport management research. Both quantitative and qualitative researchers are concerned about establishing creditability in terms of reliability and validity during data collection.

The results or findings of a study are only as trustworthy as the accuracy and consistency of its measures. Inaccuracy and inconsistency in measurement will undercut the value of a research project and call its findings into question. Reliability and validity are both dynamic in nature because they are neither static components of an instrument, nor automatically generalizable; therefore, they should be recalculated each time an instrument is used (Waltz, Strickland, & Lenz, 1984). However, if an instrument is adjusted or is applied to a new target population, it is crucial that its validity and reliability be reexamined to ensure the new instrument assesses with the same degree of accuracy in the new target population. The construct validation and the reliability should be checked after a measure is adapted to a new linguistic context (Geisinger, 1992; Lin, Chen, & Chiu, 2005).

Reliability as a Measurement Criterion in Sport Management Research

The reliability of an instrument refers to its level of being stable, being equivalent, and possessing internal consistency over time (Gall, Gall, & Borg, 2003). Internal consistency refers to the degree of cohesion among items that form a scale. Even though reliability has been conceptualized in several different dimensions, in this chapter, three major categories—stability, equivalence, and internal consistency (McMillan & Schumacher, 2006, p. 185)—will be reviewed (see Exhibit 13.2).

Reliability as stability is determined by giving the same test to the same individuals at two different times and is reported in two procedures: test-retest and intra-rater reliability. An instrument is consistent (reliable) when it behaves the same way under different circumstances and/or when it yields similar results when different people use it.

Reliability as equivalence is determined by giving two equivalent forms of a test to the same individuals on two separate occasions. Reliability as equivalence is usually reported in two forms: the parallel form and inter-rater (also known as inter-rater agreement or inter-coder agreement). The third major category of reliability is internal consistency. It is determined by the extent to which all items on an instrument measure the same variable. Internal consistency is usually reported as a Kuder-Richardson (KR20), Spearman-Brown split-half, odd-even coefficient, or Cronbach's alpha

Exhibit 13.2. Major Dimensions of Instrument Reliability

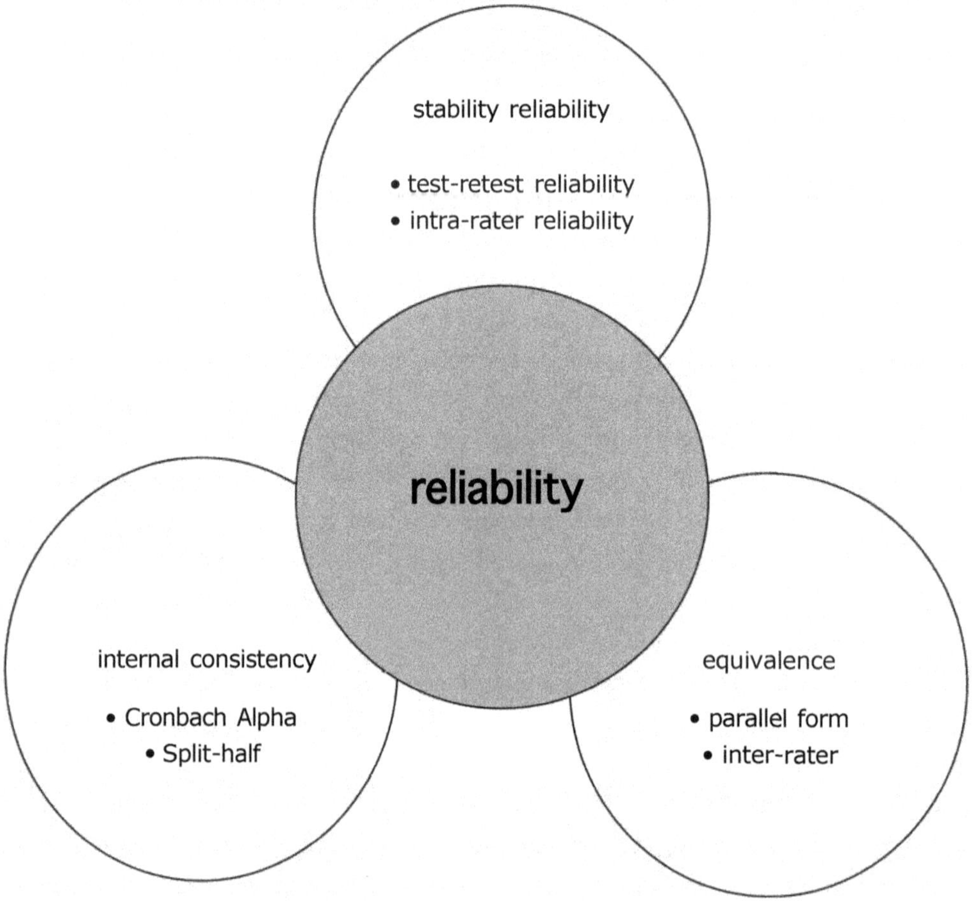

coefficient. Overall, reliability is concerned with instrument dependability (DePoy & Gitlin, 1994; Gall, Gall, & Borg, 2003; McMillian & Schumacher, 2006; Neuman, 2003). Cronbach's alpha coefficient, which measures the extent to which performance on any one item in an instrument indicates performance on any other item in that instrument, is the most preferred measure of internal consistency.

Reliability as Stability

Reliability as stability is captured in two forms: test-retest reliability and intra-rater reliability. The test-retest reliability is the correlation between scores from the same subjects tested at two different times. It involves giving a test twice to the same subjects and statistically comparing the two sets of scores by calculating the correlation coefficient. Exhibit 13.3 shows an illustration of the test-retest reliability. The closer the two sets of scores are to each other, the greater the test reliability. One of the major concerns for researchers is the time interval between the first and second test sessions. It is important that the time interval between the assessments be long enough that the subjects are not able to recall the items and inflate the reliability coefficient. For example, Li (1993) examined job satisfaction and performance of coaches of spare-time sports schools in China. He tested the test-retest reliability of 14 variables in the instrument using a pilot study. A pretest was administered, followed by a post-test 15 days later. The correlation coefficients for the variables ranged

Exhibit 13.3. A Symbolic Illustration Test-Retest Reliability

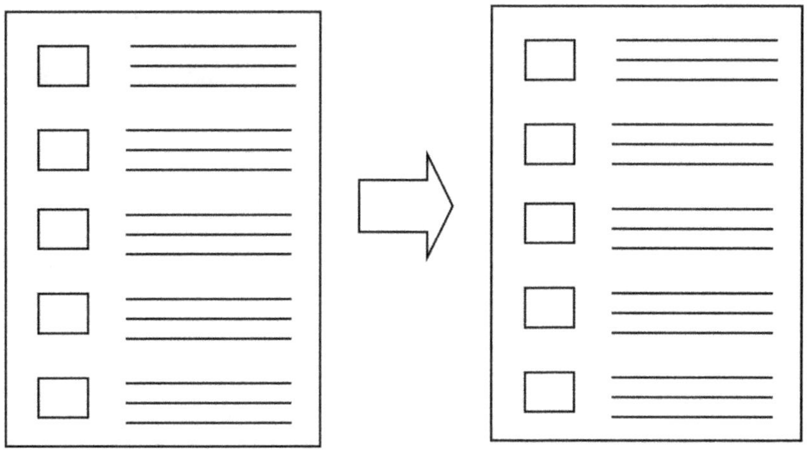

from .68 to .88. The intra-rater reliability refers to the correlations of scores by a single observer, rating the same behaviors on two or more occasions. It is also known as intra-coder reliability and intra-rater agreement.

Reliability as Equivalence

Reliability as equivalence has two forms: parallel forms of reliability and inter-rater reliability. Parallel forms refer to the correlations of scores of two different instruments given to the subjects at the same time. Parallel forms extend from the development of two different tests that measure the same items in the same way. The correlation coefficient is based on the scores of the same individual taking two equivalent forms of the test, and the scores on the first form of the tests are compared to the scores on the second form of the tests. Exhibit 13.4 shows an illustration of the alternative form reliability test. After which, the correlation coefficient scores of each form are calculated. The higher the correlation coefficients between the two forms, the greater the test's reliability. The biggest concern for this type of reliability is that forms are difficult to construct. Parallel forms are used mostly for psychological constructs, self-esteem, self-image, and self-concepts.

Exhibit 13.4. An Illustration of the Alternative Form Reliability

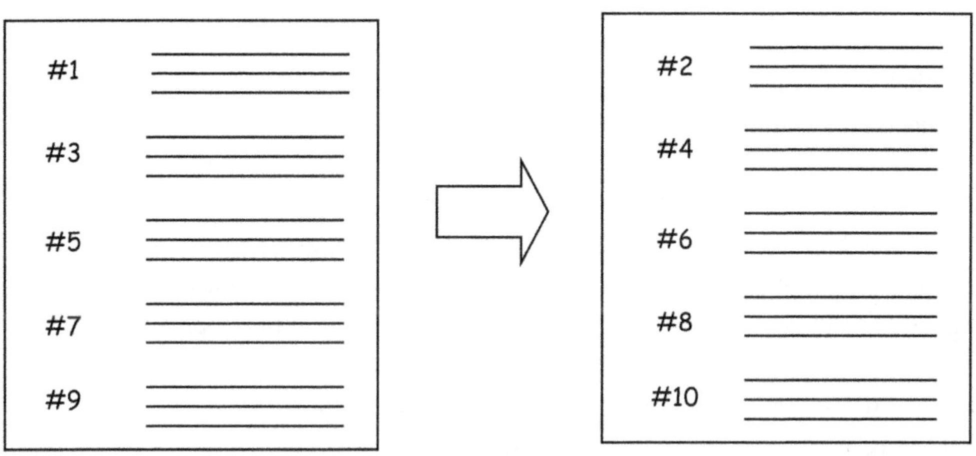

Inter-rater reliability is also known as inter-coder reliability, inter-judge reliability, and inter-rater agreement. It refers to the correlations between two or more observers that independently rate the same behaviors on a single occasion of a particular subject or group of subjects. Inter-rater reliability can be assessed through the I_r coefficient developed by Perreault and Leigh (1989). It is established by comparing coding sheets for the same data completed by the same coder 24 hours apart, ranged from 96% to 100%. The 24-hour duration was chosen as an acceptable length of time for coders to lose awareness of their previous coding schemes on the same material (Stacks & Hocking, 1998). The inter-rater reliability is commonly used in sport management research. Exhibit 13.5 shows examples of recent studies in sport management using inter-rater reliability as reliability as equivalence.

Internal Consistency

Of the three forms of reliability, internal consistency is the most widely used in sport management research. Unlike stability and equivalence, it is the measure of "the consistency with which the items on a test measure a single construct" (Johnson & Christensen, 2014, p. 169). For example, if a researcher developed a questionnaire of 30 items to measure job satisfaction among NCAA Division I athletic directors, the researcher would analyze the items on the instruments to evaluate the extent to which all 30 of the items measured the concept of job satisfaction.

The empirical approach most commonly used by researchers to estimate the internal consistency of instruments is known as Cronbach's alpha (Cronbach, 1990; Nunnally, 1978; Nunnally & Berstein, 1994). Cronbach's alpha measures the extent to which performance on any one item in an instrument indicates performance on any other item in that instrument. In addition to Cronbach's alpha, there are two other forms of reliability coefficients used for reliability consistency: Kuder-Richardson (KR) 20 and 21, and the Spearman Brown techniques.

Among sport management studies, McGuire and Trail (2002) investigated the extent to which three groups of university presidents (NCAA Divisions I-A, I-AA, I-AAA and III) differed on level of importance attributed to various intercollegiate athletic department goals and the level of satisfaction with such goals. The views of the presidents in the NCAA member institutions in each division were tested with the Scale of Athletic Department Goals (SADG). Internal consistency of the scale was evaluated using Cronbach's alpha. The mean score of each subscale was calculated to determine the president's satisfaction and important levels for each goal.

Interpretation of Reliability Coefficient Correlation

Reliability assessments are typically reported as correlation coefficients. It must be kept in mind that no universally accepted correlation values have been established for reliability; however, several schemes have been proposed. There is no simple answer as to how high the reliability of a research instrument should be. According to Nunnally (1978) and Nunnally and Bernstein (1994), the judgment of how high the reliability should depend on the type of variable being measured and the stage of the development of the research instrument. The lowest generally agreed upon limit for Cronbach's alpha coefficient is .70, although a number between .60 and .69 is also considered an acceptable moderate range for a scale in its beginning stage of development (Hair, Anderson, Tatham, & Black, 1998; Nunnally, 1978; Nunnally & Burnstein, 1994). However, .80 or higher is acceptable for well-developed instruments, and 0.90 or higher is acceptable for important decision making (Nunnally, 1978).

Exhibit 13.5. Examples of Inter-Rater Reliability by Levels of Coefficient Correlations in Sport Management Research

Source of Research Article Example	Estimated Reliability Coefficient Correlation	Purpose of Study and Description of the Validity Technique Used in the Article
Kelley, S. W. , & Turley, L. W. (2004). The effect of content on perceived affect of Super Bowl commercials. *Journal of Sport Management, 18*(4), 398–420	The I_r values for the variables used in this study ranged from .77 for the slogan variable to .98 for the variables focusing on product type and price claims.	Kelly & Turley (2004) analyzed the content of the national advertisements that were telecast during the Super Bowl over a seven-year period from 1996 to 2002. Interjudge reliability was assessed through the calculation of the I_r Coefficient developed by Perreault & Leigh (1989). The I_r value for each of the qualitative nominally scaled variables included in our study is provided. The relatively high I_r coefficients associated with the variables included in this study are indicative of a high degree of inter-judge reliability for each of the variables encoded through our content analysis procedures.
Cuneen, G., & Sidwell, M.G. (1998). Gender portrayals in *Sports Illustrated* for kids' advertisements: A content analysis of prominent and supporting models. *Journal of Sport Management, 12*(1), 39–50.	Overall percentage of agreement between all subcomponents of analysis was .85. Reliability coefficient between coders for each subscale (.88 for advertisements 1989 issues, .84 for 1990 issues, .86 for 1991 issues, .87 for 1992 issues, .80 for 1993 issues, .82 for 1994 issues).	Cuneen & Sidwell (1998) analyzed gender portrayals in *Sports Illustrated* for kids' advertisements. Inter-coder reliability coefficients were examined to provide evidence of the agreement of the coder. All advertisements on which coders disagreed on prominent and/or supporting models were deleted from analyses.
Peterson, P. M., Whiserent, W. A., & Schneider, R. G. (2003). Using a content analysis to estimate the gendering of sports newspaper personnel and their coverage. *Journal of Sport Management, 17*, (3), 376–393.	The agreement between the coders resulted in percentages that ranged from 89.1 to 97.2 with articles and from .845 to .965 with photographs.	Peterson, Whisement, & Schneider (2003) examined the gendering of sports newspaper personnel and then coverage. Inter-coder reliability coefficients were used to examine the percentage of agreement of the coders. Two trained individuals, independently of each other, coded the daily issues. After precoding sessions and a pilot test, the two coders independently examined the same 120 selected newspaper sport sections (20% of the sample) to test inter-coder reliability. According to the authors "high intercoder reliability percentages . . . for this study was manifest content . . . confirmed that the two coders had become thoroughly familiar with the coding protocol . . . by the time this study was conducted" (p. 383).

Validity as a Measurement Criterion in Sport Management Research

Validity addresses whether or not the researchers are measuring what they think is supposed to be measured. In other words, it is the degree or extent to which items on a scale or research instrument measure what they are intended to measure. According to Nunnally and Bernstein (1994), no amount of statistical elegance or a sophisticated measurement strategy can be substituted for validity. If an instrument lacks validity, there is no point in establishing its reliability. This implies that reliability and validity are not independent characteristics of an instrument. An instrument that is not reliable cannot possibly be valid. There are different ways to obtain validity evidence, as "empirical evidence and theoretical rationales that support the interpretations and actions that we take on the basis of the score or scores we get from an assessment procedure" (Johnson & Christensen, 2014, p. 172). There are three ways to obtain evidence-based validity: content, internal structure, and relations to other variables (Johnson & Christensen, 2014).

Evidence Based on Content

Content validity is "based on a judgment of the degree to which the item, task, or questions on an instrument adequately represent the construct domain of interest, such as burnout or self-esteem" (Johnson & Christensen, 2014, p. 174). Evidence for content validity is based on item content but also based on the wording, administration, formatting, and scoring of the test. Judgments of content validity must be assessed by experts in the domain of interest.

Let us use examples in sport management research to illustrate the application of these approaches. To assess the economic and fiscal impacts generated by a state high school sport championship, Turco (1997) established an expert panel to generate content validity for the questionnaire used to gather the needed data in visitor group expenditures. Several key event officials and tourism researchers were asked to serve on the panel to develop all possible categories of visitor spending. Based upon the recommendations of the panel, a number of minor revisions were made to the survey instrument. According to Turco (1997), the content validity of the instrument was enhanced through the critique provided by the panel of experts. For another example, Ridpath's (2006) study on the perceptions of college athletes about the degree of emphasis that their coaches placed on academic progress and graduation. "The content validity of the survey instrument used in the study was developed first through an extensive review of past and present related literature, then by the approval by a panel of eight experts in the higher education and intercollegiate athletic fields, and through a pilot test of a like population" (Ridpath, 2006, p. 21). Cianfrone and Zhang (2006) examined the differential effectiveness of television commercials, athlete endorsements, venue signage, and combined promotions as assessed by Generation-Y consumers. The authors created a panel of experts that included professors in business, sport management, and measurement. Following the content validity guidelines suggested by Safrit and Wood (1989), the panel members were specifically instructed to look into content relevance, representatives, and clarity. The improvements to the instrument were made with the input from the two panels.

Content validity can be measured with the application of a rigorous statistical approach developed by Lawshe (1975), who insisted the content validity of a measurement item can be reflected by the agreement among the members of an expert panel regarding how essential the item is in measuring the proposed construct. To use the method proposed by Lawshe (1975), the researcher should ask the members of the panel if the item is essential to measure the construct under investigation. The following formula was suggested to use to determine the content validity of the item:

$$CVR = \frac{n_e - \frac{N}{2}}{\frac{N}{2}}$$

Where
CVR = content validity ratio
n_e = number of panel members indicating the item as essential
N = number of members on the expert panel

According to Lawshe (1975), an item will exhibit some content validity if more than half the panel members felt the item was essential. The higher the number of panel members who indicated the item to be essential, the higher the content validity of the item.

Evidence Based on Internal Structure

Some instruments are developed to measure a single construct, but others are designed to measure multiple dimensions of a construct. Construct validity is regarded as the most theoretical form of validity. Leary (1991) referred to it as subjective judgment of entities (hypothetical constructs) that cannot be directly observed but are inferred on the basis of empirical evidence by a panel of experts on whether a test performance or result is related to the hypothetical construct(s) or set of related variables (p. 60). Babbie (2000) considered construct validity as the logical relationships within the questionnaire and whether a particular measure relates to other measures consistent with the content being examined. The principal aspect of construct validity is the determination of whether the items in an instrument or a test have each been derived mainly from their underlying theoretical construct. More specifically, it measures whether each item contributes to the underlying constructs of the instrument.

There are two procedures for assessing construct validity: factor analysis and test of homogeneity. Factor analysis is a statistical procedure that computes correlations among items and presents the number of factors within. When assessing the internal structure of an instrument, you can also examine a measure of homogeneity (i.e., how well the different item in an instrument measure the same construct). It can be tested by the item-to-total correlation or coefficient alpha. The purpose of assessing convergent validity and discriminate validity is to determine whether the measures are indeed measures of the constructs they aim to assess, and whether the measures are isolated to the construct that they are said to gauge and not to other constructs as well (Hair et al., 1998).

The study conducted by Koo, Quarterman, and Flynn (2006) is a good example of determining the convergent validity of an instrument. The purpose of the study was to assess the effect of perceived sport event and sponsor image fit on consumers' cognition, affect, and behavioral intentions. Evidence of convergent validity was sought by examining each construct's average variance extracted (AVE) (Hair et al., 1998). A construct was considered to exhibit convergent validity if the average variance extracted was .50 or greater (Fornell & Larcker, 1981). In their study, the

AVE values for all constructs were between .55 and .84. Therefore, they concluded that the measurement scale possessed convergent validity.

Evidence Based on Relations to Other Variables

Validity evidence can also be collected by relating your scores to scores on other variables. The crucial type of evidence in this category is criterion-related validity. Criterion-related validity refers to the correlation of validating a new scale of an instrument with some other measure of a trait or interest. An instrument is said to possess criterion-related validity when there is a high degree of correlation between two measures of the same concept as measured (1) by the instrument being assessed and (2) by another established and presumed valid instrument or scale.

For instance, if you measure self-esteem, for content validity you will attempt to describe the things that self-esteem represents and translate them into a measure. You might describe that self-esteem refers to how good you feel about yourself, including your self-confidence. For criterion-related validity, you will attempt to distinguish people with high self-esteem with ones with low self-esteem. You might claim that high self-esteem people will tend to volunteer more for jobs that require self-confidence, such as speaking in public. Hence, you validate your self-esteem measure by showing that it is correlated with some other independent indicator, such as volunteering for public speech (Trochim & Donnelly, 2008). There are two types of validity within this category: predictive validity and concurrent validity. The distinction between them pertains to the time period when data are gathered. In concurrent validity, both the test and criterion are measured about the same time. Concurrent validity is reflected by the correlation of scores between a measurement instrument or scale and the criterion instrument when administered at the same time. On the other hand, there is a time lapse between the two measures when determining the predictive validity. Predictive validity is indicated by the correlation of scores between a measurement instrument or scale and the criterion instrument or scale when administered at different times or for measuring future performance.

Criterion-related validity can be reported as a correlation coefficient. The correlation coefficient is an index for summarizing the degree of relationships between variables, usually ranging between +1.00 (for a perfect positive correlation) through 0.0 (for no relationship) to -1.00 (for a perfect negative correlation). The value of criterion-related validity is considered inadequate if below .70 and desirable if above 0.80.

Relationships Between Reliability and Validity

Reliability is a precondition for validity because unreliable measurement cannot be valid. However, reliability does not guarantee validity. A metaphor of a marksman's shots on a target can be used to illustrate the relationships of reliability and validity (see Exhibit 13.6). In using this metaphor, reliability can be thought of as consistency, stability, or repeatability in measurement, and validity can be described as accuracy for the claims to be measured. In situation 1, the shots are widely scattered. Thus, there is little or no consistency or accuracy. In situation 2, the shots are clustered around the most outer ring of the target, illustrating that consistency has been accomplished, but not accuracy. In situation 3, the shots are clustered around the bull's eye. As such, the marksman's shots are considered highly consistent and highly accurate. Although reliability and validity are

Exhibit 13.6. A Metaphor of Reliability and Vailidity

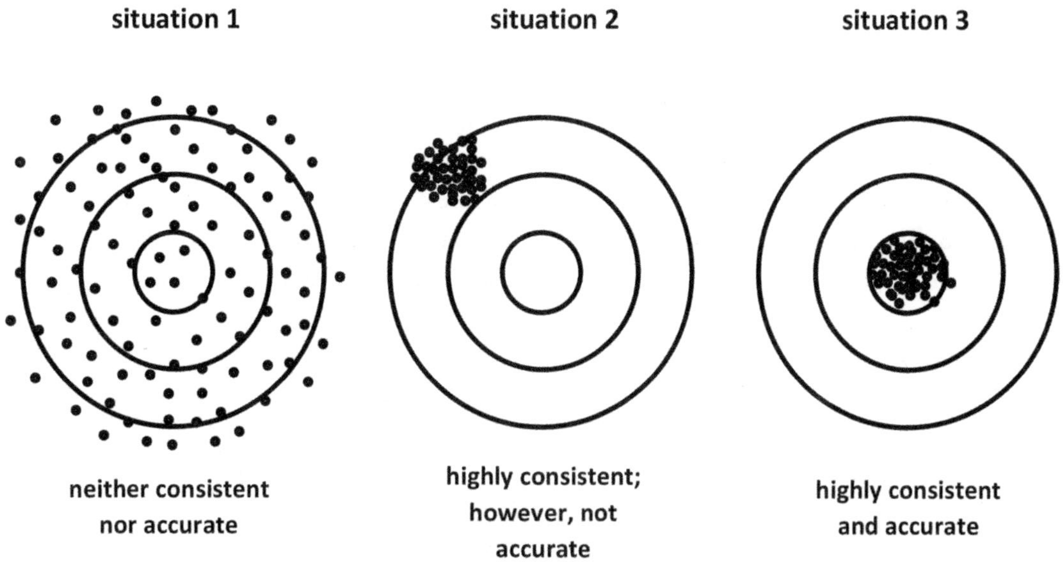

two different procedures, each one is needed to compliment the other for credibility of research in sport management.

A Case in Sport Management

An example in sport management that demonstrates the efforts of the researchers in establishing the credibility of the survey instrument is a study conducted by Fink, Pastore, and Riemer (2001). The study examined a framework of diversity initiatives as a basis for exploring top management beliefs and diversity management strategies of Division I-A intercollegiate athletic organizations. The researchers used a number of procedures, such as a panel of experts, a field test, and a pilot test, as well as internal consistency measures (Cronbach's alpha) and item-to-total correlations to evaluate the validity and reliability of the survey instrument.

First, the survey was sent to a panel of experts (14 individuals) who were provided directions for judging the instrument. They were asked to review the overall survey and provide suggestions for improvement. Most importantly, they were asked to look at the second portion of the instrument and rate each item on how well it represented the dimension it was intended to represent (i.e., compliance, reactive diversity management, proactive diversity management). Statements that were rated as good (as opposed to poor or neutral) by at least eight of the 14 judges were kept in the survey. After the initial survey was revised based on the panel of experts' comments, it was field tested by 25 sport management graduate students. They commented on the survey's clarity, wording, thoroughness, appropriateness, and ease of use. Again, the survey was revised based on these comments.

The survey was pilot tested as it was sent to athletic directors, senior women administrators, middle managers, baseball coaches, and softball coaches randomly selected from Division I-A and Division III institutions (n = 100). Thirty people responded to the pilot test. A Cronbach's alpha was calculated for the beliefs and the three subscales of the diversity practices sections of the survey. The Cronbach's alphas were: 0.88 for the beliefs section, 0.96 for compliance, 0.77 for reactive diversity management, and 0.73 for proactive diversity management. Item-to-total correlations were also analyzed. Based upon information from the pilot test, the survey was again revised (seven items were reworded and two were broken down into two items to avoid double-barreled questions). The final version of the survey was then sent to the sample noted previously.

Results from the analysis of the final survey indicated that it was a valid and reliable instrument. Cronbach's alpha for the beliefs section of the survey was 0.89. Cronbach's alpha for the three subscales of the diversity practices section of the survey ranged from 0.83 (reactive diversity management) to 0.91 (proactive diversity management), while compliant diversity management had an alpha of 0.88. The results from the item-to-total correlation analysis indicated that 26 of the 29 items correlated at least 0.25 with the sum of the other items in its own dimension.

Summary

The purpose of this chapter was to provide students with an understanding of the different levels of measurement, and concepts of reliability and validity of measurement.

There are typically four classical levels of measurement, known as nominal, ordinal, interval, and ratio levels. The nominal level of measurement is considered the lowest degree of quantification, showing differences by names or categories. The ordinal level of measurement indicates differences in categories; however, it also shows differences by rank order without any specific distances between the ranks or order. The interval level of measurement has an additional property for determining specific distances between the order and ranks of the data. The ratio level of measurement has a zero starting point and is considered the highest degree of quantification.

The criteria for determining how accurate or consistent repeated measurements are to describe phenomena in sport management are known as reliability (consistency) and validity (accuracy). Without validity and reliability, there would be no criteria by which the findings of a study could be judged as credible.

Reliability can be thought of as consistency, stability, or repeatability in measurement, and validity as accuracy. The reliability of an instrument refers to its level of being stable, being equivalent, and possessing internal consistency over time. Reliability has several different dimensions, which include stability, equivalence, and internal consistency. The concept of validity addresses whether or not a measurement scale measures what it is intended to measure.

Study Questions

1. What are the differences among the four levels of measurement?
2. What is reliability, and what are the differences of the three types of reliability?
3. What is validity, and what are the different types of validity discussed in this chapter?
4. Why are validity and reliability critical attributes in evaluating a quantitative research report?

Learning Activities

1. Evaluate the reliability and validity of the measurement scale used in an article published in a refereed sport management journal.

2. Review a nonexperimental research article that was published this year in one of the journals in sport management and argue for or against the type of reliability technique(s) used in the article.
3. Review a nonexperimental research article that was published this year in one of the journals in sport management and argue for or against the type of validity technique(s) used in the article.

References

Aaker, D. A., Kumar, V., & Day, G. S. (1998). *Marketing research* (6th ed.). New York: John Wiley & Sons, Inc.

Aczel, A. (1995). *Statistics: concepts and applications.* Chicago, IL: Irwin.

Babbie, E. (2000). *The practice of social research* (9th ed.). Belmont, CA: Wadsworth/Thomson Learning.

Cianfrone, B., & Zhang, J. (2006). Differential effects of television commercials, athlete endorsement and venue signage during a televised action sports event. *Journal of Sport Management, 20*(3), 322–344.

Cronbach, L. J. (1990). *Essentials of psychological testing* (5th ed.). New York, NY: Harper & Row.

Cuneen, G., & Sidwell, M. G. (1998). Gender portrayals in Sports Illustrated for kids' advertisements: A content analysis of prominent and supporting models. *Journal of Sport Management, 12*(1), 39–50.

DePoy, E., & Gitlin, L. N. (1994). *Introduction to Research: Multiple strategies for health and human services.* St. Louis, MO: Mosby-Year Book, Inc.

Fink, J., Pastore, D. L., & Riemer, H. A. (2001). Do differences make a difference? Managing diversity in Division IA intercollegiate athletics. *Journal of Sport Management, 15*(1), 10–50.

Fornell, C., & Lacker, D. (1981). Evaluating structural equation models with unobservable variables and measurement error. *Journal of Marketing Research, 18*, 30–50.

Gall, M. D., Gall, J. P., & Borg, W. R. (2003). *Educational research and introduction* (7th ed.). Boston, MA: Allyn and Bacon.

Geisinger, K. F. (1992). The metamorphosis of test validation. *Educational Psychologist, 27*, 197–222.

Hair, J. F., Anderson, R. E., Tatham, R .L., & Black, W. C. (1998). *Multivariate data analysis* (5th ed.). Upper Saddle River, NJ: Prentice Hall.

Johnson, R. B., & Christensen, L. (2014). *Educational research: Quantitative, qualitative, and mixed approaches* (5th ed.). Thousand Oaks, CA: SAGE Publications, Inc.

Kelley, S. W., & Turley, L. W. (2004). The effect of content on perceived affect of Super Bowl commercials. *Journal of Sport Management, 18*(4), 398–420.

Koo, G., Quarterman, J., & Flynn, L. (2006). Effect of perceived sport event and sponsor image fit on consumers' cognition, affect, and behavioral intentions. *Sport Marketing Quarterly, 15*, 80–90.

Lawshe, C. H. (1975). A quantitative approach to content validity. *Personnel Psychology, 28*, 563–575.

Leary, M. (1991). *Introduction to behavioral research methods.* Belmont: Wadsworth Publishing Company.

Levine, D. M, Ramsey, P. P., & Berenson, M. (1995). *Business statistics for quality and productivity.* Englewood Cliffs, NJ: Prentice Hall, Inc.

Li, M. (1993). Job satisfaction and performance of coaches of the spare-time sports schools in China. *Journal of Sport Management,* 1(2), 134–135.

Lin, Y., Chen, C., & Chiu, P. (2005). An overview on issues on cross-cultural research and back-translation. *The Sport Journal, 8*(4). Retrieved from http://www.thesportjournal.org/article/cross-cultural-research-and-back-translation

McGuire, R., & Trail, G. (2002). Satisfaction and importance of athletic department goals: The views of university presidents. *International Journal of Sport Management, 3*(1), 53–60.

McMillian, J. H., & Schumacher, S. (2006). *Research in education* (6th ed.). Boston, MA: Pearson Education, Inc.

Neuman, W. L. (2003). *Social research methods: Qualitative and quantitative approaches.* (5th ed.). Boston, MA: Pearson Education, Inc.

Nieswiadomy, R. M. (2002). *Foundations of nursing research* (4th ed.). Upper Saddle River, NJ: Prentice Hall.

Nunnally, J. C. (1978). *Psychometric theory* (2nd ed.). New York, NY: McGraw Hill.

Nunnally, J. C., & Bernstein, L. H. (1994). *Psychometric theory* (3rd ed.). New York: McGraw Hill.

Perreault, W. D., & Leigh, L. E. (1989). Reliability of nominal databased on qualitative judgments. *Journal of Marketing Research, 26*(2), 135–148.

Peterson, P. M., Whiserent, W. A., & Schneider, R. G. (2003). Using a content analysis to estimate the gendering of sports newspaper personnel and their coverage. *Journal of Sport Management, 17*(3), 376–393.

Ridpath, B. D. (2006). College athletes' perceptions of the emphasis their coaches place on academic progress and graduation. *The SMART Journal, 3*(1), 14–27.

Safrit, M. J., & Wood, T. M. (1989). *Measurement concepts in physical education and exercise science*. Champaign, IL: Human Kinetics.

Schwab, D. P. (2005). *Research methods for organizational studies*. New York, NY: Routledge.

Shao, A. T. (2002). *Marketing research: An aide to decision making* (2nd ed.). Cincinnati: South-Western.

Stacks, D. W., & Hocking, J. E. (1998). *Communication research*. New York: Longman.

Triola, M. F., & Franklin, L. A. (1994). *Business statistics*. Reading, MA: Addison-Wesley Publishing Co.

Trochim, B., & Donnelly, J. (2008). *Research methods knowledge base* (3rd ed.). Mason, OH: Cengage Learning.

Turco, D. M. (1997). Measuring the economic and fiscal impacts of state high school sport championships. *Sport Marketing Quarterly, 6*(3), 49–53.

Waltz, C. F., Strickland, O. L., & Lenz, E. R. (1984). *Measurement in nursing research*. Philadelphia, PA: Davis Company.

14

Attitude Measurement

After completing the study of this chapter, you should be able to:

- Understand the nature of attitude
- Identify the three components of attitude and understand what each component entails
- Identify the two main measurement techniques in attitude research
- Identify the techniques used to measure the affective component of attitude
- Apply any noncomparative scale to design a questionnaire used to measure the attitudes of sport consumers
- Apply any comparative scale to design a questionnaire used to measure the attitudes of sport consumers
- Apply either the categorical or the behavioral differential scale to evaluate sport costumers' behavioral intentions

Most of the research in sport management is conducted for the sake of obtaining information about the *attitudes* of those people who are involved in sport, such as sport consumers, sport participants, and employees or employees of a sport organization. With the information, sport marketers are able to increase their sales with the adoption of effective strategies of sales and promotion. The information allows sport event organizers to make changes so as to enhance the experience of the sport participants, and it enables sport managers to identify and address the areas of concern expressed by the employees so as to improve their job satisfaction. To effectively measure attitude, it is imperative that students understand what it is, what its components are, and how it can be measured. This is the main purpose of this chapter.

Nature of Attitude

There is a universally accepted definition of the term attitude. Zikmund (2000) referred to attitude as "an enduring disposition to respond consistently in a given manner to various aspects of the world, including persons, events, and objects" (p. 288). Bradburn, Seymour and Wansink (2004, p. 121) defined it as "a bundle of opinions that are more or less coherent and are about some complex object." Attitude is also defined as "a complex mental state involving beliefs and feelings and values and dispositions to act in certain ways" (Attitude, 2017). A close review of these three definitions seems to suggest that attitudes are predispositions to act in a certain way (Smith & Albaum, 2010) and about consistent and lasting evaluative feelings toward an object. Examples of attitudinal statements in sport include

- "The 2006 World Cup is boring due to the domination of European teams";
- "Professional baseball players are overpaid"; and
- "Professional athletes should be excluded from the Olympic Games to preserve amateurism."

In sport management, consumer attitude research typically looks at attitude-related issues from several angles. A common one is to examine consumers' attitudes toward a product (e.g., Taylor Made's new R10 to determine if the product is well received by golf enthusiasts because of its new design and new shaft technology). Some research also focuses on consumers' attitudes toward a company. For example, the heated debate a few years ago regarding practices by Nike in Southeast Asia may affect consumers' perception of Nike, consequently affecting its bottom line. Sometimes, the researcher may be interested in consumer attitudes toward a retailer, such as the Foot Locker, Sports Authority, and Dick's Sporting Goods. Another angle of consumer attitude research is on the attitudes toward the attributes of a product, particularly, toward various types of brand associations of a product, such as logos (their design), symbols (their meanings, such as the five rings used for the Olympic Games), and product endorsers.

Components of Attitudes

It is generally believed by scholars (McDaniel & Gates, 1991; Zidmund, 2000) that attitudes are composed of an *affective* (feelings) *component*, a *cognitive* (beliefs) *component*, and a *behavioral* (actual actions) *component*. In sport management research, each of the components could be used for the researcher to gain an understanding of how persistent sport consumers evaluate, feel about, and react toward a specific object (e.g., products, organizations, advertisements, sponsorships) or issue. A detailed description and analysis of each of the attitudinal components with an example is provided in the following section.

Cognitive Component of Attitudes

The cognitive component represents a person's understanding of or knowledge and beliefs about an object. For instance, in the 2006 World Cup, Adidas aired several commercials, one of which is Adidas +10 (including choose your team and dream team) in which kids in a sandlot pick a team of soccer stars for a game. Tiffany Cox, a sport management researcher, conducts a study to determine viewers' attitudes toward the commercial. Specifically, she is interested in knowing what the beliefs of viewers are about Adidas. The results of her studies may reveal that viewers seem to hold such collective beliefs:

- Adidas is a brand that is widely endorsed and used by world class athletes.
- Adidas is a brand that is full of imagination reflected in the slogan of the commercial, impossible is nothing.
- Adidas brings fun to life.
- Adidas is innovative and dares to dream.

Even though the above belief statements could be true or false, they may represent the viewers' general evaluation of Adidas as a brand. Their positive evaluations may result in favorable impressions of Adidas.

A belief is a psychological association between an object and its attributes or features (and associated benefits). Beliefs are cognitive (based on knowledge, experience, perception, etc.). The stronger the association of features or attributes (and associated benefits) is with the object, the stronger the individual's belief.

Affective Component of Attitudes

"What are the feelings or emotional reactions of viewers toward Adidas?" is one of the research questions Tiffany wants to answer. The answers reflect the affective component of viewers' attitudes. Thus, the affective component of attitudes is a person's feeling or emotional reactions to an object. As attitudes are related to people's evaluations of an object, the affective component then elicits people's feelings. Tiffany's research may find that viewers generally feel Adidas is a lovable brand and that the commercial makes them feel good.

It must be pointed out that the cognitive component and the affective component are not always consonant with each other. For example, many consumers may be aware that some of the sporting goods or athletic apparel companies have been accused of wrongfully exploiting workers in their foreign factories, but still demonstrate a strong affectional attachment to the products made by those companies. That is, they may hold a negative cognitive evaluation of those companies, but a positive affective evaluation of the products of those companies.

The affective component of attitude reflects the way in which we feel (e.g., like, dislike, are neutral) in response to external stimuli. It is emotive rather than cognitive (i.e., beliefs). It is comprised of both a person's knowledge of stimuli and that individual's evaluations of the stimuli.

Behavioral Component of Attitudes

The intention of an individual to purchase a particular product reflects the behavioral component of the individual's attitude toward the product. The answer to such a question used in Tiffany's study, the likelihood for viewers to purchase Adidas branded products, may reveal the viewers' predisposition to act. People may try to obtain an object about which they have a favorable impression; they may try to avoid or get rid of an object that they dislike or about which they have negative feelings.

People's intentions to act (e.g., purchase a particular product) positively, negatively, or neutrally toward an object are based on their affective evaluations of the object. As such, people's purchase decisions are often influenced by their affective response to the stimuli of an object. To obtain a comprehensive understanding of an individual's attitude toward an object or an issue, the researcher should approach it from the above-discussed three components.

Measurement Techniques in Attitude Research

In sport management research, *rating* and *ranking* are the two most common measurement techniques used by sport management researchers to measure attitudes. Even though both the sorting and choice techniques are among the options adopted by business professionals in measuring attitudes (Zikmund, 2000), they have been rarely appeared in sport management-related literature.

The rating technique is used when respondents are asked to rate an idea, concept, individual, program, or product by providing their estimates on a measurement scale that typically has a number of choice points or response categories. The Likert scale, the most frequently utilized measurement scale, is a typical rating scale, as respondents are presented with a series of statements and asked to indicate their degree of agreement to the statements. The rating technique is used to elicit responses from respondents about their options or feelings.

Ranking is a technique in which respondents are asked to rank a set of ideas, issues, or services, based on their relative importance or preference as perceived by the respondents. For example, in the study conducted by Li and Burden (2002) to explore the status of outsourcing of marketing operations in Division I intercollegiate athletics, they provided the respondents with a list of keys to maintaining an effective relationship between the outsourcing agency and an athletic department, and then asked them to rank the keys in order of importance. The following is an actual question used in the study, as well as the keys to maintaining an effective outsourcing relationship:

In your opinion, what is the major key to ensuring that the outsourced program will be successful? (Please rank each of the following keys in terms of their importance, with one being the most important)

- ❑ Support from the administration of the institution
- ❑ A cooperative partnership between the institution and the contracted marketing company
- ❑ A well-thought-out strategic plan in outsourcing
- ❑ A long-term partnership between the institution and the contracted marketing company

In this research, it was more appropriate to use ranking than rating because of the nature of the issue at hand. The use of the ranking technique forced the respondents to prioritize the different choice options that were provided, in an attitude scale.

There are many types of scales using the ranking technique. The typical ranking scale is one that asks the respondent to rank given options in terms of either the importance or preference, such as the example given above about asking athletic administrators in NCAA Division I athletic departments to rank keys to maintaining an effective relationship between the outsourcing agency and their departments. Each rank can be used only once. The researcher sometimes also asks the respondent to compare a pair of ideas, issues, or objects. A detailed description about attitude ranking scales is provided in the following section.

Exhibit 14.1. Commonly Used Category Scales and Their Response Choices

Quality									
Excellent	Good	Fair	Poor	Very Good	Fairly Good	Neither Good Nor Bad	Not Very Good	Not Good at All	
Importance									
Very Important	Fairly Important	Neutral	Not So Important	Not at All Important					
Interest									
Very Interested	Somewhat Interested	Not Very Interested							
Satisfaction									
Very Satisfied	Somewhat Satisfied	Neither Satisfied Nor Dissatisfied	Somewhat Dissatisfied	Very Dissatisfied	Not At All All Satisfied				
Frequency									
All of the Time	Very Often	Often	Sometimes	Hardly Ever	Rarely	Never	Most of the Time	Some of the Time	Just Now and Then
Truth									
Very True	Somewhat True	Not Very True	Not at All True	Definitely True	More True Than False	More False Than True	Definitely Not True		

Source. Zikmund, W. G. (2000). Business research methods (6th ed.). Fort Worth, TX: Harcourt College Publishers (p. 292); and Wrenn, B., Stevens, R., & Loudon, D. (2002). Marketing research: Text and cases. Binghamtom, NY: Best Business Books (pp. 132–133).

Techniques of Measuring the Affective Component

In general, the measurement techniques used in attitude research include two types of scales: *noncomparative* and *comparative scales*. In responding to questions designed with a noncomparative scale, respondents are asked to provide their evaluations of one object without reference to another object (McDaniel & Gates, 1991). The category scales, the graphic rating scales, the semantic scale and semantic differential scale, the Likert scale, and the staple scale are the noncomparative scales commonly used in sport management research. The comparative scale, on the other hand, is the type of scale that requires respondents to provide their judgments based on the results of comparison among or between objects. Examples of comparative scales include the rank order scale, the paired comparison scale, and the constant sum scale.

Noncomparative Scales

Category scales are the most commonly used scales in sport management research. They are designed in such a way that the respondent is provided with a question usually having a set of 3 to 5 response choices. The response choices are listed in order of intensity. The commonly used category scales and their responses choices are shown in Exhibit 14.1.

The following are two examples of category scales. The first one is used to determine how frequently the respondents watch sports on TV. A scale used shows the five response choices of frequency from very often to never.

How often do you watch sports on TV?
- ❑ Very Often (every day)
- ❑ Often (2–3 times a week)
- ❑ Sometimes (once a week)
- ❑ Rarely (once a month)
- ❑ Never

In the second example, the researcher desires to find out the level of interest in golf among college students. A set of three response choices are provided from which the respondents may choose.

Please place a check mark to the response choice that best reflects your level of interest in golf.
- ❑ Very Interesed
- ❑ Somewhat Interesed
- ❑ Not Very Interested

The Importance-Performance Analysis (IPA) technique, a well-documented marketing technique used mostly in the analysis of (1) consumer expectations (importance) of attributes of a product and (2) consumer satisfaction (performance) with those attributes, combines two category scales in retrieving useful information. It has also been applied in other disciplines, such as recreation and sport management, since its introduction in 1977. Exhibit 14.2 shows an example using the IPA technique in designing a survey instrument to be used in identifying the areas of a bicycle path to which the management of the facility needs to pay attention.

To design a questionnaire using the IPA technique, the researcher first needs to review related literature and conduct focus groups to identify the basic design attributes of a facility (e.g., the bicycle path), such as accessibility, layout, length, and lighting. The identified design attributes or characteristics are then incorporated into a questionnaire that includes two category scales, the Importance Scale and the Performance Scale. The people who complete the questionnaire are asked to rate each attribute in terms of its importance (e.g., how important is this attribute in a bicycle path relative to participation in physical activity?) and performance (how satisfied are they with this attribute in the bicycle path they are currently using for participation in physical activity?). The mean values of each attribute in the two scales of the questionnaire are cross-tabulated and plotted on a four-quadrant grid according to its rated importance and performance as shown in Exhibit 14.2.

The horizontal axis of the grid presents bicycle path users' perceptions of the quality of performance, and the vertical axis indicates the importance of the attributes of the bicycle path. As indicated in Exhibit 14.3 and Exhibit 14.4, the labels for the four quadrants are Keep up the Good Work, Concentrate Here, Possible Overkill, and Low Priority. The attributes in the Keep up the Good Work areas are those that users consider to be important and are satisfied with in the path they use. The Concentrate Here quadrant represents attributes that receive low satisfaction scores from users but are perceived by them to be important. The Possible Overkill items are those with

Exhibit 14.2. Example of the IPA Technique in a Survey

Please rate the importance of each of the following physical environmental characteristics of a bicycle trail in Column I. Record your rating by circling one of the numbers listed in Column II, with "1" being "very unimportant" and "5" being "very important."

Column I	Column II
1. Accessibility (to residential areas, schools, shopping, work).	1 2 3 4 5
2. Scenery enjoyability.	1 2 3 4 5
3. Opportunities to observe wildlife.	1 2 3 4 5
4. Level of crime.	1 2 3 4 5
5. Connection between residential areas and schools.	1 2 3 4 5
6. Connection between residential areas and shopping malls.	1 2 3 4 5
7. Connection between residential areas and work.	1 2 3 4 5
8. Connection between parks and other recreation facilities.	1 2 3 4 5
9. Connection between communities.	1 2 3 4 5
10. Connection to transit stops.	1 2 3 4 5
11. The trail width accommodating various users (e.g., bikers, roller skaters).	1 2 3 4 5
12. Lights.	1 2 3 4 5
13. Hills in the trail.	1 2 3 4 5
14. The trail surface's suitability for various users (bikers, runners, etc.)	1 2 3 4 5
15. Signs for traffic control, warning, and information.	1 2 3 4 5
16. Devices that slow or restrict traffic or speed.	1 2 3 4 5
17. Maintenance of the trail (e.g., its surface is even and no holes, cracks).	1 2 3 4 5
18. Availability of parking along the trail for trail users.	1 2 3 4 5
19. Roofed shelters/protected seating areas.	1 2 3 4 5
20. Kiosks displaying trail information.	1 2 3 4 5
21. Pavement markings warning trail users before intersections.	1 2 3 4 5
22. Availability of restroom facilities.	1 2 3 4 5
23. Fitness/exercise stations.	1 2 3 4 5
24. Drinking fountains.	1 2 3 4 5
25. Accessibility for people with disabilities.	1 2 3 4 5

Please indicate the level of your satisfaction toward each physical environmental attribute of the bicycle trail that you are currently using in Column I. Record your rating by circling one of the numbers listed in Column II, with "1" being "very unsatisfied" and "5" being "very satisfied."

Column I	Column II
1. Accessibility (to residential areas, schools, shopping, work).	1 2 3 4 5
2. Scenery enjoyability.	1 2 3 4 5
3. Opportunities to observe wildlife.	1 2 3 4 5
4. Level of crime.	1 2 3 4 5
5. Connection between residential areas and schools.	1 2 3 4 5
6. Connection between residential areas and shopping malls.	1 2 3 4 5
7. Connection between residential areas and work.	1 2 3 4 5
8. Connection between parks and other recreation facilities.	1 2 3 4 5
9. Connection between communities.	1 2 3 4 5
10. Connection to transit stops.	1 2 3 4 5
11. The trail width accommodating various users (e.g., bikers, roller skaters).	1 2 3 4 5
12. Lights.	1 2 3 4 5
13. Hills in the trail.	1 2 3 4 5
14. The trail surface's suitability for various users (bikers, runners).	1 2 3 4 5
15. Signs for traffic control, warning, and information.	1 2 3 4 5
16. Devices that slow or restrict traffic or speed.	1 2 3 4 5
17. Maintenance of the trail (e.g., its surface is even and no holes, cracks).	1 2 3 4 5
18. Availability of parking along the trail for trail users.	1 2 3 4 5
19. Roofed shelters/protected seating areas.	1 2 3 4 5
20. Kiosks displaying trail information.	1 2 3 4 5
21. Pavement markings warning trail users before intersection.	1 2 3 4 5
22. Availability of restroom facilities.	1 2 3 4 5
23. Fitness/exercise stations.	1 2 3 4 5
24. Drinking fountains.	1 2 3 4 5
25. Accessibility for people with disabilities.	1 2 3 4 5

Exhibit 14.3. Cross-Tabulating and Plotting on the IPA Grid

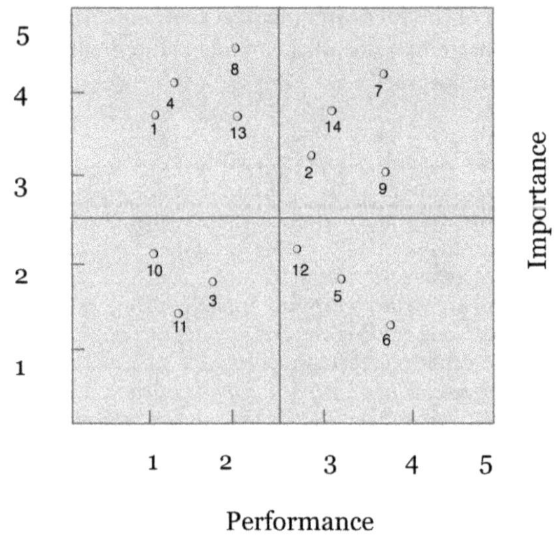

Exhibit 14.4. Interpretation of the Plotting and Labels of the IPA Grid

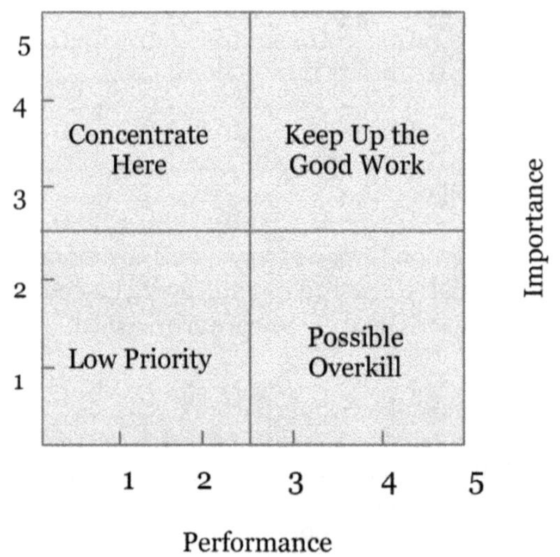

which users are happy but that they do not consider essential. The Low Priority quadrant includes path characteristics that receive low satisfaction evaluation and importance ratings.

In sport management, just like in many other social science-based disciplines, such as business, category rating scales are most commonly used to measure the affective domain of attitudes. Researchers often label them as Likert-type scales and treat them as interval scales so that they can use parametric statistics to analyze the data collected. Such practice is commonly acceptable; some scholars, however, insist that category scales measure only ordinal variables, not interval variables. Consequently, they believe that it is inappropriate to use interval-level statistics with ordinal-level data. The debate on whether or not category scales are interval scales has gone on for a long time. The purpose of mentioning such controversy here is to remind users of this type of scale that they should use them with care. In the following section, in which the Likert scale is described, a set of protocols is recommended for researchers to observe in order to collect interval data with the Likert scale.

Exhibit 14.5. Graphic Rating Scale as a Tool of Visual Communication

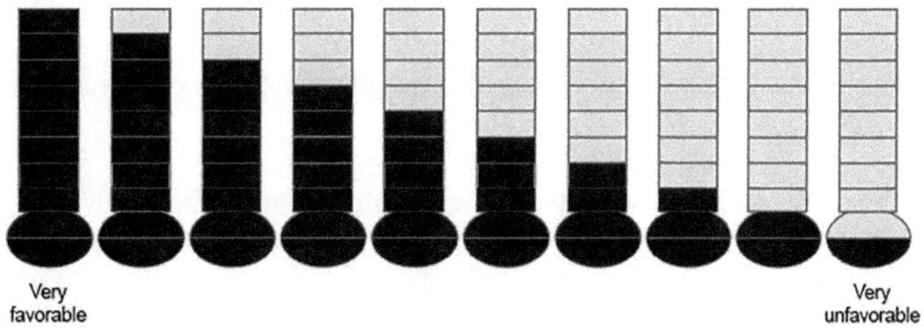

Graphic Rating Scales

The graphic rating scales are the type of scales in which respondents are asked to indicate their feelings by placing a check mark at any point on a graphic continuum that has two extreme adjectives anchored at each end. The most common form of the graphic rating scale is called the continuous rating scale. Once a check mark is made, the researcher measures the length of the checked point to one end of the continuum to determine the score of the respondent. Let us use the following example to illustrate how it works.

In a study conducted to investigate viewers' reactions to one of the commercials shown in the recent Super Bowl, the surveyed viewers were asked to indicate how favorable they were in terms of the commercial with the following graphic scale:

Assuming that the line is seven inches long, the measurement of the check mark placed by one of the respondents is about 2.1 inches away from favorable. So, this respondent had a favorable reaction (scored 4.9 = 7 – 2.1) toward the commercial provided the starting point at the favorable end is coded as 7.

There are graphic rating scales other than the continuous rating scale. An example of such an alternative is shown below. The respondent's score is determined either by dividing the line into as many categories as the researcher desires and assigning the respondent a score based on the category into which the respondent's mark falls, or by measuring the distance from either end of the scale as is done with the continuous rating scale.

Favorable									Unfavorable
☐	☐	☐	☐	☐	☐	☐	☐	☐	☐
100	90	80	70	60	50	40	30	20	10

The graphic rating scales do not involve just a continuous line or continuum. One of the alternatives is the use of other forms of visual communication, as shown in Exhibit 14.5. The respondents are provided with a series of measuring tubs that are filled with liquid to a variety of levels. The respondents are asked to indicate the tubs that reflect their feelings about an object or issue. Whichever of these forms of the graphic rating scale is used, the results are normally analyzed as an interval scale.

Semantic Differential Scale

Using the semantic differential scales, the researcher uses a set of bipolar scales with semantic labels or adjectives placed at the end point of each of the scales to measure the respondents' reactions toward or feelings about an object or issue. The semantic differential scales use words rather than numbers. The following are examples of an attitude measurement scale with the use of the semantic differential scale.

The XYZ professional baseball team management conducted a survey to measure the fans' reactions to the new team logo:

	Extremely	Quite	Slightly	Neither	Slightly	Quite	Extremely	
Innovative	☐	☐	☐	☐	☐	☐	☐	Conventional
Modern	☐	☐	☐	☐	☐	☐	☐	Old Fashioned
Motivating	☐	☐	☐	☐	☐	☐	☐	Discouraging
Interesting	☐	☐	☐	☐	☐	☐	☐	Boring
Pleasant	☐	☐	☐	☐	☐	☐	☐	Unpleasant

The semantic differential scales are used more commonly in business and sport management research. Each semantic differential scale usually has seven response choices or positions. A weight is assigned to each position: 3, 2, 1, 0, –1, –2, and –3. The middle one with 0 weight symbolizes the neutral position. The 1 positions (both positive and negative) are labeled slightly, the 2 positions quite, and the 3 positions extremely. A semantic differential scale measures directionality of a reaction (e.g., good versus bad), intensity (slight through extreme), and activity (active—passive). Ratings are combined in various ways to describe and analyze the respondent's feelings. So the researcher should attempt to identify adjectives along these three major factors or dimensions of judgment.

In a study designed to test a causal model revealing the relationship between consumer satisfaction and a number of such variables as pleasure, arousal, loyalty, and emotions, Caro and Garcia (2007) employed a semantic differential scale to measure emotions. The following shows the actual scale (p. 81):

Please circle the number that best reflects your emotions during the race based on a series of adjectives.

Pleased	1 2 3 4 5	Angry
Happy	1 2 3 4 5	Unhappy
Delighted	1 2 3 4 5	Undelighted
Glad	1 2 3 4 5	Sad
Hopeful	1 2 3 4 5	Disillusioned
Amused	1 2 3 4 5	Bored
Lively	1 2 3 4 5	Down
Excited	1 2 3 4 5	Calm
Active	1 2 3 4 5	Passive
Surprised	1 2 3 4 5	Indifferent

Likert Scale

The Likert scale is a special type of the category scale. It is known as a summated scale, which means that the attitudes making up a Likert scale are summed up to produce an index score. In that sense, a Likert scale is a composite of itemized attitude statements or scales. Typically, each statement has five categories, with scale values ranging from -2 to +2 with 0 as the neutral response, or from 5 to 1 with 3 as the neutral response. In other words, subjects are asked to express agreement or disagreement on a five-point scale. Each degree of agreement is given a numerical value. Thus a total numerical value can be calculated from all the responses. The following is an example of a questionnaire that is designed with the Likert scale technique.

For each of the statements below, please indicate the extent of your agreement or disagreement by placing a check mark in the appropriate box.	Strongly Agree	Agree	Neutral	Disagree	Strongly Disagree
1. A decrease in the number of television exposures will negatively affect ticket sales.					
2. A decrease in the number of television exposures will negatively affect boosters' contributions.					

Li and Burden (2002) conducted a study to investigate the perceptions of college athletic administrators toward outsourcing. The Likert scaling technique was used in constructing the section of the questionnaire that was drafted for assessing the perceived benefits on outsourcing of sport marketing operations in Division I-A athletic programs. The surveyed athletic administrators were asked to indicate their agreement or disagreement to eighteen statements on benefits of outsourcing along four sub-scales. Exhibit 14.6 shows that particular section of the questionnaire.

As mentioned previously, it is still controversial as to whether or not the category scales, including the Likert scale, can be treated as interval scales. Therefore, certain precautions need to be taken while using parametric statistics in analyzing the data obtained from the use of the Likert scale: follow the required procedure to construct the Likert scale, acquire a large sample size, and ensure the sample is normally distributed. Specifically, to use the Likert scale in collecting interval data, the sport management researchers should follow a set of protocols:

- A scale should include at least five, preferably more, categories. The more the number of points, the more likely the data collected meet the assumption of normal distribution. As Friedman and Amoo (1999) indicated, an 11-point scale may produce more valid results than a 3-, 5-, or 7-point scale. If the sample is large enough, the data collected could also meet the assumption of normal distribution.
- The rating scale should be balanced. That is, it must offer the same number of favorable and unfavorable response choices (Friedman & Amoo, 1999).
- To follow the recommended procedure for constructing a Likert scale, the index score should approximate the interval level of measurement (Sirkin, 1999). When scores from several items are summed to create an index, the process overwhelms minor differences and yields interval data (Sirkin, 1999).

According to Likert (1974, pp. 233–235), there are a few basic rules for constructing a Likert scale. Each statement included in a Likert attitude scale should be of such a nature that persons with

Exhibit 14.6. Section of Division I-A Athletic Programs Questionnaire

The following are some statements describing the possible benefits of the outsourcing of sport marketing operations in Division I-A athletic programs. Please indicate your agreement or disagreement. For each statement, please check the appropriate box to indicate whether you:

1 = Strongly Agree (**SA**)
2 = Agree (**A**)
3 = Neutral (**N**)
4 = Disagree (**D**)
5 = Strongly Disagree (**SD**)

Circle only one answer for each statement. There are no right or wrong answers to these questions.

1. Outsourcing can broaden the media coverage of your athletic programs so as to enhance the public awareness of them. SA A N D SD
2. Outsourcing can help you build sustainable competitive advantage. SA A N D SD
3. It is less expensive for you to outsource your marketing operations than keep them in house. SA A N D SD
4. It is more efficient for you to outsource your marketing operations than keep them in house. SA A N D SD
5. Outsourcing of your marketing operations can bring in greater financial return than keep them in house. SA A N D SD
6. Outsourcing can help reduce bureaucracy and red tape in your marketing operations. SA A N D SD
7. Outsourcing can help you enhance the quality of your marketing operations. SA A N D SD
8. Outsourcing enables you reallocate resources to other more critical areas within your department. SA A N D SD
9. Outsourcing allows you to access external specialized talent and expertise in marketing and promotion. SA A N D SD
10. Outsourcing enables you to cut down overhead expenses. SA A N D SD
11. Outsourcing can help you avoid the administrative complexity of in-house employment. SA A N D SD
12. Outsourcing enables you to use the contracted company as an extension of your overall marketing efforts. SA A N D SD
13. Outsourcing allows you to handle all aspects of your marketing operations in a more timely fashion. SA A N D SD
14. Outsourcing may cause loss of control of your marketing operations. SA A N D SD
15. Outsourcing may result in loss of opportunities to bring in more revenues for you athletic program. SA A N D SD
16. Outsourcing can provide you with opportunities to expand marketing efforts to nonrevenue sports. SA A N D SD
17. Outsourcing can help you reduce capital expenses related to marketing operations. SA A N D SD
18. Outsourcing allows your department to be part of a marketing network and synergy. SA A N D SD

different points of view, so far as to the particular attitude is concerned, will respond to it differently. It is essential that all statements be expressions of desired behavior and not statements of fact. Each statement needs to be clear, concise, and straight-forward. It would seem desirable to have each statement be worded such that the modal reaction to it is approximately in the middle of the possible responses. It seems desirable to have the different statements so worded that about one-half of them have one end of the attitude continuum corresponding to the left or upper part of the reaction alternatives and the other half have the other end of the attitude continuum corresponding to the right or lower part of the reaction alternatives. It is necessary to assign a value to the undecided positions on each statement.

Staple Scale

The staple scale is designed in such a way that a criterion that the researcher wants to assess is placed in the middle of the scale with typically five response choices or alternatives anchored on each side. The staple scale is used as an alternative to the semantic differential scale when it is difficult to find bipolar adjectives to meet the needs of the researcher. The following is a section of a questionnaire designed to assess spectators who are attending a professional football game in terms of the quality of food served by a concessionaire.

The following questions concern your ratings of the XYZ concessionaire, who sells food in the facility. You are asked to circle one number that best reflects your opinion in each of the quality measures.

-5	−4	−3	−2	−1	Taste	1	2	3	4	5
-5	−4	−3	−2	−1	Service Quality	1	2	3	4	5
-5	−4	−3	−2	−1	Cleanliness	1	2	3	4	5
-5	−4	−3	−2	−1	Price	1	2	3	4	5
-5	−4	−3	−2	−1	Variety	1	2	3	4	5

The following is another example using the staple scale in leadership assessment.

Please evaluate the leadership style demonstrated by the athletic director with the following scales:

+3	+3
+2	+2
+1	+1
Acceptability	Effectiveness
-1	-1
-2	-2
-3	-3

Comparative Scales

Rank-Order Scales

Rank-order scales require the respondent to make a judgment on one item against another. The biggest strengths of this type of scales are that it is easy to use and form an ordinal scale of items, and the respondent is forced to evaluate the object, idea, and product in a realistic manner. Results could be misleading if the choices provided are inclusive. Assuming that a researcher wants to know the brand preference of racquets by tennis enthusiasts, a rank-order scale can be applied to the survey instrument, which is shown as follows:

Rank the following brands of tennis racquets from 1 (best) to 5 (worst) according to your taste preference:

Dunlop ________
Head ________
Prince ________
Prokennex ________
Wilson ________
Yonex ________

Paired Comparison Scale

In a paired comparison scale, the researcher provides the respondent with a set of paired objects, and asks the respondent to pick one of the objects in each pair. The data collected with the use of the paired comparison scale are ordinal in nature. Compared to the ranking order scale, the biggest advantage of the paired comparison scale is the ease with which respondents can pick one of the two choices and then rank all the items provided in a rank order scale. The disadvantage of this type of scale, on the other hand, is that it is very cumbersome to complete the list if there are a large number of paired items in the list. If the researcher who wants to obtain information about the brand preference of racquets by tennis enthusiasts uses the paired comparison scale in the survey, the question and the response choices will look like the following:

For each of the following pairs, which brand of tennis racquet do you think is better (please check one brand for each pair).

____ Dunlop	____ Head
____ Head	____ Prince
____ Prince	____ Prokennex
____ Prokennex	____ Wilson
____ Wilson	____ Yonex
____ Yonex	____ Dunlop
____ Dunlop	____ Prince
____ Prince	____ Wilson
____ Wilson	____ Dunlop
____ Dunlop	____ Prokennex
____ Prokennex	____ Yonex
____ Yonex	____ Head
____ Head	____ Prokennex
____ Prince	____ Yonex

Constant Sum Scale

The constant sum scales refer to the scaling method that requires the respondent to divide a given number of points, typically 100, among two or more attribute options based on their importance to the respondents. For example, in the study above, designed to determine the brand preference of racquets by tennis enthusiasts, instead of using the rank-order scale to design the survey instrument, the researcher employs the constant sum scaling technique. The questionnaire looks like this:

Allocate a total of 100 points among the following brands of tennis racquets depending on how favorable you feel toward each brand; the more highly you think of each brand, the more points you should allocate to it. (Please check that the allocated points add to 100.)

Dunlop	________
Head	________
Prince	________
Prokennex	________
Wilson	________
Yonex	________
	100 points

In another example, the purpose of a study is to evaluate how women runners view the nine characteristics of a newly designed women's jogging outfit. Each respondent is given 90 points and asked to allocate points to each of the characteristics based on its importance to her. The more important a characteristic, the more points she should allocate to it.

Characteristics	**Number of Points**
Comfortable to wear	________
Durable	________
Endorsed by famous athletes	________
Made in the United States	________
Made by a well-known manufacturer	________
Latest styling	________
Freedom of movement	________
Not too expensive	________
Authentic, like the pros wear	________
	90 points

Techniques of Measuring the Behavioral Component

The behavioral component of attitude can be assessed with either the category scales or the behavioral differential scales. The category scales are used to inquire a respondent's likelihood of action or intention to perform some action. Statements used in these scales often include action phrases, such as "I would recommend," "I would buy," and "I would do," to indicate action tendencies. The following example illustrates how to use a category scale in assessing respondents' intention to renew their season tickets.

Please indicate how likely it is for you to renew your season ticket.

- ❑ I definitely will purchase.
- ❑ I probably will purchase.
- ❑ I might purchase.

Sometimes, the researcher can use a question, such as "how likely are you," instead of a statement to reflect the intention to act. As the likelihood to act is explicitly stated in the question itself, those action phrases as mentioned above do not have to be included in the scale, as exemplified in the following example:

How likely are you to renew your season ticket?

- ❑ Extremely likely
- ❑ Very likely
- ❑ Somewhat likely
- ❑ Likely, about 50-50 chance
- ❑ Somewhat unlikely
- ❑ Very unlikely
- ❑ Extremely unlikely

The sport management researcher can also use the behavioral differential scale to measure the behavioral intentions of subjects toward any object or category of objects. The behavioral differential scale, similar to the semantic differential scale that has a pair of bipolar objectives placed in the two ends of the scale, has "would" and "would not" anchored in both sides of the scale. The researcher who wants to assess people's intention to renew their season tickets could use the behavioral differential scale to achieve the same purpose, as shown in the following example:

You would : : : : : : : would not renew your season ticket.

Selection of an Attitude Measurement Scale

The most appropriate scale for a research project is the one that can yield the most useful information. Therefore, to select a right attitude measurement scale, the sport management researcher should closely review the following factors.

What are the preferred measurement techniques, the rating or ranking scale or the scale used to measure behavioral intention?

Each measurement technique has one intended purpose. The rating scale is widely used in attitude-related research because of its applicability for use of parametric statistics. However, it usually does not provide the researcher with information about relativity of importance or preference among items. On the other hand, the ranking technique forces respondents to give their preferences about items evaluated. But the data collected can be analyzed only with the statistical procedures that are appropriate to ordinal data. If the main goal of a research project is to evaluate people's intention to act, the researcher should choose either category scales or the behavioral differential scale that are appropriate to measure behavioral intention.

Are the collected data nominal, ordinal, or interval?

The type of data collected dictates the type of statistical procedures (parametric or nonparametric) to be used in data analysis. The researcher should use the objectives of the study to select the type of statistical procedures to apply and, accordingly, the type of scale to use. For example, if the purpose of the study is to simply find out the education levels and occupations of the spectators attending a golf tournament, the researcher can just use the category scale technique to derive some nominal information (e.g., frequency and percentage) about them.

Does the issue to be investigated require a unidimensional or multidimensional scale?
A unidimensional scale refers to the scale used to only measure one attribute of an object. For example, the following scale designed to measure the service quality of athletic clubs is a unidimensional scale, as the question and all provided response choices measure the same thing, that is, service quality.

Compared with the service quality provided in Club A, the service quality in Club B (seven-point semantic differential scale):

Favorable ❑ ❑ ❑ ❑ ❑ ❑ ❑ Unfavorable

Good ❑ ❑ ❑ ❑ ❑ ❑ ❑ Bad

Pleasant ❑ ❑ ❑ ❑ ❑ ❑ ❑ Unpleasant

On the other hand, there are phenomena or objects that have multiple attributes. To measure them, a multidimensional scaling approach is required. To assess how favorable youth skateboard enthusiasts are to a number of newly designed skateboard decks produced by several manufacturers, the researcher must choose a scale that measures a number of attributes of the skateboard deck, including size, shape, material, heaviness, length, structural strength (easiness to break), and exterior design. In other words, a multidimensional scale may contain a number of sub-scales with each measuring one attribute.

Should a balanced or an unbalanced scale be used?
A balanced rating scale has an equal number of positive and negative categories; a neutral or indifference point is at the center of the scale. The following is an example of such a scale:

If you could not find a job in your chosen area in sport management, would you attempt to find a job in any other area in sport management?

Very Likely	Likely	Unsure	Unlikely	Very Unlikely
❑	❑	❑	❑	❑

On the other hand, an unbalanced rating scale refers to the scale that has more response categories piled up at one end of the scale. The positive categories do not equal negative categories in number. An example of the use of an unbalanced rating scale is shown below:

How would you rate the quality of service provided by the tournament organizer?

❑ Excellent
❑ Very Good
❑ Good
❑ Fair
❑ Poor

How many response choices or categories should be used?

As discussed in Chapter 13, one of the criteria to assess whether or not a survey instrument is a good measurement is its sensitivity, which is basically the ability of a survey instrument to catch the variability of responses (Zikmund, 2000). The more response choices provided in a scale, the more sensitive the scale will be. To answer the question of how many response choices or categories should be used, the researcher needs to decide how sensitive the scale should be. In general, five to nine response choices are appropriate. As mentioned in the section in which the Likert scale is discussed, the more response choices, the more likely the data collected will be normally distributed.

A related issue also should be addressed by researchers when deciding the number of response choices. That is, should the response choices be an even or odd number? Many scales, such as some category scales and the Likert scale, offer the respondent a neutral response choice in the middle of the scale. Typically, these scales have an odd number of response choices. The neutral option provides those respondents who may not have an opinion toward the object or issue one way or the other with the opportunity to express their opinion (i.e., neutral). Some scholars believe that having a neutral response choice could be problematic as it gives the respondent the option of hiding their opinions or not thinking hard on the question. It is commonly the case that responses tend to regress to the middle point of the scale if there is a neutral answer choice.

Is a forced choice scale a better option than a nonforced choice scale?

A forced-choice scale is a scale that does not provide such response choices as "do not know" or "can't tell" for the respondent to choose from the fixed alternatives. The semantic differential scaling technique is a typical example of a forced-choice scale, as it does not have an option for the respondent to indicate no opinion. The following are two examples, one of the forced choice scale and the other one of the nonforced choice scale.

How reliable is the treadmill produced by the XYZ sporting company?

Forced Choice Scale

Extremely Reliable	Very Reliable	Somewhat Reliable	Somewhat Unreliable	Very Unreliable	Extremely Unreliable	Can't Tell
❑	❑	❑	❑	❑	❑	❑

Unforced Choice Scale

Extremely Reliable	Very Reliable	Somewhat Reliable	Somewhat Unreliable	Very Unreliable	Extremely Unreliable
❑	❑	❑	❑	❑	❑

There is not a consensus in terms of whether or not the researcher should use or not use the nonforced choice scale in attitude research (McDaniel & Gates, 1991), but a large majority of research conducted in sport management uses the forced choice format.

Should it be a labeled or end-anchored scale?

The following is a labeled scale, in which each response choice is labeled.

Excellent	Very Good	Fair	Poor	Very Poor
❑	❑	❑	❑	❑

The semantic differential scale is an end-anchored scale, as it has only a pair of polarized adjectives placed in two ends of the scale. There are no labels given to each response choice in the middle of the scale between two anchored adjectives.

Summary

This chapter discussed the basic issues of which sport management researchers should be aware while choosing a scale or scales to measure attitudes. In general, among the four types of measurement techniques that can be used to assess attitudes, sport management researchers prefer rating and ranking more than sorting and choice techniques.

All the measurement techniques used in attitude research can be categorized as either the noncomparative scales or comparative scales. The noncomparative scales that are commonly employed by sport management researchers include category scales, graphic rating scales, semantic scale and semantic differential scale, the Likert scale, and the staple scale. The rank-order scale, the paired comparison scale, and the constant sum scale are examples of comparative scales.

Sport management researchers also use either category scales or behavioral differential scales to measure the behavioral component of attitude. While assessing the behavioral component of attitude, the researcher typically asks the respondent to indicate the likelihood or intention of doing something. "I would recommend," "I would buy," and "I would do," are some of the common phrases provided in the answer choices.

To choose an appropriate scaling technique to yield the most meaningful information, the sport management researcher must carefully consider a number of factors. Such factors include the following:

- What are the preferred measurement techniques, the rating or ranking scale, or the scale used to measure behavioral intention?
- Are the collected data nominal, ordinal, or interval?
- Does the issue to be investigated require a unidimensional or multidimensional scale?
- Should a balanced or an unbalanced scale be used?
- How many response choices or categories should be used?
- Is a forced choice scale a better option than a nonforced choice scale?
- Should it be a labeled or an end-anchored scale?

Study Questions

1. What are the three components of attitudes? Explain each of them with an example.
2. What are the four basic measurement techniques in attitude research? Which technique is the most common one used in sport management research?
3. What are the differences between the noncomparative scales and the comparative scales?
4. What are advantages and disadvantages of the following pairs of scaling techniques?
   a. The category scales and the semantic differential scales
   b. The Likert scale and the staple scale
   c. The rank-order scale and the constant sum scale

5. What are the scaling techniques that can be used to measure the behavioral component of attitudes?
6. What are the factors that need to be considered while choosing a particular type of scaling technique in questionnaire design?

Learning Activities

1. Assume that the Marketing Department of the Red Devil, an arena football team, wishes to understand how fans react to the half-time show the team just unveiled, specifically around four attributes: excitement, creativity, attractiveness, and satisfaction. You are asked to design a questionnaire to be used as a survey to obtain the information. What will the questionnaire look like:
   a. if it is designed with one of the category scales?
   b. if it is designed with the Likert scale?
   c. if it is designed with the semantic differential scale?
2. Assume that you were just hired as assistant ticker manager in a sports facility. One of the things you have been assigned to do is to design a short questionnaire, using either the categorical or the behavioral differential scale, to evaluate customers' behavioral intentions to purchase season tickets and tickets to other events to be held in the facility in the near future. What will the questionnaire look like? Be sure to include at least four questions.

References

Attitude. (2017). *WordWeb Online*. Retrieved from http://www.wordwebonline.com/search.pl?w=attitude

Bradburn, N. M., Seymour, S., & Wansink, B. (2004). *Asking questions: The definitive guide to questionnaire design—for market research, political polls, and social and health questionnaires* (rev. ed.). San Francisco, CA: Jossy-Bass.

Caro, L. M, & Garcia, J. A. M. (2007). Consumer satisfaction with a periodic reoccurring sport event and the moderating effect of motivations. *Sport Marketing Quarterly, 16*(2), 70–81.

Friedman, H. H., & Amoo, T. (1999). Rating the rating scales. *Journal of Marketing Management, 9*, 114–123.

Li, M., & Burden, W. (2002). Outsourcing sport marketing operations by NCAA Division I athletic programs: An exploratory study. *Sport Marketing Quarterly, 11*(4), 227–232.

Likert, R. (1974). The method of constructing an attitude scale. In G. M. Maranell (Ed.), *Scaling: A sourcebook for behavioral scientists* (pp. 233–243). Chicago, IL: Aldine Publishing Company.

McDaniel, C., & Gates, R. (1991). *Contemporary marketing research*. St. Paul, MN: West Publishing Company.

Sirkin, R. (1999). *Statistics for the social sciences* (2nd ed.). Thousand Oaks, CA: Sage Publications.

Smith, S. M., & Albaum, G. S. (2010). *An Introduction to Marketing Research*. Retrieved from http://cloudfront.qualtrics.com/q1/wp-content/uploads/2012/02/IntrotoMarketResearch.pdf

Wrenn, B., Stevens, R., & Loudon, D. (2002). *Marketing research: Text and cases*. Binghamtom, NY: Best Business Books.

Zikmund, W. G. (2000). *Business research methods* (6th ed.). Fort Worth, TX: Harcourt College Publishers.

15

Sampling

After completing the study of this chapter, you should be able to:

- Explain the difference between a sample and a census.
- Define the key terms used in sampling (e.g., sample, generalize, statistic, and parameter).
- Differentiate random and nonrandom sampling techniques.
- Explain the advantages and disadvantages of specific random and nonrandom sampling techniques.
- Compare and contrast the four basic methods of probability sampling.
- Identify the sampling procedure used in a sport management study.
- Discuss the factors that influence the determination of sample size.

Sport management researchers often collect data from a representative *sample* of the *population* in which they are interested. There are numerous sampling methods that are available for the researchers to choose. In this chapter, these sampling techniques will be reviewed, and examples in sport management will also be provided to illustrate how each procedure is used in the selection of a proper sample. In the last section of this chapter, a method that can be used to determine sample size in sport management research will be discussed briefly.

Nature and Scope of Sampling

Sampling is the process of choosing a sample from a population. A population refers to an entire group of units, elements, or individuals to be studied. If you study every individual in a certain population, you are conducting a *census*, not a survey (Johnson & Christensen, 2014). The makeup of the population may consist of the observations of events, places, individuals, or groups. For example, Reese and Mittelstaedt (2001) used the population of ticket operators to investigate ticket pricing strategies in the National Football League. "Subjects consisted of the entire population of individuals responsible for ticket operations in the NFL (n=31). This population was selected due

Exhibit 15.1. Advantages of Conducting a Sample Versus a Census

Advantage	Comment
	Conducting a Sample
Lower cost	Selecting some, but not all, of the entities usually costs less.
Current information	Enhances opportunity to collect more current data.
Less time involvement	Samples can be calculated more rapidly than populations.
Larger scope of information	By evaluating smaller groups, it is sometimes possible to collect more extensive data from each unit or element being investigated.
	Conducting a Census
Eliminates uncertainty	An inferential analysis based on a sample always involves risk. Conducting a census removes this doubt. For example, information from the census conducted every 10 years in the United States is used to apportion federal funds and allocate Congressional seats to states.
Promotes quality	Conducting a census, especially on consumer goods that lend themselves to inspection or testing, projects an image of quality. For instance, a slip of paper stamped "Inspected by No. 3" in an article of clothing implies that someone examined the good and declared it suitable. As another example, more manufacturers of floppy disks certify one or both sides before shipping to guarantee all tracks and sectors are not damaged.

to the expertise of each subject in the areas of establishing ticket prices and identifying variables that affect the decision-making process" (p. 225). Academically, a census refers to an investigation of all the units, elements, or individuals of a population. Exhibit 15.1 provides a comparison of the advantages of using a sample versus a census.

In general, there are two types of populations: a *target population* and an *accessible population.* The target population refers to the population to which the investigators want to generalize the major finding of the investigation. For example, a target population is all of the NCAA Division I head women volleyball coaches in the US from both private and public universities. An accessible population refers to the sector of the target population that is available to the researcher during the time of the investigation.

Unlike a population, a sample refers to a subset of units, elements, or individuals selected to represent a population of interest. There are several advantages of using a sample, such as low cost, less time, increased accuracy (fewer opportunities for errors), feasibility of having the core of the population of interest, and the possibility of gathering more extensive information (Levine, Rampsey, & Berenson, 1995). It is important for the investigator to ensure that the sample is representative—referring to the extent the sample can stand for the population from which the sample was drawn—of the population under study. Other terms related to sampling that may be encountered:

- *Population parameter*—a numerical measurement that describes or summarizes a particular characteristic of a population, census, or universe;
- *Sample statistic*—a numerical measurement that describes or summarizes a particular characteristic of a sample;

- *Sampling error*—"reflects the heterogeneity or 'chance differences' from sample to sample based on the probability of subjects being selected in the particular samples" (Levine, Berenson, & Stephan, 1998, p. 20); and
- *Generalizability*—the extent to which the results of a study can be applied to groups other than the one studied.

The list of all units, elements, or subjects in a population is referred to as a *sampling frame*. For example, if Florida State University (FSU) students were defined as the population, the sampling frame would be the university registrar's student records that show all the students at FSU. An example of a sampling frame was used by Kwon and Trail (2001) when they compared the differences between American and international students' motivation for attending intercollegiate sports events. The sample was collected from a Midwestern university in the US. The Office of the Registrar of that university provided the list of all the students. The sample was then drawn from the list or the sampling frame. The survey was sent to 600 students, divided into two strata made of 300 American students and 300 international students. The sample for each stratum was created using systematic *random sampling*. If the Office of the Registrar had omitted a student's name from the list, the list would not be representative of all students of the university as a sampling frame.

Basic Approaches to Sampling

There are two basic approaches to sampling: random sampling techniques (simple random sampling, systematic sampling, stratified random sampling, and cluster random sampling) and *nonrandom sampling* techniques (convenience sampling, quota sampling, purposive sampling, and snowball sampling).

Random Sampling Techniques

In random sampling, every person, unit, or element of the population has an equal chance to be selected for an investigation. The major advantages of these sampling techniques are that the techniques allow for more accurate representation and generalization of the population being studied. It is free of sampling bias because no population members are favored, and each unit, element, or individual has an equal chance of being selected for a study. It is not subjected to the conscious bias of the researcher. The major disadvantage is that it may be time consuming. Cost also may be a factor. To execute the random sampling procedure, the researcher must specifically define the population, list all units or elements of the population, and then randomly select samples of units from the population. There are four types of random sampling techniques.

Simple Random Sampling

A simple random sample refers to a sample selected by a procedure in which every member of the population has an equal chance of being selected. In order to draw a simple random sample, you can use a table of random numbers or a random number generator. A table of random numbers provides a list of random numbers in a random order so no number appears more frequently than any other number. For a random number generator, a researcher can access websites such as www.

random.org or www.randomizer.org. In the context of sport management, Danylchuk (2000) used simple random sampling to survey spectator perceptions of tobacco sponsorship at an LPGA event.

Systematic Sampling

A systematic sample is a sample drawn by determining the sampling interval—the population size divided by the desired sample size, which is symbolized by *k* (Johnson & Christensen, 2014)—selecting a random starting point between 1 and *k*, and then selecting every *k*th element. Let us assume that there are 50 sport management researchers in the United States, and we have a list of these 50 researchers, numbering them one through 50. You want to select five researchers to be on your panel. If you use systematic sampling, your sampling interval is 10 (50/5). Now, you need to choose one number between 1 and 10 for a starting point. You should use a table of random numbers or a random number generator. Let's say that your randomly selected number is 3. Hence you will choose researchers numbered 3, 13 (3+10), 23 (13+10), 33 (23+10), and 43 (33+10).

Stratified Random Sampling

Stratified sampling is a procedure where a population is divided into mutually exclusive groups called strata and then a simple random sample or a systematic sample is drawn from each stratum. For instance, you can divide the population into males and females and take a random sample of males and a random sample of females using gender as the stratification variable. Within stratified random sampling, *proportional stratified sampling* refers to the procedure that the sample proportions are made to be the same as the population proportions on the stratification variable. For example, let's assume that your stratification variable is gender. Then the proportion of males and females in your sample are made to be the same as the proportion of males and females in the population. In other words, if the population consists of 60% females and 40% males, then 60% of your samples should be randomly chosen from the female subpopulation (i.e., all females in the population), and 40% of your samples should be randomly chosen from the male subpopulation. On the contrary, *disproportional stratified sampling* refers to where the sample proportions are made to be different from the population proportions on the stratification variable. Let's assume that you are interested in studying a traditionally female college of 5,000 students that recently (this academic year) began accepting male students. In the first academic year of accepting male students, assume that 90% of the students (4,500) were female and only 10% (500) are male. You are interested in comparing the academic preparation of male and female students in this college and you have resources to gain 300 samples. In this case, you should have the same number of males and females (150 and 150), which results in obtaining a disproportional stratified samples.

As an example, Stoldt, Miller, and Comfort (2001) used stratified random sampling for their study. In this study, the samples included random samples of athletic directors at NCAA institutions stratified based on their institutional level of affiliation (Division I, II, III). For this, a list of NCAA members was obtained from the organization's website (NCAA, 2000). There were a total of 973 institutions that were active NCAA members. A total of 32.6% completed in Division I, 26.9% competed in Division II, and 40.4% competed in Division III. Based on the proportions, the sample was stratified accordingly and athletic directors at one of every three institutions at each level were randomly selected for their study. Names and addresses were retrieved from the National Directory of College Sports (1999).

Cluster Random Sampling

Cluster random sampling refers to a form of sampling that uses clusters (a collective type of elements, such as schools, classrooms, universities, sport teams) rather than individual elements (e.g., individual student, individual fan, individual coach). For instance, a sport team is a cluster because it is composed of many individual athletes, coaches, and administrators. If you want to randomly select sport teams with cluster sampling, you will use basic simple random sampling except that rather than taking a random sample of individuals, you take a random sample of clusters at the team level (e.g., Cincinnati Bengals, Pittsburgh Steelers). Cluster sampling requires a larger sample size than the other random sampling techniques.

Nonrandom Sampling Techniques

Unlike random sampling, nonrandom sampling does not require the sample to be selected randomly. There are four different types of nonrandom sampling techniques: convenience sampling, quota sampling, purposive sampling, and snowball sampling.

Convenience Sampling

For convenience sampling, a researcher collects the information of individuals who are available, who volunteer, or can be easily obtained in the sample. Hence, when convenience samples are used, it is vital for researchers to describe the characteristics of the people participating in the study. Convenience samples are not an optimal way to collect data, especially when you want to generalize a population on the basis of your study. Nevertheless, many researchers are frequently forced to employ convenience sampling due to practical constraints such as money and time.

For example, Park (2001) used convenience sampling to explore the involvement profiles in selected recreational activities in Korea. Boyd and Shank (2004) used convenience sampling to examine how the gender of athlete endorsers and type of product that they endorse (sport-related versus non-sport related) is related to respondents' perceptions of source expertise, trustworthiness, and attractiveness. Swanson, Gwinner, Larson, and Janda (2003) used convenience sampling procedures to explore four individual psychological motivations of college student game attendance and word-of-mouth behavior for attending sporting events. A table was set up in a high-traffic area of the university student center during dining hours for four consecutive days. Individuals walking by the data collection table were invited to participate in the survey and were offered either a button or pen that promoted the university's football team. A total of 537 usable surveys were collected.

Quota Sampling

In quota sampling, a scholar identifies the major group or subgroups of interest, decides on the number of people to be included in each group, and then chooses a convenience sample of people for each group. It is analogous to stratified random sampling in that the population is divided into subgroups or strata. It differs from stratified random sampling because it obtains subjects through convenience samples, whereas stratified sampling obtains its subjects through a random sample method. The researcher must specify in advance that the sample will include certain percentages or a particular kind of subject. Like convenience samples, quota samples are relative easy to administer and are not expensive.

Matsuoka, Chelladurai, and Harada (2003) assessed the direct and interaction effects of team identification on spectators' intent to attend future games using the quota sampling method. Each data collector selected subjects from each seating block of the stadium according to the approximate proportions of age and sex of spectators in that block. Before the game, the selected individuals were administered the first questionnaire containing items on demographic characteristics (e.g., age, gender, occupation, and city/town of residence), team identification, and intent to attend future games This procedure yielded a sample of 2,845 respondents.

Purposive Sampling

Purposive sampling is often called judgmental sampling. In this sampling, a researcher handpicks the units or subjects, choosing the group of subjects who best serve the purpose of study. Purposive sampling is well suited for exploratory studies. For instance, if you are interested in adult females over 55 years old who are season ticket holders of an NFL team, you can solicit individuals with specific characteristics to participate in your research. A number of sport management studies employed the purposive sampling technique (e.g., Danylchuk & Chelladurai, 1999; Kropp, Lavack, Holden, & Dalakas, 1999). Kwon and Armstrong (2002) designed the study to examine factors influencing impulse buying of sport team licensed merchandise by a sample of college students. They used purposive sampling by selecting subjects from related disciplines in sport/physical education and textile classes:

> *One hundred forty-five students (48 male and the rest female) at a large mid-western university participated in the study. Among the 145 students, 82.8% were white, and 10 subjects were African-American. The average number of years the participants had been enrolled in the university was 2.39, with approximately 41% of the subjects having been enrolled for more than 2 years. (p. 155)*

Snowball Sampling

For snowball sampling, a researcher will ask each research participant to identify other potential research participants. If you can only identify a few individuals in the beginning of a research study, you will ask each participant to bring additional participants so that you can recruit larger samples.

Determining the Sample Size

When you design a research project, you are expected to determine how many subjects need to participate in your research. The process to determine an appropriate sample size is quite complicated. To determine how big a sample must be, the researcher should first determine the sample mean and the population mean. The sample mean is typically different from the population mean. The difference between them is viewed as an error. The margin of error (E) is the maximum difference between the observed sample mean and the value of the population mean. The margin of error can be calculated as

$$E = Z\frac{S}{\sqrt{n}}$$

in which

Z is the standardized value corresponding to a confidence level,
S is the standard deviation of the population, and
n is the sample size.

If we rearrange the formula to yield the sample size:

$$n = \left(\frac{ZS}{E}\right)^2$$

For instance, let us assume the manager of Red Tigers Sports Authority, a specialized store for winter sport gear and equipment, wants to use a 95% confidence level (which is equivalent to a Z value of 1.96) and a range of error of less than $15 to estimate the average price of skis. The estimated standard deviation is $20. How many different types of skis should the manager survey?

$$n = \left(\frac{ZS}{E}\right)^2 = \left(\frac{1.96 \times 20}{15}\right)^2 = \left(\frac{39.2}{15}\right)^2 = (26.1)^2 = 601$$

There are some implications tossing this formula. First, the more precise (less error) the estimate must be, the bigger the sample size must be. Second, doubling sample size increases the degree of precision by one fourth (Zikmund, 2000).

Although researchers can use the above-mentioned equation to determine the sample size for their studies, there are some practical issues that need to be considered. In general, the bigger the sample size is, the more accurate the results will be. However, researchers should be aware of the so-called law of diminishing returns, which means additional units may become less and less meaningful. Time and cost factors should also be considered in the determination of how big the sample needs to be. The type of statistical procedures to be used in data analysis is another consideration. Multivariate statistics usually require a large sample. The researcher must keep in mind that to a certain extent, quality of the sample is more important than its size.

Choosing Sampling Design: Practical Implications

A number of factors may determine the appropriate sampling design for a particular study. These factors include the degree of accuracy needed, resource and time availability, degree of homogeneity present, prior knowledge of the population, whether it is a national versus local project, and the need for statistical analysis (McDaniel & Gates, 1991; Zikmund, 2000). As mentioned previously, the accuracy of the research outcome correlates with the size of the sample and the sampling method used in the study. However, the more sophisticated the sampling design, the more resources required; the more sophisticated the sampling design, the more time needed for the researcher to execute the sampling plan. There is always some tradeoff between sampling design, and accuracy, resources, and time.

If the population to be investigated is somewhat homogeneous, a small sample size and relative simple sampling design could suffice to yield accurate and reliable results. Also, whether the population elements are clearly defined and available to study or not are some of the issues that researchers should consider while choosing their sampling plan. Obviously, it will require a complex and multi-staged sampling design if a research project is to be conducted nationally. The last factor that may influence the selection of a sampling design is the needs of the researcher for statistical analysis. A sample taken with a nonrandom sampling technique does not give the freedom for the researcher to make projection beyond the sample (Zikmund, 2000).

The Sampling Process

Before attempting to select a sample, the researcher must first develop a sampling plan that outlines the steps that should be taken in order to acquire a sample appropriate for the research design. Recommended by scholars (McDaniel & Gates, 1991; Zikmund, 2000), the following are the steps constituting a sampling plan:

1. Define the population of interest.
2. Choose a data collection method.
3. Choose a sampling frame.
4. Select a sampling method.
5. Determine the sample size.
6. Develop operational procedures to select the sample.
7. Execute the sampling plan.

Summary

This chapter reviews the key concepts of sampling in the context of sport management research. An understanding of the terminology will facilitate the comprehension of various sampling procedures discussed in this chapter. Basic terminology includes population, target population, accessible population, sample, census, sampling frame, data, sample data, parameter, statistic, sampling error, statistical relationship, representative, and generalizability. Sampling can be categorized under two major groupings: random sampling and nonrandom sampling. The process of random sampling requires that every person, unit, or element of the population have an equal chance of being selected for the investigation. Simple random sampling, stratified random sampling, systematic random sampling, and cluster random sampling are four types of random sampling procedures that sport management researchers can use in their studies. The process of simple random sampling allows for all elements or participants of a population to have an equal chance of being selected for a study. On the other hand, the primary goal of stratified random sampling is to guarantee that all groups in the population are adequately represented in proportion to the size of each stratum. The major advantages of systematic random sampling include an easily-obtained sample, it is economical, and it requires less time than simple or stratified random sampling techniques. The primary purpose of cluster random sampling is to assist the researcher in gathering data in the most meaningful way for large-scale studies when the population is geographically spread out.

Sample size could be determined based on three parameters: the margin of error, the estimated standard deviation, and the standardized value corresponding to a confidence level. There are several issues that can affect the determination of an appropriate sample size.

When determining the appropriate sampling design for a particular study, the researcher should consider several factors such as degree of accuracy, resources, time, degree of homogeneity, prior knowledge of the population, whether it is a national versus local project, and the need for statistical analysis.

The researcher should follow these seven steps to execute the sampling plan: (1) define the population of interest, (2) choose data collection method, (3) choose sampling frame, (4) select a sampling method, (5) determine the sample size, (6) develop operational procedures to select the sample, and (7) execute the sampling plan.

Study Questions

1. What is a population? What is a sample? What is a sampling frame?
2. What are the differences between these terms used in sampling?
   a. population versus sample
   b. target population versus accessible population
   c. random sampling versus nonrandom sampling
4. What are the advantages and disadvantages of each random and nonrandom sampling procedure?
5. What are the four basic methods of probability sampling?
6. How is a table of random numbers used to select a sample?
7. What are the factors that influence the determination of sample size?

Learning Activities

1. Review an article that was published this year in one of the journals in sport management that has used the nonrandom sampling technique(s) in selecting the sample for the study. Comment on its sampling design.
2. Review an article that was published this year in one of the journals in sport management and provide your critique on the random sampling technique(s) used in the study in terms of its appropriateness.
3. Discuss the sampling designs employed in the following studies. Which sampling methods are used for the journals? How would you classify these?
   a. O'Rourke, S. M., & Chelladurai, P. (2006). Effectiveness of the national collegiate athletic association: Perceptions of intercollegiate athletic administrations. *Internal Journal of Sport Management, 7*(1), 82–101.
   b. Cunningham, G. B., & Sagas, M. (2005). Diversified dyads in the coaching profession. *Internal Journal of Sport Management, 6*(4), 305–323.
   c. Mahony, D. F., Riemer, H. A., Breeding, J. L., and Hums, M. A. (2006). Organizational justice in sport organizations: Perceptions of college athletes and other college students. *Journal of Sport Management, 20*(2), 159–188.
   d. Kahle, L., Duncan, M., Dalakas, V., and Aikon, D. (2001). The social values of fans for men's versus women's university basketball. *Sport Marketing Quarterly, 10*(2), 36–42
4. Find a study that utilizes one of the following sampling techniques in a sport management related journal published in a 2007 or 2008 issue:
   a. Cluster random sampling
   b. Stratified random sampling

References

Boyd, T. C., & Shank, M. D. (2004). Athletes as product endorsers: The effect of gender and product relatedness. *Sport Marketing Quarterly, 13*(2), 82–93.

Danylchuk, K. E. (2000). Tobacco sponsorship: Spectator perceptions at an LPGA event. *Sport Marketing Quarterly, 9*(2), 103–111.

Danylchuk, K. E., & Chelladurai, P. (1999). The nature of managerial work in Canadian intercollegiate athletics. *Journal of Sport Management, 13*, 148–166.

Johnson, B., & Christensen, L. (2014). *Educational research: Quantitative, qualitative, and mixed approaches.* (5th ed.). Thousand Oaks, CA: SAGE Publications, Inc.

Kropp, F., Lavack, A. M., Holden, S. J. S., & Dalakas, V. (1999). Attitudes toward beer and tobacco sports sponsorships. *Sport Marketing Quarterly, 8*(3), 49–58.

Kwon, H. H., & Armstrong, K. L. (2002). Factors influencing impulse buying of sport team licensed merchandise. *Sport Marketing Quarterly, 11*(3), 151–162.

Kwon, H., & Trail, G. (2001). Sport fan motivation: A comparison of American students and international students. *Sport Marketing Quarterly, 10*(2), 27–35.

Levine, D. M., Berenson, M. L., & Stephan, D. (1998). *Statistics for managers using Microsoft Excel.* Upper Saddle River, NJ: Prentice Hall Inc.

Levine, D. M., Ramsey, P. R., & Berenson, M. L. (1995). *Business statistics for quality and productivity.* Englewood Cliffs, NJ: Prentice Hall Inc.

Matsuoka, H., Chelladurai, P., & Harada, M. (2003) Direct and interaction effects of team identification and satisfaction on intention to attend games. *Sport Marketing Quarterly, 12*(4), 244–253.

McDaniel, C., & Gates, R. (1991). *Contemporary marketing research.* St. Paul, MN: West Publishing Company.

McMillian, J. H., & Schumacher, S. (2006). *Research in education* (6th ed.). Boston, MA: Pearson Education, Inc.

Park, S. (2001). A further exploration of the involvement profiles in selected recreational sport activities: Results from a study in Korea. *Sport Marketing Quarterly, 10*(2), 77–82.

Reese, J. T., & Mittelstaedt, R. D. (2001). An exploratory study of the criteria used to establish NFL tickets prices. *Sport Marketing Quarterly, 10*(4), 223–230.

Stoldt, G. C., Miller, L. K., & Comfort, P. G. (2001). Through the eyes of athletics directors: Perceptions of sports information directors, and other public relations issues. *Sport Marketing Quarterly 10*(3), 44–52.

Swanson, S. R., Gwinner, K., Larson, B. V., & Janda, S. (2003). Motivation of college student game attendance and word-of-mouth behavior: The impact of gender differences. *Sport Marketing Quarterly 12*(3), 151–162.

Zikmund, W. G. (2000). *Business research methods* (6th ed.). Fort Worth, TX: Harcourt College Publishers.

16

Data Editing, Coding, and Transforming

After completing the study of this chapter, you should be able to:

- Define and explain editing and coding.
- Explain the tasks involved in editing.
- Explain the reasons or causes of missing data.
- Explain how to deal with the "I don't know" answer.
- Elaborate on the four principles that researchers should use in data coding.
- Use the content analysis technique to conduct a research project.

The purpose of this chapter is to present information about how to prepare the data collected so as to facilitate the data analysis process. Specifically, this chapter will cover two major topics: data preparation and analysis and *content analysis*. In discussing the topic on data preparation and analysis, attention will be given to the review of the activities in which the researcher is engaged at different stages in the data preparation and analysis process. The primary aim of those activities, such as *validating*, *editing*, *coding*, and *data entry*, is to ensure the accuracy of the data while transforming them from the raw data form to a new arrangement that can be analyzed easily.

The other topic to be covered in this chapter is content analysis. It is introduced in the hope that students will find a useful technique to create a new set of data while analyzing available documents or other materials.

An Overview of the Stages of Data Preparation and Analysis

Data preparation and analysis is the act of transforming data with the aim of extracting useful information and facilitating conclusions (McDaniel & Gates, 1991). Depending on the type of data and the question, this might include the application of statistical methods, curve fitting, selecting or discarding certain subsets based on specific criteria, or other techniques. In general, the data preparation and analysis process involves four stages: validating, editing, coding, and data entry and analysis.

Validating

Data validating has been seen as an integral step of the data process. It is used to make sure that the data collection process, such as interviews (including those conducted by phone, by intercepting respondents at a sport facility, etc.), in fact took place. The main goal is to determine whether or not the survey instrument did in fact measure what it was intended to measure or whether or not "each of the questionnaires to be processed represents a valid interview" (McDaniel & Gates, 1991, p. 482). The manner of how an interview is conducted affects the validity of an interview. For example, an unskilled interviewer may forget to provide proper instruction to the interviewees in terms of how to complete the survey instrument. Interviewer fraud or cheating, a common phenomenon in research involving interviews, also affects the validity of an interview. Unethical interviewers may complete the survey instruments themselves by filling in fabricated numbers and answers to the questionnaires.

To validate the data, it is recommended that 10% to 20% of participants interviewed be contacted again to provide information about (1) whether or not they were interviewed, (2) whether or not the interviewees were in fact qualified to be interviewed, (3) whether or not the interviewer followed a proper interview protocol to conduct the interview, (4) whether or not the interviewee was presented with all the questions included in the survey instrument, and (5) whether or not the interviewee believed the interviewer was courteous, etc. (McDaniel & Gates, 1991).

Editing

In order for the data collected to be useful information, they should be edited (Statistics Canada, 2013). "Editing is the process of ascertaining that questionnaires were filled out properly and completely" (McDaniel & Gates, 1991, p. 482). Editing involves two stages. The first stage is called field editing, as it is often completed by the researcher who personally conducts the data collection with an interview or field survey. Its purpose is to examine the collected raw data in order to be sure that it is accurate, complete, and usable. When interviews are used in data collection, the main task of field editing is to check for incomplete sentences and to make sense of the abbreviations used by the interviewer.

Central editing follows the field editing process. Its aim is to complete the remaining editing tasks. Specifically, the researcher finishes the final editing of the questionnaire to ensure that there will be no ambiguity in each question before coding and data entry. Central editing often occurs in a central location.

The aims of editing include the verification of legibility of entries, completeness of entries, consistency of entries, and accuracy of entries (Green & Tull, 1978; Statistics Canada, 2013). The following is a brief description of each of the tasks.

Legibility of Entries

The first step of field editing is to make sure the recorded data are clear and legible. This process is especially important in research that involves data recorded by human beings, such as interviews and observations. It is very common that the recorded information was drafted in a hurry so that the handwriting is illegible, because the researcher tried to note quickly every piece of information during the interview or observation. In order for the data to be useful, they must be reviewed and clarified if there is any doubt about their clarity. The researcher who recorded the data should be the individual involved in the first order of editing for legibility.

Completeness of Entries

It happens frequently that not all the questions are completed if a questionnaire is used in data collection in a mail or online survey or in a personal interview. The incompleteness could be attributed to a number of reasons, such as the unwillingness of the participant to provide the information, the unintentional omission of the respondent, or the failure of the interviewer to record the data.

Consistency of Entries

Discrepancies may exist between the answers provided by the respondent to the questions included in a survey or an interview. For example, in a questionnaire designed to examine consumer purchasing behavior related to golf, the following question was included: What is the amount of discretionary income (i.e., the amount of your income available for spending after the essentials, such as food, clothing, and home mortgage payment, have been taken care of)? A respondent indicated a discretionary income was less than $500. However, while responding to the question, "please provide your estimate in terms of how much money you spend monthly on average on playing golf or purchasing golf-related products," the same respondent said an amount usually spent on golf was at least $600 a month on playing golf or purchasing golf-related products. Such inconsistency should be detected during the editing phase.

Accuracy of Entries

One of the major sources of inaccuracy is caused by the respondent's selection of more than one answer in a question with exclusive answer choices. The following is an example to illustrate this commonly committed error.

What is your marital status?

- ❑ Single
- ❑ Widow
- ❑ Separated
- ❑ Divorced
- ❑ Married
- ❑ Cohabited

When a situation like this occurs, the researcher should determine which answer choice is more logical and accurate to the respondent. Scholars in research methods in social sciences have identified a number of reasons why data might be missing in survey instruments (Huisman & van der Zouwen, 1999; McDaniel & Gates, 1991). These reasons:

- *Missing by logic.* The missing data is caused by the inapplicability of the survey question to the respondent. For example, one of the questions in a questionnaire distributed at the entrance of a professional football stadium reads: "to what extent has the commercial influenced your decision to purchase a season ticket?" A respondent skipped the question as this person was not a season ticket holder. Asking the participant to respond to all the questions included in a questionnaire regardless of their background is one of the most commonly committed errors by new or unskillful questionnaire designers. To avoid missing data caused by this reason, the researcher could use the routing or branching technique or with a response choice of "not a season ticket holder."
- *Missing by design.* Often, the interviewer can decide not to ask the respondent a particular subset of questions because the interviewer feels that the characteristics of the subject do not match the intended purpose of those questions. For instance, there is a particular section in the exit survey on Discover Hong Kong Year Campaign that is devoted to those tourists who have visited any of the new tourist attractions during their trips. If a respondent interviewed did not visit the new attractions, the interviewer will skip to another part of the survey instrument.
- *Item omission.* There are many reasons causing item omission in interviews or in a mailed or online survey. A respondent may omit a certain section of the questionnaire unintentionally if that section was laid out in such a way that it could be skipped easily. Sometimes, the omission results from a missing page in the questionnaire. The illegibility of some questions in the survey instrument caused by poor printing quality could also irritate the respondent, who then decides not to respond to those questions. The inappropriate use of the branching technique can lead to item omission as well (discussion of the branching technique can be found in Chapter 12).
- *Refusal to respond.* The willingness of the respondent to answer a survey question is influenced by several factors, including the topic of the question, the way the question was phrased and presented, and the presence of the interviewer. A respondent may refuse to respond to a question under the following circumstance: (1) the question asked is threatening or invasive of privacy; (2) the question is testing the respondent's knowledge of certain issues to which the respondent would be embarrassed not to know the answer; (3) the question asked is cognitively too difficult for the respondent to answer; (4) the question asked is perceived to be either unclear or irrelevant; or (5) the answer provided to the question could be socially undesirable and the presence of an interviewer leads to the decision of not answering the question.

The following is an example illustrating the type of question that could be considered too threatening. A survey conducted to determine alcohol consumption behavior in tailgate parties before college football games may include a question such as, "do you frequently use alcoholic beverages such as liquor, wine, or beer during tailgating?" This type of question could be perceived as threatening if the respondent is either a faculty member or a university administrator, which can influence a respondent's decision to use the phrase "I don't know."

As Huisman & van der Zouwen (1999) stated, "[I don't know] is the abbreviation representing the group of nonsubstantive responses like 'no idea,' 'no opinion,' or 'I don't know'" (p. 6). Sometimes, it also means it is hard to say or it depends. The researchers must decide how to treat them. Specifically, they need to decide whether or not the "I don't know" answer in fact means that the respondent truly did not know the answer or, rather, that the question asked was irrelevant to the respondent so the respondent chose not to answer (i.e., missing by logic) or the respondent wanted to hide a true feeling (equivalent to the effect of refusal to respond).

As mentioned in Chapter 12, the researcher must understand the pros and cons of using this type of response choice in a mailed or online survey. The basic reason for its usage is to give the respondent an opportunity to indicate their true feeling (i.e., "I don't know") about a particular issue when the respondent in fact does not have sufficient information to formulate an opinion. The following are some questions that could potentially be answered by the respondents in an interview.

A. Who is the current NFL commissioner who succeeded Paul Tagliabue?
B. Do you think the new drug policy imposed recently by the MLB is acceptable?
C. Do you like your current job as a sales representative?
D. What brand of athletic shoes do you like the most?
E. How many times did you attend professional sport games last year?

To question A, the "I don't know" answer is logical, as the intention of the question is to determine if the respondent knows the name of the new NFL commissioner. There is no reason why respondents would give the "I don't know" answer unless they were not aware of the switch in power in that association. If the "I don't know" answer is given to question B, however, it is difficult to judge whether or not the respondent really either did not know the MLB's new drug policy or was aware but had not formed a particular opinion yet. To the remaining questions, C to E, the "I don't know" answer could imply the inappropriateness in drafting those questions. The meaning of ""I don't know" may be multifaceted: "I don't want to answer the question," and "I don't want to answer the question, as the question is not important to me."

Inadequate Response

"An inadequate response is a response that does not fall within the accepted range of values" (Huisman and van der Zouwen, 1999, p. 7). For example, one of the questions included in a mailed survey asks if the respondent plays golf and a dichotomous set of answer choices (yes or no) is provided. A respondent might ignore the answer choices and write sometimes in the space between yes and no. The effect of inadequate response is similar to that of item omission. The researcher should decide if it is legitimate to choose one of the two answer choices immediately next to the mark or to simply treat it as a missing answer. Choosing more than one answer to a question that requires only one of the response choices provided is another type of inadequate response. The same treatments suggested to deal with the problem mentioned previously can also be applied to edit this type of inadequate response.

The above discussion implies that the causes of missing data or the "determinants of item nonresponse" as referred to by Huisman and van der Zouwen (1999) are multidimensional. The researcher, the respondent, and the individual who conducts the interview can all contribute to the problem. First, the researcher can be a cause of missing data. Asking the respondents to complete a questionnaire that contains questions that are utterly irrelevant to them is one of the errors that could be committed by the researcher, which will inevitably lead to many questions not being answered adequately. The way the questionnaire is designed may affect the response behavior of the respondent. Huisman and van der Zouwen (1999) noted the areas in a questionnaire that can potentially cause the respondent not to answer certain questions:

> *The structure of the questionnaire (e.g., the incorrect use of the branching technique may confuse the respondent and increase item nonresponse), the position of the item in the questionnaire (e.g., items placed early in the survey instrument are more likely to be completed), the item wording (the terminology used in a question is unknown to the respondent), the topic and questions in the scale (e.g., questions concerning income, sexual behaviors, etc. have a higher rate of item nonresponse), the task involved in the questions (e.g., the difficult questions are more likely to be unanswered than easier ones), and the format of the questions (e.g., the number of answer choices provided in a close-ended question may affect the response behavior of the respondent). (p. 9)*

The respondent can be another cause of missing data. The demographic characteristics and the level of motivation and interest of the respondents influence their response behavior. Previous research found that less educated respondents are more likely to commit item nonresponse (Huisman and van der Zouwen, 1999, p. 11). When respondents are approached with topics with which they are not familiar or in which they have little interest, they are more likely to be apathetic and less willing to complete the questionnaire.

The interviewer can also be a cause of missing data. Unskillful interviewers tend to have a higher rate of missing data as they often make mistakes, including skipping questions, noting answers in the wrong place, presenting questions incorrectly, and forgetting to record responses. To prevent interviewers from making these mistakes, it is important to provide adequate training for them.

Simply put, the purpose of editing is to prepare the survey instrument for the next stages of analysis, coding and tabulating. Even though "the editing process is extremely tedious and time consuming … [it] is an important step in the processing stage" (McDaniel & Gates, 1991, p. 488).

The use of online survey or an electronic questionnaire can present many aforementioned data errors from occurring. There are no more issues related to legibility of entries while the problem of completeness of entries or missing data could also be prevented by the design of the survey instrument. Basically, the respondents won't be able to advance to the next question or page if they fail to complete all the answer choices.

Coding

Coding refers to the process of grouping and assigning numeric codes to the various responses to a particular question (McDaniel & Gates, 1991; Ghauri, Gronhaug, & Kristianslund, 1995). As mentioned in Chapter 12, questions in a questionnaire are presented either in an open-ended format or a closed-ended format. Most of the survey instruments in sport management research use the closed-ended format.

Coding for closed-ended questions is a relatively simple process. The researcher often precodes the questionnaire prior to putting it into use (see Exhibit 16.1 for an example of a precoded questionnaire). The precoding practice simplifies the coding process, as the researcher can directly enter the data from the questionnaire that has been through the editing process. If a respondent circles two in response to question A, the researcher can input the number two on the computer directly, without the need to go through the coding process.

The purpose of coding the responses collected with an open-ended format is to make them processable and analyzable with statistic procedures. The process of coding open-ended questions involves four steps (McDaniel & Gates, 1991, pp. 489–491):

Exhibit 16.1. Example of a Pre-Coded Questionnaire

Please check the box in front of the answer choice that best represents your situation.

A. What is the division designation of your conference?
1. ❑ Division I Bowl Subdivision
2. ❑ Division I Championship Subdivision
3. ❑ Division I Non-Football Schools
4. ❑ Division II
5. ❑ Division III
6. ❑ Other (explain)

B. If your conference competes at the Division I level, is your conference commonly referred to as:
1. ❑ Equity Conference
2. ❑ Mid-Major Conference
3. ❑ Other (explain)

C. How many institutions are affiliated with your conference?

D. How many conference tournaments are currently sponsored by your conference?

E. How many years has your conference been in existence?
1. ❑ Fewer than 5
2. ❑ 6–10
3. ❑ 11–15
4. ❑ 16–20
5. ❑ More than 20

F. Has your conference outsourced any part of its marketing operations and rights?
1. ❑ Yes
2. ❑ No

1. The researcher lists all the responses provided by the respondents to each of the open-ended questions on the same piece of paper. If the number of respondents is large and makes the task cumbersome, the researcher may randomly select a manageable number of responses to complete the coding process.
2. The researcher reviews all the responses and consolidates those responses that essentially have the same meaning into a particular category. A name is given to each of the newly created categories.
3. After the consolidation, the researcher assigns a numeric code to each of the categories.
4. To complete the tasks in the final step, the researcher first associates all the responses that belong to the same category created in Step 2. Based on the results of Step 3, the researcher then assigns a numeric code to all of them by writing the code in the questionnaire.

Following is an example illustrating how to employ this four-step process to provide coding to open-ended questions. In a recent survey conducted with a group of several senior sport facility executives, the following open-ended question was asked: "in your opinion, what skills are important for a sports facility manager to get their job done effectively?" The 26 skills identified are shown in Exhibit 16.2. The questions were consolidated and accordingly coded.

Exhibit 16.2. Responses from Sport Facility Executives Important Skills

Question: In your opinion, what skills are important for a sports facility manager to do the job effectively?

Responses:

1. Develop and manage standard operating procedures for all operational functions of the facility
2. Develop organizational structures and personnel staffing requirements
3. Develop marketing, advertising, and public relations plans
4. Manage booking and scheduling
5. Generate accurate projections for attendance and revenue
6. Develop contracts with lessees and promoters/performers
7. Develop and manage emergency preparedness procedures
8. Manage security personnel and develop crowd management plans
9. Handle procurement of products and services
10. Develop contracts and agreements with vendors for products and services
11. Oversee novelty sales and contract negotiations
12. Oversee equipment and supply purchases
13. Operate the facility in accordance with applicable local, state, and federal regulations
14. Manage personnel assigned to various facility functions
15. Develop job descriptions, duties, responsibilities, and performance standards for employees
16. Oversee the compliance of state and federal regulations regarding personnel practices
17. Negotiate with unions for all outside services
18. Assess the quality of services and the facility's effectiveness
19. Handle and settle grievances of employees
20. Develop financial reports and manage audit activities
21. Oversee innovative improvements in facility and facility services
22. Communicate effectively with clients, subordinates, and the public
23. Manage catering activities for food and alcoholic beverages to ensure quality of service
24. Oversee food preparation and concession sales
25. Develop management plans for safety of the facility and equipment
26. Develop operational budgets

Exhibit 16.3. Consolidated Response Categories and Codes for the Open-Ended Responses to Important Skills in Exhibit 16.2

Response Categories	Response Number from Exhibit 16.2	Assigned Numeric Code
Policies and Procures	1, 7, 8, 13	1
Accounting, Budgeting, and Purchasing	4, 5, 9, 12, 20, 26	2
Sales and Service Contracts	6, 10, 11, 23, 24	3
Marketing, Promotion, and PR	3, 22	4
Personnel Management	2, 14, 15, 16, 17, 19	5
Facility Management	13, 18, 21, 25	6

As Exhibit 16.3 indicates, the 26 responses can be consolidated into six categories: policies and procedures; accounting, budgeting, and purchasing; sales and service contracts; marketing, promotion, and public relations; personnel management; and facility management. Each category is given a numeric code from 1 to 6. Based on the code assignment sheet, the researcher can then assign a numeric code to each of the 26 responses to facilitate data entry.

It is suggested that a number of principles be followed while coding all questions, regardless of whether or not they are closed-ended or open-ended, to achieve optimal outcomes (Luck & Rubin, 1987):

- *Sufficiency.* The number of categories should be sufficient to include all the themes represented by the responses but not too many to hamper the statistical process. It is impractical to have fewer than three categories, as much important information could be hidden in so few categories.
- *Appropriateness.* The created categories should fit well with the purpose of the research. In other words, the responses should be categorized to facilitate the approval of the research hypothesis or causal relationship between variables.
- *Exclusiveness.* The researcher must make sure that the categories created are mutually exclusive, meaning that the categories should not overlap in the themes and meanings they represent. Mutual exclusivity also means that a response can be placed appropriately into only one category, but also different responses are grouped into different categories (interclass heterogeneity, Luck & Rubin, 1987, p. 348). For example, in a survey, the respondent was asked to indicate an occupation. The following occupation categories were provided: (1) professional, (2) manager, (3) sales representative, (4) priest/pastor/minister, (5) artist, (6) technician, and (7) unemployed. Due to the nature of the categories, the respondent could easily have chosen more than one response. Those who thought artists were professionals, for instance, would have chosen both (1) and (5).
- *Single-dimensionality.* Each category should represent only one dimension conceptually. All the homogeneous responses placed in a specific category should be similar (intra-class homogeneity) (Luck & Rubin, 1987, p. 348) to cluster around one single dimension. The occupation case discussed previously can be used to illustrate the single-dimensional rule in coding. Asked to indicate an occupation, a respondent who used to be a manager of a sporting goods

store but is currently unemployed might have chosen both manager and unemployed, as these two answer choices represent two concepts that are in different dimensions. The unemployed category shows the current status of the respondent, but it is not an occupational category.

If the desire of the researcher is to prepare and code the open-ended responses so that they can be analyzed with parametric statistical procedures, the following three rules, as recommended by Luck and Rubin (1987) should be observed: (1) avoid open-ended class intervals; (2) choose class intervals of the same width; and (3) deal with midpoints of class intervals.

There is a consensus that the coding process should be handled by one person for each open-ended question to ensure consistency if the coding is manually accomplished (Huisman & van der Zouwen, 1999; Statistics Canada, 2015). When more than one coder is needed to code a particular question, it is strongly encouraged that they consult with each other during the coding process, particularly during the step involving consolidation of responses that essentially have the same meaning into a particular category.

In large-scale projects, and especially in cases in which the data entry is performed by a subcontractor, researchers utilize a data code book. The data code book contains every variable used in the research, including its variable name, and the code number associated with each possible response to each question that makes up the data set. Exhibit 16.4 shows an example of the data codebook for a project that examines the attitudes of American adults toward sport participation.

Data Entry and Analysis

Data entry refers to the creation of a computer file that holds the raw data taken from all of the questionnaires deemed suitable for analysis. It is the last step before data analysis in which the researcher keys in the coded data on the computer. The online survey has shortened the data entry process considerably. In some cases, it eliminates it entirely as the researcher can simply download the answers or data sheet to a preformatted spreadsheet and process it with a specific type of statistical computer software, such as the SPSS, SAS, etc. The process of data entry is no longer needed with the use of the online survey technique.

Using Content Analysis on Coding Open-Ended Questions

Content analysis is one of the methods that sport management researchers have used in analyzing the collected information to achieve their research objectives. For example, Pedersen and Pitts (2001) used content analysis as a method to analyze all the articles published in *Sport Marketing Quarterly* to determine the frequency of various types of research methods and statistical procedures utilized by sport marketing researchers. To determine how gender was portrayed in *Sports Illustrated Kids*, Cuneen and Sidewell (1998) conducted a content analysis of the advertisements included in the magazine over a six-year period.

Content analysis is defined as "the systematic, quantitative description of the composition of the object of the study" (Gay & Diebl, 1992). The object described here may include a whole array of things, including the minutes of meetings, letters, memoranda, diaries, speeches, newspapers, magazine or journal articles, films, television programs, and photographs (Robson, 2002).

Exhibit 16.4. A Section of a Data Code Book for a National Survey of Adult Attitudes to Sport Participation

Question	Variable Number	Code Description	Variable Name
	1	Record Number	RECNUM
	2	Respondent Number	RESPID
1	3	5-Digit Zip Code 99999 = Missing	ZIP
2	4	2-Digit Birth Year	BIRTH
3	5	Gender 1 = Male 2 = Female 9 = Missing	GENDER
4	6	Marital Status 1 = Married 2 = Widowed 3 = Divorced 4 = Married but Separated 5 = Unmarried 9 = Missing	MARITAL
5	7	Household Income 1 = Below $20,000 2 = $20,001–$40,000 3 = $40,001–$60,000 4 = $60,001–$80,000 5 = $80,001–$100,000 6 = Above $100,000	INCOME

Executing a Content Analysis

As mentioned above, content analysis is a systematic process that involves several steps (Fraenkel & Wallen, 2000). The steps will be described briefly in the following section.

Determination of Objectives

It is critically important for sport management researchers desiring to use the content analysis technique to determine what is to be accomplished. For example, Pederson, Whisenant, and Schneider (2003) yielded some descriptive information about the gender of the journalists (reporters, photographers, editors, etc.) who provided the coverage of interscholastic athletics in the state of Florida. Exhibit 16.5 shows the percentages of male and female journalists working in the newspaper sport departments during a certain period of time. In general, content analysis is adopted to "obtain descriptive information about a topic" (Fraenkel & Wallen, 2000, p. 472).

Exhibit 16.5. Percentages of Male and Female Journalists Working in the Newspaper Sport Departments During a Certain Period of Time

	Male	Female	Other	Total
Editors				
Executive Sports Editors	43 (100%)	0 (0%)	0 (0%)	43 (100%)
Prep Sports Editors	41 (95.3%)	2 (4%)	0 (0%)	43 (100%)
Written Articles				
Editorial Guidance	1636 (91.3%)	156 (8.7%)	0 (0%)	1792 (100%)
Journalists	276 (91.4%)	26 (8.6%)	0 (0%)	302 (100%)
Articles	953 (53.2%)	93 (5.2%)	746 (41.6%)	1792 (100%)
Photographs				
Editorial Guidance	721 (87.2%)	106 (12.8%)	0 (0%)	827 (100%)
Photographers	132 (78.6%)	36 (21.4%)	0 (0%)	168 (100%)
Photographs	396 (47.9%)	90 (10.9%)	341 (41.2%)	827 (100%)

Source. Pedersen, P. M., Whisenant, W. A., & Schneider, R. G. (2003). Using a content analysis to examine the gendering of sports newspaper personnel and their coverage. *Journal of Sport Management, 17*(4), 376–393.

Other reasons to use the content analysis technique are to verify other research findings and to test hypotheses. Sport management researchers can use the content analysis technique to validate the results of studies utilizing other research methodologies, such as observation and survey. In the aforementioned study, Pedersen, Whisenant, and Schneider (2003) developed four hypotheses, one of which is, "Female reporters will give greater coverage to girls' interscholastic athletics than will male reporters" (p. 382). Some scholars also believe that content analysis can be used to identify a new research idea as an exploratory technique when large numbers of documents are available for the researchers to review and study (Robson, 2002). Through the review process and the application of the content analysis method, the researchers may hope to generate an actionable research theme.

Definition of Terms of Interest

It is also essential for the sport management researcher to operationally define the concepts or terms that will be under investigation. For example, in the study conducted by Cuneen and Sidewell (1998) to investigate how gender was portrayed in *Sport Illustrated Kids*, they operationalized several key terms, such as prominent models and supporting models.

Identification of the Unit of Analysis

In this step, the sport management researcher must decide what is to be analyzed: words, phrases, sentences, abstracted meanings, etc. "The unit most commonly used is probably the individual word" (Robson, 2002, p. 351). The unit of analysis could be syntactical, referential, or thematic. A word or phrase is a syntactical unit. Robson (2002) maintained that the syntactical unit could be the number of stories on a topic, the column inches, or the size of headlines in analyzing newspaper, magazine, or journal articles. On the other hand, a referential unit could be something that is referred to with different expressions, but the meanings of those expressions are the same. For example, expenses and expenditure are terms used interchangeably by sport marketers. In using the thematic unit, the researcher attempts to extract a number of themes from the content.

Selection of a Sampling Plan

After the units of analysis have been identified, the sport management researcher should determine how to obtain the needed sample(s) for analysis. The samples could be the publications of a particular journal or newspaper over a certain period of time (e.g., five years) or all the commercials aired during a particular TV program. For example, all the advertisements appearing during Super Bowl games from the years 1996 to 2002 were included in the sample for a study conducted by Kelley and Turley (2004) that was designed to investigate the relationship between advertising content and the effect of the advertisement. All of the sampling techniques discussed in Chapter 15 can be employed to select samples for studies using the content analysis technique.

Data Recording and Analysis

Counting is the basic method of data recording. Each time a unit of analysis appears, it is counted. When the counting process is complete, the collected data is tabulated and analyzed. The most commonly used statistical data analysis techniques in content analysis are descriptive statistics, particularly the frequency (e.g., the number of times a particular unit of analysis appears) and percentage and/or proportion that a particular term appeared in the total occurrence of all the units of analysis. χ^2 test is also commonly used by sport management researchers to determine the degree of association between two units of analysis.

Advantages and Disadvantages of Content Analysis

Content analysis can be used to collect information that is often difficult or impossible to collect through the use of other traditional research methods because of its unobtrusive nature. This is its major advantage. The other advantage is that it can be used to obtain information about events that happened in the past so as to enable the researcher to make comparisons between the historical information and data collection from present events. Another reason why content analysis is advantageous over other means of data collection is that the needed data (journals, newspapers, books, photos, videotapes, etc.) are readily available so that it is relatively economical to acquire the data. Content analysis can be repeated and replicated (Fraenkel & Wallen, 2000).

A major disadvantage of content analysis is that it is extremely time consuming, as the systematic nature of the method requires the commitment of human resources to record the data carefully. The other disadvantage is that "there is a temptation among researchers to consider that the interpretations gleaned from a particular content analysis indicate the causes of a phenomenon rather than being a reflection of it" (Fraenkel & Wallen, 2000, p. 481).

A Content Analysis Case

Green, Costa, and Fitzgerald (2003) conducted a content analysis on 2002 NCAA Women's Final Four telecasts. The purpose of the study was to determine the nature and extent of the television exposure generated for the city of San Antonio by hosting the 2002 NCAA Women's Final Four basketball tournament. The ESPN coverage of the tournament was content analyzed for any verbal mentions of San Antonio or its associated images and for the variety and duration of San Antonio imagery that appeared in the ESPN broadcast coverage. Two research questions were raised for the study: (1) How extensive is the host city's exposure during the telecast of a large and nationally significant sport event? and (2) What kinds of city mentions and city images occur during the telecast?

All coders involved in the study engaged in code training for both verbal mentions and imagery. During this stage, the coders developed lists of the types of verbal mentions and visual images appearing in the broadcasts. The coders shared their lists and discussed the emerging categories and assignment rules with the other coders. This initial process allowed for the clarification of some ambiguous rules, the development of more precise definitions, and the addition of new categories. Definitions for each coding category were agreed upon and compiled in a coding manual. Coding matrices were developed for each analysis, both visual and verbal. Specific image or verbal types were listed down the left side of the matrix, and each occurrence was listed along the top. Coders then entered the duration of each occurrence by its image type.

The second phase consisted of pilot coding. During this phase, coders individually analyzed a sample of the telecasts. Inter-coder reliability assessed at the pilot coding phase yielded a coefficient of reliability greater than .90 for all variables.

In the final analysis, two researchers independently coded each telecast twice. First, the telecast was coded for the number and type of verbal name mentions of San Antonio or San Antonio-related words (e.g. Alamodome, Alamo, River Walk). In the second analysis, coders recorded the types of San Antonio-related images (i.e., what was shown), and the duration of those images (i.e., how many seconds the image appeared on the screen). The achieved coefficient of reliability was greater than .90 for all but two variables. In both cases, the broadcast was then analyzed by a third researcher. The two scores that correlated at .90 or higher were used to calculate aggregate scores.

Ten programs (11 hours and 46 minutes of coverage) were analyzed: (1) the Selection Show; (2) ESPN Sports Center broadcasts on the Friday, Saturday, and Sunday of the tournament weekend; (3) pregame show for each of the semifinal games and the championship game; (4) both semifinal games; and (5) the national championship game.

Overall, the findings of the study question the value of the exposure generated by events as a tool for place marketing.

Summary

Data editing, coding, and transforming play key roles in ensuring the accuracy and quality of a research project. The importance of this research phase should not be overlooked.

Editing, whose purpose is to ensure the completeness of a survey instrument, involves two stages: field editing and central editing. Regardless of where editing takes place, the researcher involved in the editing process typically has to complete four tasks: (1) ensure the legibility of entries, (2) check the completeness of entries, (3) ascertain the consistency of entries, and (4) ensure accuracy of entries.

Coding is the process immediately following editing, and it involves grouping and assigning numeric codes to the various responses to a particular question. The coding procedures used to process open-ended or closed-ended questions are different. Coding for closed-ended questions is a relatively simple process, but the coding process for responses collected with an open-ended format is somewhat time consuming and tedious. The goal of coding is to make data processable and analyzable with statistical procedures. A four-step model is commonly used by sport management professionals in coding open-ended questions. The four steps include reviewing and listing responses, consolidating responses, setting codes, and entering codes.

It is suggested that four basic rules be observed while coding questions, regardless of whether or not they are closed-ended or open-ended. These four rules are: sufficiency, appropriateness, exclusiveness, and single-dimensionality. While handling a large-scale project, researchers often utilize a data codebook, which contains every variable used in the research, including variable names and the code number associated with each possible response to each question that makes up the data set.

Data entry, the last step before data analysis, is the process of creating a computer file that holds the raw data taken from all of the questionnaires deemed suitable for analysis.

Study Questions

1. Can you define and explain the following two terms in your own words?
   a. Editing
   b. Coding
2. What are the tasks involved in editing? Explain them.
3. According to scholars in research methods, what are the reasons or causes of missing data?
4. How can researchers deal with the "I don't know" answer?
5. What are the four principles the researcher should use in data coding? Explain each.
6. What are the steps or stages in executing a study using the content analysis technique?

Learning Activities

1. In a study conducted by Li and Burden (2001), they asked the athletic directors in all NCAA Division I institutions the question, "what are the benefits of outsourcing of sport marketing operations in intercollegiate athletics?" The following are 18 reasons the surveyed athletic directors provided. You are asked to use the four-step coding process for open-ended questions to complete the coding task. The end product of your coding should be a table like Exhibit 16.3 that shows the consolidated response categories and codes, and the response associating to each category.
   a. Outsourcing can help broaden the media coverage of your institution so as to enhance the public awareness.
   b. Outsourcing can help your institution build a sustainable competitive advantage.

c. It is less expensive for your conference to outsource your marketing operations and rights than to keep them in-house.
d. It is more efficient for your institution to outsource your marketing operations and rights than to keep them in-house.
e. Outsourcing of your marketing operations and rights can bring in a greater financial return than keeping them in-house.
f. Outsourcing can help reduce bureaucracy and red tape in your marketing operations.
g. Outsourcing can help your institution enhance the quality of its marketing operations.
h. Outsourcing enables your institution to reallocate resources to other more critical areas.
i. Outsourcing allows your institution to access external specialized talent and expertise in marketing and promotion.
j. Outsourcing enables your institution to use the contracted company as an extension of your overall marketing efforts.
k. Outsourcing allows your institution to handle all aspects of your marketing operations in a more timely fashion.
l. Outsourcing may cause loss of control of your marketing operations and rights.
m. Outsourcing may result in loss of opportunities to bring in more revenues for your institution.
n. Outsourcing can provide your institution with opportunities to expand marketing efforts to nonrevenue sports.
o. Outsourcing can help your institution reduce capital expenses related to marketing operations.
p. Outsourcing allows your institution to be part of a marketing network and synergy.

2. Ask 10 of your classmates this question: "where would you like to see yourself 10 years from now?" Using one of the basic steps of content analysis, analyze their answers and then describe the frequency of the answers.

References

Cuneen, J., & Sidwell, M. J. (1998). Gender portrayals in Sports Illustrated for Kids advertisements: A content analysis of prominent and supporting models. *Journal of Sport Management, 12*(1), 39–50.

Fraenkel, J. R., & Wallen, N. E. (2000). *How to design and evaluate research in education.* (4th ed.). Boston, MA: McGraw Hill.

Gay, L. R., & Diebl, P. L. (1992). *Research methods for business and management.* New York: Macmillan Publishing Company.

Ghauri, P. N., Gronhaug, K., & Kristianslund, I. (1995). *Research methods in business studies: A practical guide.* New York: Prentice Hall.

Green, B. C., Costa, C., & Fitzgerald, M. (2003). Marketing the host city: Analyzing exposure generated by a sport event. *International Journal of Sports Marketing and Sponsorship, 4*, 335–353.

Green, P. E., & Tull, D. S. (1978). *Research for marketing directions* (4th ed.). Englewood, NJ: Prentice Hall.

Huisman, M., & van der Zouwen, J. (1999). Item nonresponse in scale data from surveys: Types, determinants, and measures. In M. Huisman (Ed.), *Item nonresponse: Occurrence, causes, and imputation of missing answers to test items* (pp. 63–90). Leiden: DSWO-press.

Kelley, S. W., & Turley, L. W. (2004). The effect of content on perceived affect of Super Bowl commercials. *Journal of Sport Management, 18*(4), 398–420.

Luck, D. J., & Rubin, R. S. (1987). *Marketing research* (7th ed.). Englewood, NJ: Prentice Hall.

McDaniel, C., & Gates, R. (1991). *Contemporary marketing research.* St. Paul, MN: West Publishing Company.

Pedersen, P. M., & Pitts, B. G. (2001). Investigating the body of knowledge in sport management: A content analysis of *Sport Marketing Quarterly. The Chronicle of Physical Education in Higher Education, 12*(3), 8–9, 22–23.

Pedersen, P. M., Whisenant, W. A., & Schneider, R. G. (2003). Using a content analysis to examine the gendering of sports newspaper personnel and their coverage. *Journal of Sport Management, 17*(4), 376–393.

Robson, C. (2002). *Real world research: A resource for social scientists and practitioners-researchers* (2nd ed.). Oxford, UK: Blackwell Publisher.

Statistics Canada. (2015). Data collection, capture and coding. Retrieved from http://www.statcan.gc.ca/pub/12-539-x/2009001/collection-collecte-eng.htm

Statistics Canada. (2013). Data editing. Retrieved from http://www.statcan.gc.ca/edu/power-pouvoir/ch3/editing-edition/5214781-eng.htm

17

Communicating Research Results

After completing the study of this chapter, you should be able to:

- Appreciate the importance of effective communication in reporting research findings.
- Understand the methods of communicating research results.
- Understand the three most popular writing styles used in communicating research results.
- Understand the use of bias-free language in research.
- Appreciate the factors that affect the communication process in sport management.
- Understand the various types of research reports.
- Understand the components of research reports.
- Understand the requirements for poster and oral presentations.
- Understand the general rules in the development and use of tables, figures, and charts.

When the research project has concluded, the next step is to prepare a written report of the project, the study results, conclusions, and recommendations. This may involve a written report or an oral presentation. This chapter will present the different formal writing style manuals, using bias-free language guidelines, and the types of written reports and their specific components. In the next sections, this chapter presents how to prepare different types of presentations with effective graphics, such as tables and figures.

Effectively Communicating Research Results

An important aspect of reporting research is to remember that it is a form of communication. After spending weeks or even months on a research project, the researcher might feel that writing a report is a waste of time because all of the important work has been accomplished. However, communicating the project and its results is just as important as conducting the research. The researcher is communicating the arrangement of the project, the purpose, its objectives, the methods used, the logistics of collecting the data, how the data were analyzed, what the analysis means, what conclusions can be drawn, and what recommendations can be made based on the results. Therefore, the researcher should make every attempt to communicate effectively and efficiently.

There are different formats that can be used for communicating the research. This chapter explains how research is communicated in sport management.

Communication Methods

What communication method is best? The answer to this question is contingent upon a number of factors. Research is best communicated when consideration is given to factors that affect the communication process, including what kind of *setting* is involved, who is the *audience*, what standards of reporting should be followed, and how much time (if an oral report) or space (if a written report) is allotted for the report. Exhibit 17.1 illustrates these factors and a flowchart for making determinations about the best way to communicate the research project results.

Type of Setting. The first decision that the researcher needs to make is determine if the research is being conducted for a sport business setting or an academic setting. This decision will affect the determination of the type of communication method that will be used.

Type of Audience. Type of audience refers to who will be receiving the communication of the research. In the sport business setting, there are different groups of individuals to whom the report will be targeted. Examples of those groups or individuals are executive management, marketing, production, operations, event management, grounds, game day crew, facility management, manufacturing plant, distribution staff, sales personnel, sales management, product design, and research and development. In the academic setting, the audience might be students, faculty, administrators, or researchers.

Further determination of audience includes understanding how much experience members of the audience have. In many cases, individuals in the business setting have less experience with research than those in the academic setting. Yet, some groups in the business setting have a lot of experience. For example, those who work in the research and development area will typically have experience in research.

The reason for determining how much experience the audience has with research is to guide the researchers in how technical they can be with the report. If they will be communicating with individuals with little or no research experience, then they will need to use lay terminology and simple methods of communication. If the audience has research experience, they can use greater detail and technicality in the communication.

To guide the researcher or the individual who will be developing a written report, there are standards of writing style. Some of the more common ones used in sport business and academia are presented later in this chapter. The reason for this step is to determine if there are required standards that must be followed. Secondarily, even if standards are not required, following a writing style manual offers an organized approach to putting the report together. If a writing standard is required, the researcher must follow that standard. This is also called a writing style and is found in writing style manuals—these are presented in the next section. If not, it would be helpful to choose one to follow.

Another factor to consider before planning the report is time or space. Time allotted for an oral presentation will dictate how detailed the report can be. Space allotted for a written report will influence how much information is included and how it will be presented. If time or space is short, the presentation must be brief and to the point. If there is more time or space allotted, the presenter

Exhibit 17.1. Factors Used to Determine the Best Method for Communicating the Research Project

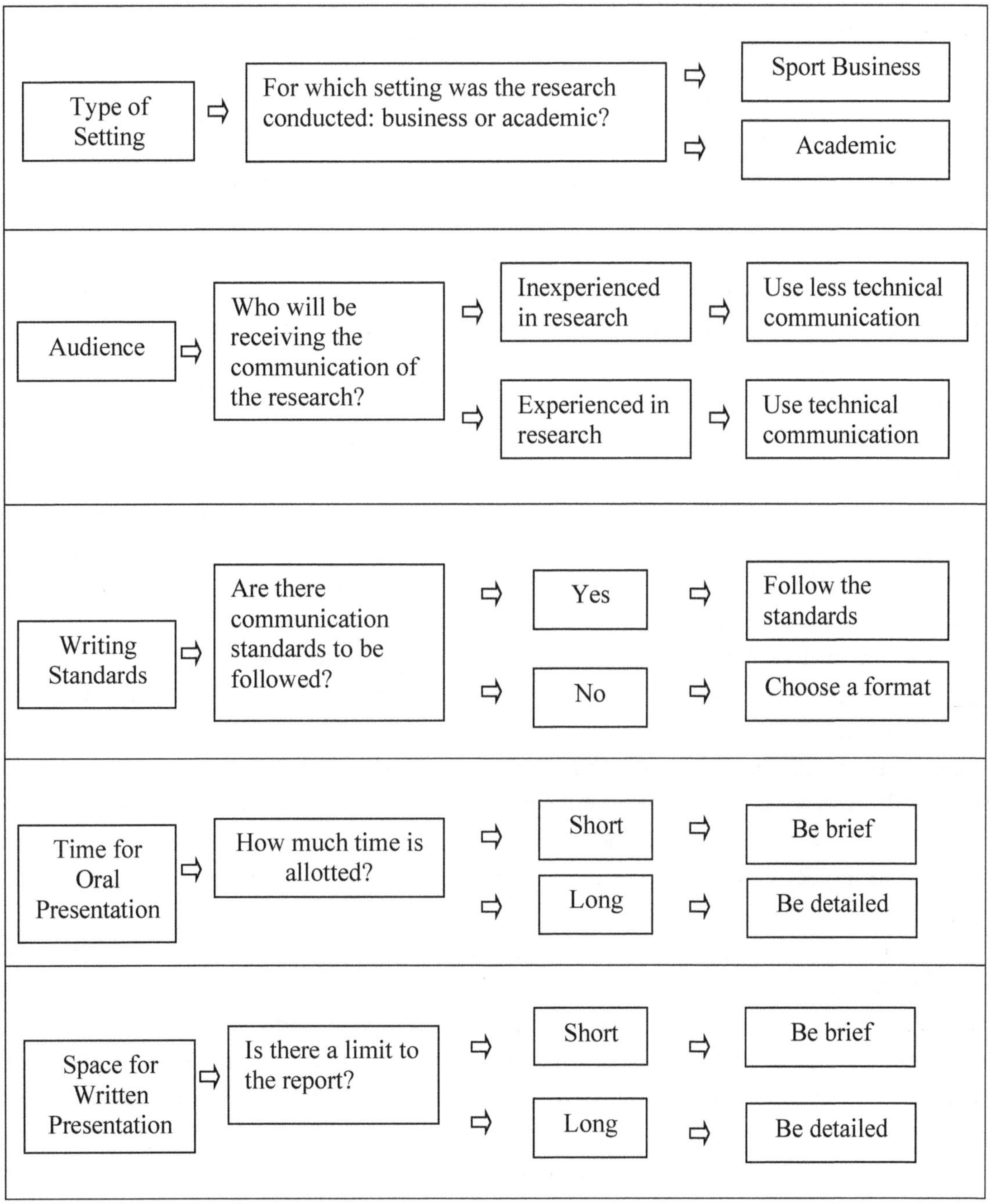

should take the opportunity to include as much information as possible about each element of the research project.

Categories of Written Reports

The categories of written reports are based on the intended audience and setting. The categories include the sport business research report, the academic professional manuscript, the thesis, the dissertation, and the poster presentation. Any category can be used in either setting, but it is most common that the business report is used in a business setting, whereas the other four categories are used in an academic setting.

The business setting might include any type of sport business in which a researcher will be presenting the results of a research project. The academic setting might include submitting the manuscript to an academic journal for publication, submission of an abstract of a research project to be presented at an academic conference as an oral presentation or as a poster presentation, a thesis manuscript developed as a graduate student's academic program requirements, or a dissertation manuscript developed as a doctoral student's academic program requirements. Either setting may require a *briefing*, *short*, or *long report.*

Some university academic programs in sport management require or allow the student to complete a thesis or dissertation. The thesis and dissertation format is much more detailed than other formats. The purpose for this detail is for the student to present every possible piece of information and material as evidence of their ability to conduct research and write a thesis or dissertation. The thesis or dissertation study is supervised by a committee of faculty, one of whom is designated as the student's major advisor for the project.

Types of Reports

The basic types of reports are briefing, short, and long. Any of these may be used in the sport business or academic setting. In the sport business setting, the approach for the report is of a more practical analytical nature. In academia, the approach is usually of a more scientific nature, but results and recommendations should be practical. In some research, either sport business or academic, the nature is of a more theoretical approach. That is, the research and its results will be primarily used for informational, or for a foundation upon which future research can be based. The report of the research is the organized presentation of the research, its results, conclusions drawn based on the findings, and recommendations for the application of the information. Exhibit 17.2 illustrates each type and their components.

Briefing

The briefing can be an oral, poster, or PowerPoint presentation, and consists primarily of three sections (preparing the oral presentation will be discussed in the latter part of this chapter). In the briefing, the lengthy material is summarized down to the shortest, but most important, points of the project.

Time is a factor always to be considered in the oral presentation. The presenter should first find out how much time will be available so that the report will fit into the amount of time allotted. A presenter should never go into a presentation without having first practiced so that the presentation is within the time limit. In the business setting, the individuals to hear the report will most likely not be interested in the logistical details of the research. They want to hear the results and recommendations for how to apply the results. One tip is to think of the presentation as containing five fifths: the introduction will take one fifth, while the other two sections should each be given two fifths of the time. For example, if the time allotted is 25 minutes, the presenter should use only five minutes for the introduction, and 10 minutes each for the remaining findings and recommendations sections. Additionally, many presentation rules require that a presenter save the last 5 minutes of the allotted time for an audience question and answer period. In this case, the presentation

Exhibit 17.2. Research Report Sections and Order of Inclusion (for Written or Oral, Business or Academic Research Reports)

Sections Included	Type: Briefing	Type: Short	Type: Long
Preface/Foreword			✓
Cover Letter			✓
Title Page	✓	✓	✓
Executive Summary		✓	✓
Abstract			✓
Introduction	✓	✓	✓
Statement of Problem	✓	✓	✓
Purpose of Research			✓
Research Objectives			✓
Review of Literature		✓	✓
Exploration for Information			✓
Analysis of Existing Literature			✓
Methodology	✓	✓	✓
Sampling Design		✓	✓
Research Design			✓
Data Collection Method			✓
Data Analysis Method			✓
Limitations			✓
Findings/Results	✓	✓	✓
Conclusions & Discussion	✓	✓	✓
Recommendations	✓	✓	✓
References		✓	✓
Tables & Figures	✓	✓	✓
Appendix/Exhibits		✓	✓

should be prepared to be just 20 minutes and the last five minutes should be saved for the audience question and answer time. The briefing should consist of the following parts:

1. *Introduction.* This should be kept to a bare minimum including coverage of the purpose and objectives of the research as a reminder of why this study was conducted.
2. *Findings and conclusions.* This can be done by presenting the findings and then conclusions afterward, or vice versa. In this part, the researcher should concentrate on presenting the results of the research along with conclusions that can be made based on these results.

3. *Recommendations.* The last part of the presentation should be spent giving recommendations about how to use the results. In the sport business, the presenter should spend this time talking about what the results mean and how they can be used as solutions to the original problem.

Short Report

The short report or executive summary includes more details and information than the briefing. The short report is appropriate when the problem is well defined, has a relatively simple methodology, and has findings, conclusions and recommendations that are uncomplicated and straightforward.

A short or full report is five to ten pages in length. It should include a brief introduction that presents the statement of the problem studied. Next a short summary of any existing information that could be used to help explain or solve the problem should be included. If answers were found in exploration, then the next section would be an explanation of how this information provides direction for solving the problem. If answers were not found, then the next section will be an overview of the research methodology utilized. The final portion of the report will contain the findings of the study, conclusions that can be formulated based on the findings, and recommendations about how to apply the information gained in the study.

Long Report

The long report is one that includes extensive details of every aspect of the research project. For the sport business setting, this report will include great detail about each of the steps of the project with heavy emphasis on the findings and recommendations. For this purpose, the report will be written with a practical management style. For the academic setting, the report will follow a specific writing manual as required by the academic publication or organization and will be written in a formal academic language style.

In the sport business setting, often the individuals to whom the report will be presented have no background in research methodology. Therefore, every effort should be taken to present the information in nontechnical language so that those individuals can understand the project and its uses. On the other hand, in the academic setting it is important to present the research according to standards set by research and scientific organizations. These standards exist to protect the integrity and validity of every aspect of research. For instance, there are specific standards and protections for research projects that will involve human subjects.

Often, practitioners (those in the business settings) believe that academics (those in academic settings) produce research that cannot be used in the real world. These individuals believe that academics produce research geared for nothing more than to be published in an academic journal. Although this is true of some research projects, it is certainly not true of all academic research. Additionally, academic research that has a theoretical or scientific basis produces information that enhances or furthers the existing information on a topic.

Further, some academic research is conducted specifically for the academic setting. For example, academics study the aspects of the young field of sport management, such as growth, curriculum development, program development, accreditation standards, and even the growth of the research literature. The information from this research is important to guide the development of the field and thus has a practical nature for those in academia.

Selecting a Writing Style

Writing style refers to guidelines and standards for a particular format for writing. The format will include guidelines for nearly everything including, but not limited to, language, spelling, grammar, organization of the paper or manuscript, and formatting for citations and references. The primary purpose of adopting a writing style is to conform to the standards of a particular field or organization. This is helpful when members and readers of a manuscript are all using the same writing style, format, referencing, and more. Additionally, it provides for a formal and professional way to write, organize a manuscript, and have it presented to be published for all the world to see.

There are several formal writing style manuals. These manuals are popular within and adopted as official manuals by different fields of study or organizational areas. For the academic field of sport management, the vast majority of academic journals and textbooks follow the American Psychological Association's *Publication Manual of the American Psychological Association* (2010)—commonly referred to as the APA Manual or APA style. Two other popular styles are Modern Language Association (MLA) and Chicago Manual of Style (CMOS). American Psychological Association (APA) style is often used for academic disciplines such as science and business, including sport management, while the Chicago Manual of Style and Modern Language Association are used primarily in the liberal arts and humanities fields.

The American Psychological Association sponsors guidelines for the preparation of formal manuscripts and presentations. These include very specific details about every aspect of writing, language, grammar, citation, referencing, tables and figures, and organizing the paper or presentation. The purpose of these guidelines is to offer recommendations to standardize the format of the paper or presentation. The *Publication Manual of the American Psychological Association* is considered the style manual of choice for writers, editors, students, and educators in the social and behavioral sciences. It provides invaluable guidance on all aspects of the writing process, from the ethics of authorship to the word choice that best reduces bias in language. Well-known for its authoritative and easy-to-use reference and citation system, the *Publication Manual* also offers guidance on choosing the headings, tables, figures, and tone that will result in strong, simple, and elegant scientific communication (American Psychological Association, 2017).

The APA style website (apastyle.org) offers help guides and provides answers for up-to-date questions regarding format. For example, as the Internet and other technology have become sources of information, many changes have been made in regards to reference citation. The website carries the most up-to-date information regarding changes and recommended writing style guidelines to accommodate those changes.

The Modern Language Association (MLA) style is most commonly used to develop papers and cite sources within the liberal arts and humanities fields. The book's title is *MLA Handbook, 8th Edition* (2016). The style recommended by the association for preparing scholarly manuscripts and student research papers concerns itself with the mechanics of writing, such as punctuation, quotation, and documentation of sources. MLA style has been widely adopted by schools, academic departments, and instructors for nearly half a century.

The MLA handbook is also used by numerous scholarly and literary journals, newsletters, magazines, small periodicals, and by some university and commercial presses. The MLA also has a website for instant and helpful information. Their website is www.style.mla.org. On the website you will also find links to other books that are targeted to helping your writing.

The Chicago Manual of Style (16th Ed.) is another of the most used writing style manuals. The Chicago Style for writing is considered to be one the oldest—the history of this well-known manual shows that it is just over one hundred years old. It began "in 1891 when the University of Chicago Press first opened its doors" (*The History of The Chicago Manual of Style,* 2017).

Published by the University of Chicago Press originally in 1906 as a guide to writers for articles and other works to be published by the Press, these guidelines also have been modified over time as techniques and technologies have changed. The *Chicago Manual* is also published as an online book. Look for information about this manual on the website http://www.chicagomanualofstyle.org.

Successful Communication with Bias-Free Language

All researchers are expected to use language free of bias in preparing their research reports. Indeed, most writing standards strongly recommend or require the use of bias-free language. Bias-free language is the use of language free from biases, discriminatory, unintended stereotypes, or labeling in communication. Most language guidelines stipulate communicating to avoid inaccurate, inappropriate, or derogatory biases. The primary purpose of communicating with bias-free language is to treat and address all people equally. The *Publication Manual of the American Psychological Association* (2010) believes authors must provide language free from all biases. As such, the researchers should make a habit of checking their work for biased language just as they check it for spelling and grammar.

The APA manual provides excellent guidelines and examples of avoiding biased language. These include such topics as labels, participation, gender, sexual orientation, racial and ethnic identity, and disabilities, as well as using language that is positive, appropriate, puts people first, and uses emotionally neutral expressions. Exhibit 17.3 provides some examples of problematic language and preferred uses.

It is imperative to take the time to learn about how to use bias-free language, practice using it, and practice writing with it. Sport management professionals come into contact with numerous diverse individuals; it is their responsibility to treat individuals as they want to be treated—with utmost respect.

Effective Presentation with Graphics: Tables, Charts, Figures, and More

Presentations of research, either written or oral, are enhanced and thus are more effective if they include graphics. These are typically tables, charts, figures, and other similar illustrations. This section of the chapter presents a general guide to the development and use of these so that the presentation will be enhanced.

Presenting research data as text is common when the report includes a very small amount of results. However, there are guidelines that should be followed for expressing research results in written language. Each writing manual, as presented in the last sections of this chapter, offers specific rules and guidelines for writing research results into the text of the document. For example, the *APA Publication Manual* (2010) states that titles of texts should be capitalized; names of factors should be capitalized; an abbreviation should be used only after the first presentation of the complete term is presented; figures should be used to express all numbers of 10 and above; and any number under 10 or that begins a sentence should be spelled out.

Exhibit 17.3. Examples of Problematic Language and Preferred Bias-Free Language

Problematic	Bias-Free
***Gender* – Make the effort to make language gender neutral, as in these 2 examples.**	
1. Let's take a look at how he/she managed the ticket sales.	Let's study how the ticket manager handles sales.
2. Our company doesn't have the manpower for that project.	Our company doesn't have the staff (or personnel, workforce, workers, human resources) for that project.
3. The girls team at the university is having a winning season.	The women's team at the university is having a winning season.
***Disability* – Put the person first.**	
1. She is a disabled individual.	She is a person with a disability.
***Age* – The age of an individual or group should be described with adjectives, and use the correct term for females and males at different ages.**	
1. As a 22-year old college girl, she has been very successful.	Use woman, not girl, for females who are 18 and older; use man, not boy, for males 18 and older.
2. Dr. Smith is an elderly man.	Dr. Smith is an older person.
***Race or Ethnicity* – Use language descriptive with specificity, not stereotype.**	
1. Study participants were 22 orientals.	Study participants were primarily Asian and most were from Thailand, Laos, and South Korea.
***Sexual Orientation* – Use language equal to both women and men; avoid using the word "homosexuals" and separate lesbian women from gay men; avoid assumption of heterosexual orientation.**	
1. People were asked about their homosexuality.	Participants were asked about experiences of the life of a lesbian woman or a gay man.

The most common method for presenting data is through the use of illustrations: tables, charts, and figures. These allow the presenter to show large amounts of data in a small amount of space. However, there are specific guidelines for the presentation of these. Writing manuals provide detailed guidelines for developing and presenting tables, charts, and figures in a written manuscript. In this section, we provide a general overview of the use of tables, charts, and figures.

A table is an illustration of data in a tabular format—see Exhibit 17.4 for two examples. All other illustrations are considered figures. Usually in the form of charts, models, diagrams, and other illustrations, figures present numerical information in a visual form. Exhibit 17.5 presents some of the many different types of charts and graphs available for the researcher to use in depicting data.

Components of the Research Report

A report is developed based on its intended audience and use. Research being reported for a sport business management team will be put together and written in a format and a language appropriate

Exhibit 17.4 Examples of Two Tables

Table A Top Ten Countries of Olympic Games Medal Count

Nation	Gold	Silver	Bronze	Total
United States	974	772	661	2,407
USSR	473	376	355	1,204
Italy	218	178	198	594
France	209	220	250	679
Germany	205	210	227	642
Great Britain	197	245	247	689
East Germany	193	167	164	524
Sweden	185	186	217	588
Hungary	156	138	161	455
Norway	149	141	125	415

Table B Factors That Affected Attendance at the Women's Final Four Championships in Atlanta, Georgia

Factors	Mean
To attend a Championship tournament	4.73
Entertainment value of the games	4.31
Support of the teams playing in the tournament	3.93
Tournament location	3.69
Availability of tickets to purchase	3.53
The Georgia Dome as a venue	2.98
Price of tickets	2.93
Dates games are scheduled	2.92
Promotion of the event	2.75
Days of the week games are scheduled	2.74
Other activities taking place in the city	2.73

Note: Likert Scale of 1 to 5 with 5 being most affect.

Source: Pitts, B.G., Lu, D., Ayers, K., Lucas, C. (2004). Factors affecting spectator attendance at the 2003 NCAA Women's Final Four Basketball Championship. Paper presented at the 2005 European Association for Sport Management, Belgium, 2004.

for that purpose. Research that will be reported in academic literature will be written and formatted in a manner appropriate for that purpose and style.

The reporting of all research should be done with a typical ordering of each section. The sections generally include introduction, literature or information, methodology, findings, conclusions, and recommendations. Exhibit 17.2 presented earlier illustrates the many possible sections of a report along with those sections that are typical in three different types of reports. Additionally, these

Exhibit 17.5 Examples of Common Charts and Graphs

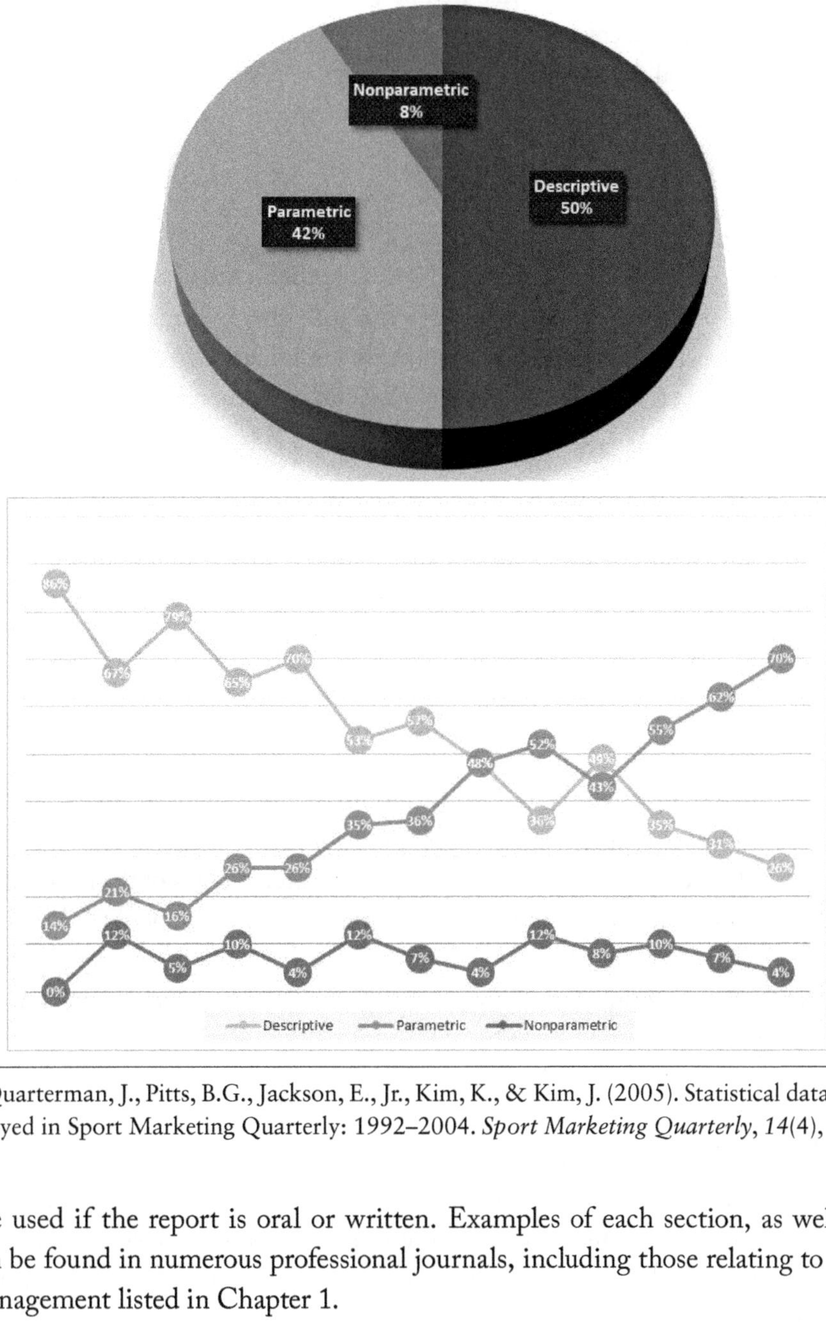

Source: Quarterman, J., Pitts, B.G., Jackson, E., Jr., Kim, K., & Kim, J. (2005). Statistical data analysis techniques employed in Sport Marketing Quarterly: 1992–2004. *Sport Marketing Quarterly*, *14*(4), 227–239.

sections are used if the report is oral or written. Examples of each section, as well as complete reports, can be found in numerous professional journals, including those relating to the discipline of sport management listed in Chapter 1.

Preface/Foreword

In a report that is extensive, there may be a short preamble to the entire document. These sections are commonly used in academic works, especially in textbooks. The preface is a brief introduction to a document that explains its scope, intention, or background. The preface is usually written by the author.

The foreword is also a brief introduction to the document, but is usually written by a person other than the author. This piece usually compares the document to other works and offers an informal view of how valuable the document is.

Cover Letter

If the research report is presented in a formal manner, a cover letter can provide an introduction for the reader from the author. The cover letter can include a brief overview of the project, an explanation of how the informational document is organized, and a note of appreciation for the opportunity.

Title Page

The title page is a one-page cover sheet. It is typically presented in a stylized format that gives the document a pleasing look. The title page should include at least four items: the title, the date, to whom the report is presented, and by whom it was prepared. Other information might include author contact information. See your selection style manual for specific information on formatting the title page of a report in a business setting and an academic paper to be submitted to an academic journal.

Executive Summary

An executive summary is a miniature report of the research project, usually two to three pages in length (single spaced). The executive summary covers all the aspects of the body of the report in abbreviated form. This section is written after the full report is completed.

Abstract

An abstract is a very brief overview of the study and usually is limited to 300 words or less. This amounts to approximately a long paragraph. It typically begins with a statement about the purpose of the study, followed immediately by selected results and conclusions. The abstract gives the reader just enough information to get an idea of what the full report is about so that the reader can decide whether or not reading the whole report will be beneficial. APA guidelines note that a well-written abstract should be accurate, self-contained, concise, and specific (American Psychology Association, 2010, pp. 12–13).

Introduction

The introduction section in a full report can be comprised of components that describe the parts of the project: the problem statement, purpose of the research, research objectives, and background information. In many projects, the introduction section may be taken directly from the proposal with some editing.

The problem statement is comprised of the stated need for the research. This is the original question that began as the reason for developing the research project. An example is, "after evaluating the annual analysis of the fans in attendance at the annual SuperCross events held at the Georgia Dome, it was determined research is needed to find out why attendance has been decreasing over a three-year period of time." It is then followed by a stated purpose of the research and research objectives.

The purpose of the research is a statement about the bottom line reason for the study. An example is, "the purpose of this research was to examine the factors that affected attendance at the 2007 SuperCross event at the Georgia Dome."

In the research objectives section, the objectives are outlined to address the purpose of the study in more specific terms. These may be a list of research questions or hypotheses. An example of research objectives is as follows. "Based on the purpose of the study, the following questions guided this study: Why has attendance been dropping? What affects the attendance of current attendees? Are there factors such as pricing or promotions that can be changed in order to have a positive effect on attendance for future events?"

Review of Literature

In this section, the results of the search for existing information are presented. In a briefing, none of this is presented unless the answer to the research question is found in existing literature that could completely answer the research question, or have an influence on it, or serve as information that could greatly add to the resulting recommendations.

Exploration for information. The information that provided answers to the research questions, or solutions to problems, will be presented in this section. It should be presented in an organized manner. There is no truly right or wrong way to organize this information, except that it should be organized based on a logical manner that presents how and where the information search took place, what was found, what information was provided from the search, and how and why this information answers the research questions or provides solutions to the research problem.

Analysis of Existing Literature. The analysis of existing literature section includes a presentation of what was found in existing literature and the researcher's analysis. This should be presented in an organized manner with attention to what information offers help in answering questions or finding solutions. This section also puts historical (old) studies or information into context. As well, it demonstrates the connections between the informational pieces (articles, books, manuscripts, presentations, etc.) that the researcher is using.

The researcher should refer to the APA manual for help in developing this section. Also, it can be helpful for the researcher to study how authors of existing manuscripts presented their literature review sections.

Methodology

In this section, a description of how the study was conducted needs be presented in detail. This description should enable the reader to evaluate the methods the researcher used and their appropriateness. This section should also present enough information about the methods used in the study so that other researchers can replicate the study. To present information in this section, the researcher should identify the sections with headings. The sections usually include sampling design, research design, data collection methods, data analysis methods, and limitations of the study.

Sampling design or subjects. Identifying the subjects in a study is critical. Identification of the target population being included in the study should be first presented in this section. In addition, the specific method used to determine an appropriate sample of this population should also be included. Furthermore, the actual logistical methods used in appropriating the sample while data collection is in process should be reviewed as well. In a study to examine the effects of time on sponsorship recognition, Pitts and Slattery (2004) describe their subjects in the following way:

> *The population selected for examination in this study was season ticket holders at a successful and nationally ranked university football organization. This population was selected because highly or*

> *nationally ranked collegiate football teams can have high interest, high involvement, and highly loyal fans (p 46)...A selective sampling technique was used to obtain a useful sample for the study (p. 47)...From the resulting list of 10,958 season ticket holders, every 22nd name was selected using a computer program to reach a sample size of 500 (p. 47).*

Research design. The research method, or design, selected for the study is presented in the research design section. The presentation should be accurate in describing each step in the execution of the plan. A description of the instrument, including how it was designed and tested for reliability and validity is a critical part of this subsection. In this section, the researcher should also justify why the design was appropriate for the study.

Data collection method. The data collection method section should give precise details about the method(s) used to collect the data, and a description of each step used and how this was controlled so that the collection was accurate and appropriate. For example, Dixon and Bruening (2007) described how they collected the data for their study designed to examine the challenges of balancing work and family by NCAA Division I female head coaches as follows.

> *Participants were notified by email when the host site was functional and each was issued a login name and password. Instructions on how to maneuver the site were included in this informational email and were also included as a link on the host site. The participants were all instructed to read the information sheet and consent letter included on the site first and then to proceed with the study (assuming they agreed with the consent letter). From there, each participant completed a brief (22-item) questionnaire to provide basic demographic information (e.g., age, number of years in the profession, number of years as a head coach, number and ages of children, spouse or partner's occupation) without taking valuable time away from the online focus group format to answer these questions. (p. 388)*

Data analysis method. In this section, the researcher should present and describe the details of the methods used to analyze the data. A justification of why these methods are appropriate for the study should also be provided.

Limitations. The limitations of the study need to be discussed in their own section. The researcher must be precise and forthcoming. The following is an example of how the limitations of a particular study are presented:

> *Three limitations for this study should be recognized. First, while explicit instructions for both sets of content analyses were provided, ambiguity and variability can still exist in the results. Set standards for what constituted being a 'critical' part of the website were not in place; thus consistency across participants can be questioned. Second, some of the content categories indicated through the SII may be extremely difficult to communicate through a website and subjective to evaluate on a website... The third limitation was the convenience samples in Studies 1 and 3 were different from the venue sample in Study 2.*

Concluding Sections

These concluding sections typically are the majority of the report. These sections present what was found in the results of the study, what conclusions can be drawn from the results, and what recommendations can be made based on the results.

Findings/results. In this section, the researcher presents the raw findings or results of the data analysis. This should be done without comment or discussion. Results should be organized in a logical manner, or in a way that follows the design of the study. Tables, charts, figures, or other illustrative aids are used in this section. The researcher should choose a medium that clearly and efficiently presents the data and other information.

Conclusions and discussion. The conclusions and discussion are the famous "so what?" sections. That is, after a study is conducted, everyone asks what does this mean and how can we use it. After presenting the results, the researcher is now in a position to interpret what the results mean and what conclusions can be formulated. In this section, the researcher is free to interpret the results in relation to the original purpose, questions, hypotheses, and objectives of the study. The researcher may begin this section with discussing whether or not the results of the study answered the research question, supported the hypotheses, or met the objectives of the study, followed by a description of how the results answered the questions.

Recommendations. Recommendations are suggestions for how the research can be applied. These should be presented with detail for the action to be taken. In academic research to be published in journals, the recommendations often include ideas for future study that will broaden, enhance, or test understanding in the research area. In sport business, the recommendations will include suggestions for business actions. Justifications should be offered. Alternatives can be suggested. The following is an example of recommendations offered after completing a study:

> *Finally, this study can also serve as a road map for future research studies. One suggestion is to replicate this investigation. Using a longitudinal approach would be useful with data collection taking place at different time intervals (for example, every two years). Conducting telephone and face-to-face interviews may assist in increasing the response rate of this type of investigation. Continuing the investigation over a 10-year period to observe changes in challenges is also recommended. Another suggestion is to utilize quantitative measures in future research by developing and implementing a scale for rating and ranking the themes and responses solicited in the current investigation. Also future investigations of challenges confronting females as athletic directors need to examine the generalizability of the current findings of more diverse groups when categorized by such dimensions as race and age. (Quarterman, Dupreé, & Willis, 2006, pp. 538–539)*

References

This section contains a list of all of the sources cited within the study. The researcher should follow a standard from a writing manual so that citations will be easy to understand. Some of the more commonly used writing manuals are presented elsewhere in this chapter.

Tables, Figures, and Charts

In a manuscript, tables and figures, as well as other illustrations, should be presented following a writing manual format. Some manuals call for tables, figures, and charts to be placed within the report closest to where they are referenced. Other writing style manuals call for them to be printed one per page at the end of the manuscript, before the appendix or exhibits. Suggestions for formulating tables, figures, and charts are in a different section of this chapter.

Appendix/Exhibits

The appendix or exhibits section is the place to put a number of pieces of information that are not placed within the body of the report. Some of those include large tables, instructions, copies of surveys and questionnaires, detailed drawings, or other material important to be included.

Thesis and Dissertation Format versus Journal Article or Business Report Format

Some university academic programs in sport management require or allow the student to complete a thesis or dissertation. The thesis and dissertation format is much more detailed than other formats. The purpose for this detail is for the student to present every possible piece of information and material as evidence of their ability to conduct research and write a thesis or dissertation. The thesis or dissertation study is supervised by a committee of faculty, one of whom is designated as the student's "major advisor" for the project.

Every section as listed in the "Long" column in Exhibit 17.2 earlier in the chapter is included in the thesis and dissertation manuscript. Each section includes great detail and information about the topic of that section.

Preparing Poster and Oral Presentations

Poster Presentation

The poster presentation is used as a method of presenting the research project by posting the results—hence, the title of poster. That is, a briefly written report using several components will be produced for posting in an area designated for the poster. This is more typical for a setting in which a number of different type of presentations are given, such as at a convention, exhibition, or conference. At an academic conference, for example, there are be oral presentations along with workshops, seminars, round-table discussion groups, symposia, and poster presentations.

Typically, a four-by-six-foot space on a professional display board is provided for the presenter. The poster presenter displays the research report within this space. The event designates a date and time for a session in which researchers display their posters. As individuals come to read and view the research projects, they may ask questions about the projects of the researchers, who are available for discussion next to their posters.

The preparation of a poster report should be done with the same attention to detail and content as any other type of report or presentation. There are a number of formats from which one can choose, or one can be creative and develop the display with a personal touch. Either way, the poster style report should include the basic components of the written report: introduction, methodology, results, conclusions and discussion, recommendations, and references. Additionally, because the poster report is a visual display, attention should be given to graphic presentation.

Oral Presentation

Researchers often present their research projects and findings in an oral presentation. These presentations have some unique characteristics that make them somewhat different from public speaking. Typically, the presentation is made to a relatively small group of people (10 to 50, depending on the setting), the presentation is limited to details about the project, the speaking time allotted can be 15 minutes to an hour, and there is time for questions and discussion following the presentation.

An oral presentation should include many of the same components as a written report. Usually, the length of each component is delimited by the amount of time allotted for the presentation. Preparation should be given as much attention as preparation of the written report, with an emphasis on specific points to be made during the presentation. Some researchers prepare an oral presentation by first writing the full report. After this, the researcher outlines the key components to include and points to be made. Then, the researcher develops the actual presentation using either brief notes or a fully written presentation.

A decision to make early in the planning stage is whether to give the presentation by memory, from notes, or by reading the paper. Giving a presentation by memorizing it can be risky. It is possible to forget certain points, or even whole parts of the report. In the end, questions could be asked that the researchers might not be able to answer because they will have no notes with them. Giving a presentation from notes is not as risky; however, it does require that the researchers still memorize some information that they include in their notes.

Giving a presentation by reading the manuscript is a good style to use if the researchers don't fall prey to the worst effect of this—putting the audience to sleep. If they choose this style, they need to be sure to practice what they have written so much that they don't have to look down at the paper constantly. They should look at the audience members so as to engage them in their presentation.

Setting can determine which style is used. It is highly advisable to use a relaxed, polished presentation with notes in a sport business setting. In an academic setting, it is more expected that a well-practiced and strictly straightforward presentation be given either from notes or by reading the paper.

Delivery and Performance

The content of a research report is the primary concern in a presentation. However, the delivery and performance of this by the presenter can affect how the audience accepts the communication. Researchers should consider the audience and setting and prepare appropriately with appearance, demeanor, speech pattern, enunciation, and gestures.

In presentations, knowing your audience comes first, so the presentation can be customized to the audience and their expectations. Outline your report and prepare the presentation accordingly.

Prepare to deliver the presentation clearly and succinctly. The written report will provide greater detail of your research. Choose graphical aids that are easy to understand and simple, not cluttered. If possible, check out the room and equipment well before the presentation day and time. Be sure to practice your presentation so that you are comfortable with the material you must cover and can deliver the presentation in the exact time allotted. Use a watch to time yourself and know where you should be in your presentation at certain points in time. Come to the room well in advance of the presentation time to set up. Have everything ready to begin on time; don't waste your presentation time getting equipment set up. Bring backup media in several formats in case your files and equipment are incompatible, or if the equipment fails. For example, bring saved presentation files on two or three flash drives and a CD. It is even advised to bring your own laptop and connector cables as a backup. In addition, bring paper copies of the presentation in case all equipment fails, or you would like the audience to have paper copies for taking notes.

On the day of the presentation, dress appropriately and arrive early to get everything set up and ready to go on your designated presentation time. Acknowledge the audience at the start of your presentation. This will get their attention immediately. During the presentation, speak so that everyone can hear you. Enunciate clearly. Don't use or overuse distracting utterances. Use a professional appearance; be aware of gestures, stance, eye contact, and other movements. Keep to the presentation without wandering around topics, and adhere to your previously practiced time limits—be at certain places in the presentation at specific times as you practiced. If you go over your time though, don't panic; adjust by speaking a bit faster to catch up, or skip extraneous examples or repeated points. Stop on time and invite questions for discussion.

Summary

Reporting research results is done in different formats contingent upon the purpose of the research. Research conducted for use in a sport business setting will be reported in a format useful to the business. Research conducted for primary use in academic settings will be reported in a format in accordance with the various academic categories for its dissemination and function.

Typical components of a research report in academia will include an abstract, an introduction to the topic, a summary of the research previously conducted, a review of the research methodology used for the study, a detailed report of the findings of the study, a conceptual discussion of the findings, and conclusions based on the findings of the study. Additionally, a list of references cited and further information provided, such as tables, figures, and appendices, will be included as the last sections of the report.

There are guidelines to be followed for writing the report. One such set of guidelines is the *Publication Manual of the American Psychological Association*—commonly known as the APA writing manual. These guidelines provide information on all aspects of the organization of the research report. The APA writing manual also outlines guidelines on appropriate language in contemporary society, including guidelines and suggestions on how to write with gender-appropriate language.

Study Questions

1. What is the importance of effective communication in reporting research findings?
2. How many different methods can sport business management professionals use to communicate the results of their research?
3. What are the factors affecting the communication process in sport management?
4. What are the different types of research reports? Explain.
5. How many components does a research report typically have and what are they?
6. What are the requirements on the preparation of a poster or oral presentation?
7. What are the general rules in the development and use of tables, figures, and charts? Explain these rules briefly.
8. Differentiate the three most popular writing styles used in communicating research results.

Learning Activity

Select an article from one of the sport management journals, critique the article, and explain what you would do to improve the quality of the study.

References

American Psychological Association. (2010). *Publication manual of the American Psychological Association, Sixth Edition*. Washington, D.C.: American Psychological Association.

American Psychological Association. (2017). *Publication manual of the American Psychological Association, Sixth Edition*. Retrieved from http://www.apa.org/pubs/books/4200066.aspx

Bennett, G., Sagas, M., & Dees, W. (2006). Media preferences of action sports consumers: Differences between Generation X and Y. *Sport Marketing Quarterly, 15*, 40–49.

Dixon, M. A., & Brunening, J. E. (2007). Work-family in coaching I: A top-down perspective. *Journal of Sport Management, 21*(3), 377–407.

Pitts, B. G., & Slattery, J. (2004). An examination of the effects of time on sponsorship awareness levels. *Sport Marketing Quarterly, 13*(1), 43-54.

Quarterman, J., Pitts, B.G., Jackson, E., Jr., Kim, K., & Kim, J. (2005). Statistical data analysis techniques employed in Sport Marketing Quarterly: 1992–2004. *Sport Marketing Quarterly, 14*(4), 227–239.

Quarterman, J., Dupree, A. D., & Willie, K. P. (2006). Challenges confronting intercollegiate female athletic directors of NCAA member institutions by division. *College Student Journal, 40*(3), 528–545.

Section

IV

Statistical Analysis in Sport Management Research

18

Introduction to the Practice of Statistics

After completing the study of this chapter, you should be able to:

- Define statistics.
- Differentiate between descriptive statistics and inferential statistics.
- Differentiate between parametric statistics and nonparametric statistics.
- Understand and differentiate between the dependent and independent variable.
- Understand the nature and scope of hypothesis and hypothesis testing.
- Properly state a null hypothesis.
- Differentiate between the two types of alternative hypotheses.
- Properly apply either type of alternative hypothesis in a research situation.
- Understand the nature and scope of research questions.
- Differentiate between difference questions and relationship questions.
- Understand the concept of statistical significance.
- Understand the difference between Type I error and Type II error.

Sport management researchers use *statistics* to answer research questions and test hypotheses so that they can contribute to a body of knowledge in their specific area or focus of study. The language of research and statistics can be cumbersome when beginners attempt to interpret statistics and statistical data analysis. The intention of this chapter is to introduce several fundamental elements that will aid beginning researchers to interpret statistics better in sport management research.

Nature of Statistics

One of the first tasks for beginning researchers is to understand clearly a working definition of the term statistics. The term has been defined by a variety of scholars in research methods and statistics. One of many examples is the one by Keller, Warreck, and Bartel (1990), who defined statistics as "a body of principles and methods concerned with extracting useful information from a set of numerical data" (p. 2). Another definition is by Kiess (1996) who defined statistics "as methods, techniques, or procedures used to summarize, analyze, and draw inferences from data of a sample" (p. 5). For this book, statistics refer to a body of rules, procedures, and standards directed to extract units of information from a set of numerical data. It is the science of collecting, organizing, and interpreting numerical data to gain understanding of phenomena or situations. Two major statistical dimensions are under the umbrella of statistics: *descriptive statistics* and *inferential statistics*.

Descriptive Statistics and Inferential Statistics

In general, descriptive statistics are procedures used to describe, organize, condense, summarize, and display numerical data for a population or a sample (DePoy & Gitlin, 2005; Gall, Gall, & Borg, 2003; Howell, 1997; Keller, Warrack, & Bartel, 1990; Leedy & Ormrod, 2001; McMillan & Schumacher, 2014; Neuman, 2003; Shiffler & Adams, 1990). Inferential statistics refer to the "use of the laws of probability to make inferences and draw statistical conclusions about populations based on sample data" (Johnson & Christensen, 2014, p. 550). Keller, Warrack, and Bartel (1990) have defined inferential statistics as "a group of statistical data analysis techniques for drawing conclusions or making inferences about characteristics of a population, based on information available in a sample taken from the population" (p. 2). In sum, the ultimate goal of inferential statistics is to make a statement about a population based on information from a sample (Levine, Ramsey, & Berenson, 1995; Shiffler & Adams, 1990).

Parametric Statistics and Nonparametric Statistics

There are two types of tests when it comes to inferential statistics—*parametric* and *nonparametric* statistics. Both parametric and nonparametric statistics allow researchers to make assumptions about random samples taken from populations. Parametric techniques are used for deriving logical conclusions about the characteristics of a population when (1) data taken from the populations are normally distributed, (2) they have equal variances on the variable(s) that is being measured, and (3) the levels of measurements are interval or ratio. Parametric statistical procedures are the most powerful in testing questions and hypotheses about the differences of groups or sets of values.

Unlike parametric techniques, nonparametric techniques are used for deriving conclusions about the characteristics of a sample; however, they do not make assumptions about the population, and the levels of measurements are primarily ordinal or normal (Gall, Gall, & Borg, 2003; McMillan & Schumacher, 2014; Neuman, 2003). Nonparametric statistical procedures are recognized as the most useful in testing questions and hypotheses about relationships and associations of groups or sets of values. Within the classifications of these two types of inferential statistics, parametric statistical tests are (1) the most powerful in testing questions and hypotheses about the differences of groups or set of values; (2) via the use of interval and ratio measurement scales; and (3) primarily via the use of the mean and median. Nonparametric statistical tests are (1) the most powerful in testing questions and hypotheses about relationships and associations of groups or sets of values; (2) via the use of nominal or ordinal measurement scales; and (3) primarily via the use of frequencies, percentages, and proportions. Exhibit 18.1 lists the different types of statistical tests for descriptive, inferential (parametric and nonparametric) statistics.

Exhibit 18.1. Grouping Statistical Analysis Procedures

Descriptive Statistics	Inferential Statistics	
	Parametric	**Nonparametric**
percentages	ANOVA	Chi Square
frequency	regression	Mann-Whitney
mean	MANOVA	Wilcoxon signed rank test
standard deviation	*t*-test	Kruskal-Wallis
range	factor analysis (EFA)	Spearman rank correlation
ranking	Pearson product-moment correlation	Fisher's exact test
median	factor analysis (CFA)	conjoint analysis
mode	discriminant analysis	
	canonical correlation	
	cluster analysis	
	structural equation model	
	Fishers *z* score	
	MANCOVA	
	ANCOVA	
	path analysis	
	trend analysis	

Independent and Dependent Variables

In order to better understand inferential statistical tests, one must understand how to interpret *dependent* and *independent variables* in a study. First, researchers must recognize that the language is different when interpreting independent and dependent variables for experimental research than for nonexperimental research (Kerlinger & Lee, 2000). For example, independent variables are referred to as predictor or treatment variables in correlational studies, and dependent variables are referred to as criterion, effect, or response variables. As shown in Exhibit 18.2, independent variables are easier to identify in experimental studies because they are the variables manipulated by the researcher to see if such manipulation causes change in the dependent (criterion) variable(s).

The variable that is influenced by the treatment variable is the dependent variable. It is the outcome variable, or the one that the researcher is trying to explain. Such a variable is determined by the researcher prior to the research investigation and is dependent on data collection. For example, a researcher may ask what effect different colors of players' uniforms will have on high school men's basketball players' attitude when playing home games. In this case the independent (or treatment) variable is the color of the players' uniforms and the dependent (outcome) variable is the players' attitudes.

Defining the Independent and Dependent Variables for Nonexperimental Designs

In nonexperimental research, the independent (predictor) variable is not manipulated or controlled as in experimental research, but is observed in its natural setting to see if it can be correlated or causally compared with the dependent (criterion) variable (Wimmer & Dominick, 2000). The independent variable describes the characteristic by which the alternative groups being studied are distinguished. Usually, the independent variable is divided into a few discrete groups distinguished from each other in meaningful ways. This is illustrated in Exhibit 18.3, as there is no manipulation of the presumed independent variable (certified and noncertified athletic trainers) on the presumed dependent variable (perceptions of motivational congruencies and discrepancies by the trainers) is inferred or speculated from our observations that the two vary together (are correlated) (Kerlinger & Lee, 2000).

One must keep in mind that dependent variables are not determined by the researcher prior to the research investigation. They are the outcome variables or the ones that the researcher is trying

Exhibit 18.2. Interpretation of Independent and Dependent Variables

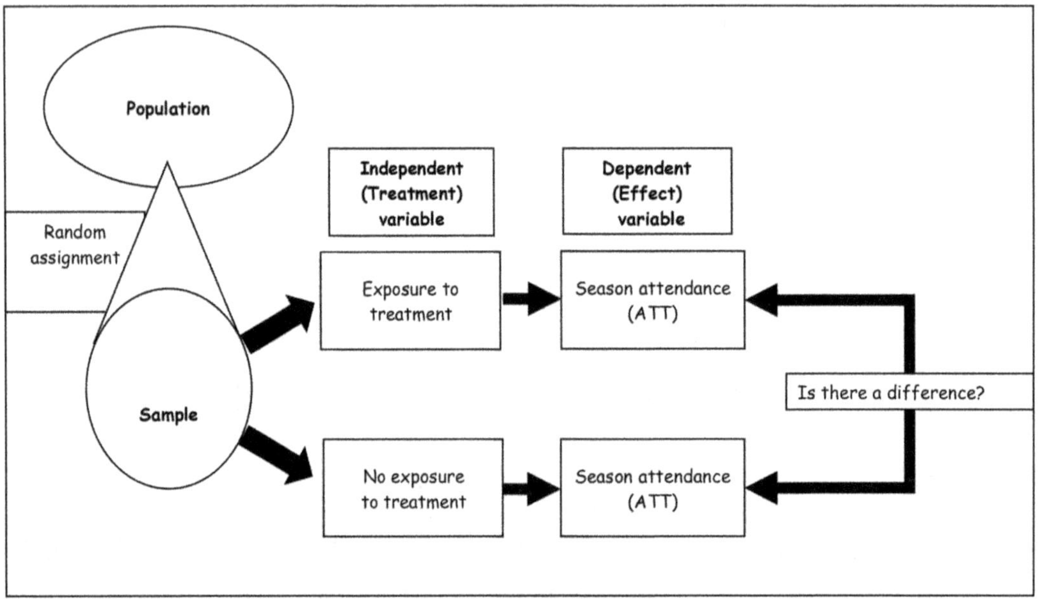

Exhibit 18.3. Independent Variables and Dependent Variables for Nonexperimental Designs

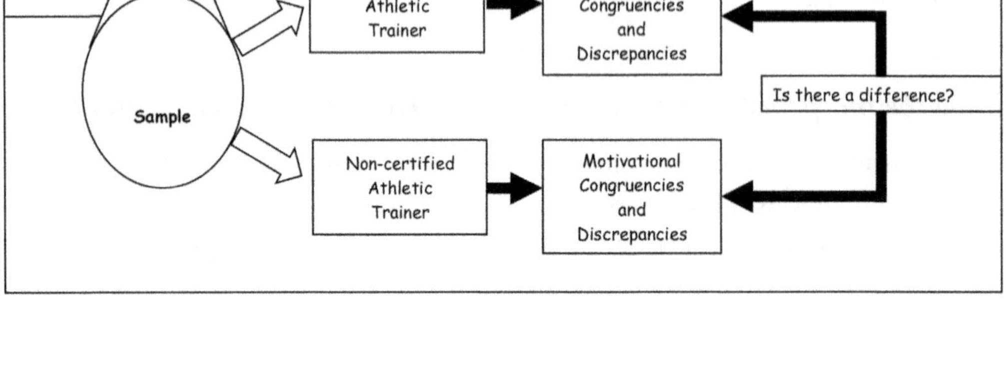

to explain. For example, Buxton, Lankford, and Noda (1992) examined the motivational congruencies and discrepancies (*dependent variables*) between certified athletic trainers and noncertified student trainers (*independent variables*) in the state of Hawaii (p. 320). The dependent variables (motivational congruencies and discrepancies) were not determined by the researcher prior the research investigation. The independent (or predictor) variables (certified status of athletic trainers) have been determined prior to and are independent of the collection of data (The athletic trainer's responses are the key factors in determining the dependent variables).

The Nature and Scope of Hypothesis and Hypothesis Testing

Hypothesis testing is a decision making process of whether or not a research hypothesis should be accepted as true or rejected as false in a statistical fashion (Lomax, 2007). In a way this is much like any other decision involving two possibilities in our lives, such as whether to go to the football game tonight or not.

Generally speaking, a hypothesis is an educated prediction or a logical assumption that is based on prior research or known facts and that can be tested by experimental or nonexperimental studies. Hypotheses are neither proved nor disproved and can only be supported or not supported by statistical analysis. As Leedy and Ormrod (2001) stated, "Hypotheses have nothing to do with proof... Their acceptance or rejection is dependent on what the data—and the data alone—ultimately reveal" (Leedy & Ormrod, 2001, p. 60). That is the function of any statistical test. The two main types of hypotheses testing require the *null hypothesis* and the *alternative hypothesis*, and these two types of hypotheses compete against one another. Alternative hypothesis may take the form of being directional or nondirectional (Bork, 1993; Leedy & Ormrod, 2001; McMillan & Shumacher, 2014).

Conceptually, the term null means without value or without significance. The null hypothesis is a statement of no difference. It is usually symbolized as H_0. This statement tells us that there will be no difference, no relationship, no association, or no pattern between the independent and dependent variables based on the results of statistical testing. If there is a statistical difference or relationship between the variables, the null hypothesis is rejected. When no differences are found between the variables, the null hypothesis is accepted. The following are a few examples of null hypothesis:

- At the $p \leq 0.05$ level of confidence, no relationship exists between organizational commitment and age.
- At the $p \leq 0.05$ level of confidence, no relationship exists between organizational commitment and gender.
- At the $p \leq 0.05$ level of confidence, no relationship exists between organizational commitment and levels of education.
- At the $p \leq 0.05$ level of confidence, no relationship exists between organizational commitment and marital status.
- At the $p \leq 0.05$ level of confidence, no relationship exists between organizational commitment and length of service.

The alternative hypothesis (H_a), or research hypothesis, generally states that there will be a statistically significant difference between the dependent and independent variables, but it does not state the magnitude of the differences. As with the null hypothesis, it is subjected to statistical analysis. Unlike the null hypothesis, it states that a statistically significant difference, pattern, or relationship exists between the independent and dependent variables, and it identifies under what conditions the differences or relationships will occur. It is the statement of hypothesis that represents the expectation of the researcher.

Alternative hypotheses can be formulated in two ways: directional and nondirectional. The directional hypothesis (H_d) not only states that there will be a statistically significant difference but also identifies where or under what conditions that difference will occur. Directional hypotheses predict differences or relationships between independent variables and dependent variables and specify the directions of such differences and relationships. For an example of a directional hypotheses, we can state: certified athletic trainers will rate the factors that motivate them in the workplace HIGHER than would the noncertified student athletic trainers. The study conducted by Kwon & Armstrong (2002) provides a good illustration of directional hypotheses. The following are the four directional hypotheses they used to examine factors influencing impulse buying of sport team licensed merchandise (p. 154–155):

- Hypothesis 1: The higher a consumer's level of shopping enjoyment, the greater the likelihood of the consumer making an impulse purchase of sport team licensed merchandise.
- Hypothesis 2: The higher the level of identification a consumer has with the respective sport team, the greater the likelihood of the consumer making an impulse purchase of the team's merchandise.
- Hypothesis 3: The more time available to sport consumers in shopping situations, the greater the likelihood of the consumer making an impulse purchase of sport team licensed merchandise.
- Hypothesis 4: The more money available to sport consumers in shopping situations, the greater the likelihood of the consumers making an impulse purchase of sport team licensed merchandise.

Like directional hypotheses, nondirectional hypotheses predict differences or relationships between the independent variables and the dependent variables; however, they do not specify the directions of such differences or relationships. As an example of nondirectional hypothesis, we can posit: there will be significant differences (or relationships) between certified athletic trainers and noncertified student athletic trainers when rating the factors that motivate them in the workplace.

In the study of Triola & Franklin (1994), the authors developed six nondirectional hypotheses:

- There are significant relationships between male and female athletic directors of NCAA Division II schools when ranking occupational affective commitment.
- There are no significant relationships between male and female athletic directors of NCAA Division II schools when ranking occupational affective commitment.
- There are significant relationships between male and female athletic directors of NCAA Division II schools when ranking occupational normative commitment.
- There are no significant relationships between male and female athletic directors of NCAA Division II schools when ranking occupational normative commitment.
- There are significant relationships between male and female athletic directors of NCAA Division II schools when ranking occupational continuance commitment.
- There are no significant relationships between male and female athletic directors of NCAA Division II schools when ranking occupational continuance commitment.

The Nature and Scope of Research Questions

The use of research questions is common where there is an absence of literature or related research studies in a particular area of interest to the researcher. One must understand the nature of research questions to better understand inferential statistical techniques. There are two types of research questions written relevant to inferential statistics: difference questions and relationship questions (McMillan & Schumacher, 2014).

Difference Questions

Difference questions are typically used in the measurements for comparing differences between two or more groups. A difference question asks, are there differences between two or more sets of data? In the example below, there are three research questions based on differences (Triola & Franklin, 1994):

- Are there significant differences between male and female athletic directors of NCAA Division II schools when rating occupational affective commitment?
- Are there significant differences between male and female athletic directors of NCAA Division II schools when rating occupational normative commitment?
- Are there significant differences between male and female athletic directors of NCAA Division II schools when rating occupational continuance commitment?

Relationship Questions

Unlike differences questions, relationship questions are raised in studies that assess the degree of relationship or association between two or more groups. The typical question is, are there significant relations or associations between two or more variables for a group of subjects? A mathematical equation is used to show how one variable is related to one or more variables and calculated through correlation coefficients. In the example below, there are three relationship-related research questions (Triola & Franklin, 1994):

- Are there significant relationships between male and female athletic directors of NCAA Division II schools when ranking occupational affective commitment?
- Are there significant relationships between male and female athletic directors of NCAA Division II schools when ranking occupational normative commitment?
- Are there significant relationships between male and female athletic directors of NCAA Division II schools when ranking occupational continuance commitment?

The Concept of Statistical Significance

Researchers should be aware that sampling is not perfect in that researchers can never select a sample of subjects that exactly matches all of those in the profile of a population. But most of the time, it is hard and impractical for researchers to collect information from the whole population due to its large size. *Statistical significance* is a term used to refer to a result that is extremely unlikely to have occurred by chance. Statistical significance helps us to determine how confident researchers are that their results are real enough not to occur by chance alone. It means that the results are not likely to be due to chance factors while indicating the probability of finding a relationship in the sample when there is none in the population.

Statistical significance estimates the probability (p calculated) of sample results from those specified by the null hypothesis for population, given the sample size (Cohen, 1992).

Statistical significance tests do not tell us what we want to know nor do they evaluate whether or not our results are important. They tell us only whether or not the results of a study were due to chance. After collecting all data, the researcher is required to state beforehand how willing the researcher is to be wrong when declaring that one group of variables is really different from another group of variables being measured. The level of significance values is symbolized as p. The p stands for probability. By rule of thumb or convention, the two most frequently used levels of significance are 0.05 and 0.01.

With a 0.05 significance level, the researcher accepts the risk that out of 100 samples, a true null hypothesis would be wrongly rejected five times. This means that the researcher feels safe in concluding that two variables are different or related when a statistical analysis shows that there is less than a 5% probability that such findings were due to chance or a fluke. Another way of saying this is that the results are significant at 0.05 level or the point of fine level. A p value of 0.05 means that the probability of the findings being caused by chance alone is 5 in 100 (McMillian & Schumacher, 2014). There is nothing magical about the 0.05 level. Other levels can be used such as 0.01, 0.005, or 0.001. Yet, it is not conventional to use a higher level than 0.05, such as 0.1.

If researchers want to have a stricter criterion, that is, they want to be even more assured that the results have not occurred by chance, then they can set the significance level at 0.01, which is a 1 in 100 chance of having occurred by chance. If the researchers wanted to be more certain that the results of their studies did not occur merely by chance, they would select a more stringent level of statistical significance.

In sport management research, p is evaluated using two levels of credence or credibility: the 5% (0.05) level or the 1% (0.01) level. The 5% (0.05) is the most commonly used level of chance employed in sport management research.

Generally, researchers report the probability level at which they were able to reject the null hypothesis. For example, if the level of significance was 0.05, the researcher rejects the null hypothesis at a probability that is less than 0.05. Instead of writing this out in words, the researcher uses the equation $p < 0.05$ (p meaning probability, the symbol $<$ meaning less than).

The level of significance is a probability that tells the researchers how unlikely the sample data must be before they can reject the null hypothesis.

Types of Decision Errors

There are two different types of decision errors that might be made in the decision-making process. Both types of errors are defined in reference to the null hypothesis. A *Type I error* is committed when the researcher rejects the null hypothesis when in fact it is true. A *Type II error* is committed when the researcher accepts the null hypothesis when in fact it is false. Exhibit 18.4 presents the statistical decision table showing the relationship between a statistical decision and the risk of committing an error.

Exhibits 18.5 and 18.6 show a graphic representation and numerical summary of the statistical data analysis methods used in three sport management journals—*Journal of Sport Management,*

Exhibit 18.4 Statistical Decision Making: Type I and Type II Errors

		Null Hypothesis	
		True	False
Decision	Accept the null	Type A Correct decision (1-α)	Type II error (false acceptance, α)
	Reject the null	Type I error (false rejection, β)	Type B (power) Correct decision (1-β)

Source: Cozby, 1997; DePoy & Gitlin, 2005; Leary, 1991.

Exhibit 18.5. A Summary of the Statistical Data Analysis Methods Used in *JSM, SMQ,* and *IJSM,* 1987–2004

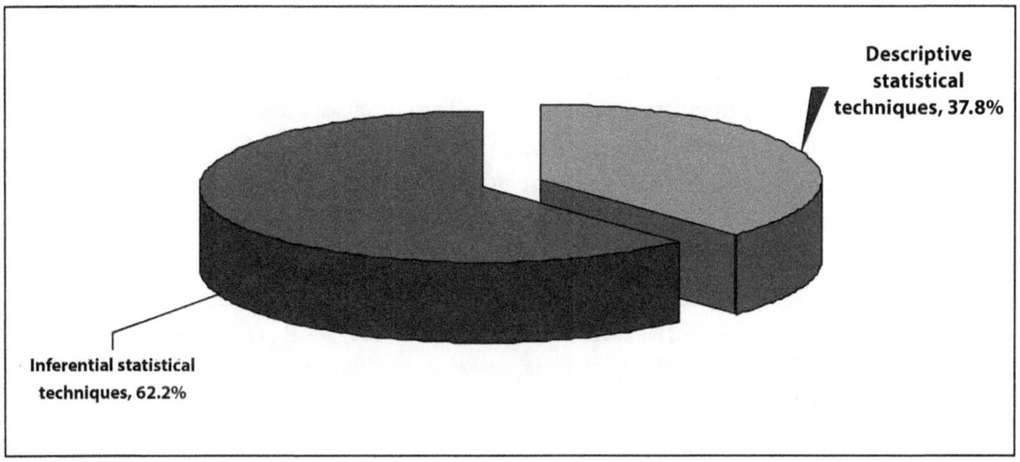

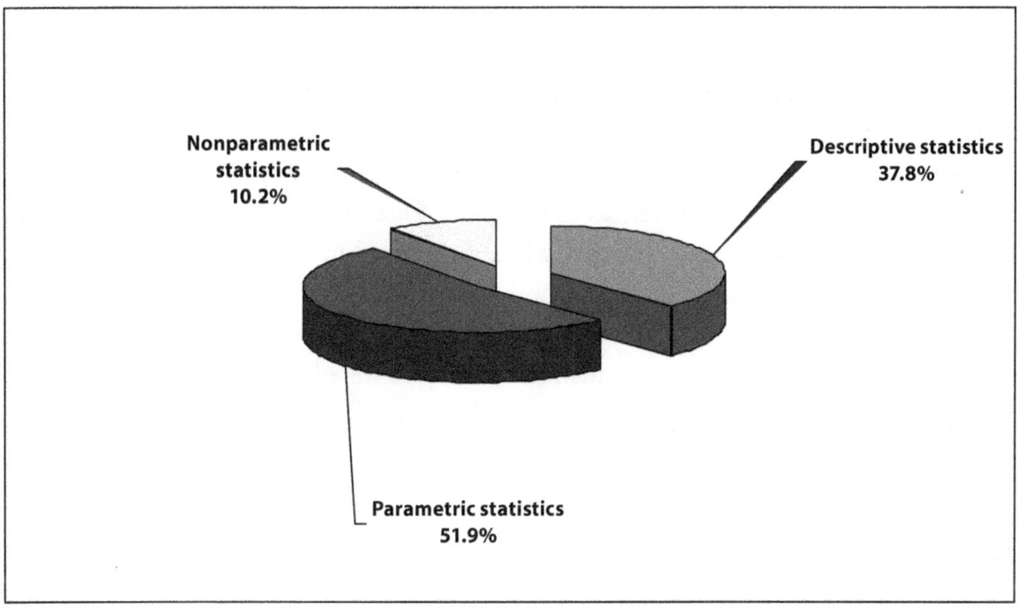

Source. Quarterman, J., Jackson, E. N., Kim, K., Yoo, E., Koo, G. Y., Prugger, B., & Han, K. (2006). Statistical data analysis techniques employed in the *Journal of Sport Management* January 1987 to October 2004. *International Journal of Sport Management*, 7(1), 13–30.

Exhibit 18.6. A Numerical Summary of the Types of Statistical Data Analysis Techniques Used in *JSM, SMQ,* and *IJSM* 1987–2006

Technique	Operational Definition	N	%
Descriptive statistics	Techniques for describing, organizing, and summarizing numerical (mean, mode, median, variance, standard deviation, range, percentage, frequency, and rank)	381	37.8
Parametric statistics	Techniques for deriving logical conclusions about the characteristics of a population from corresponding characteristics of a sample or the population itself when the level of measurement of data is primarily interval or ratio (*t* test, ANOVA, MANOVA, ANCOVA, MANCOVA, exploratory factor analysis, confirmatory factor analysis, the Pearson product-moment correlation, regression analysis, trend analysis, discriminant analysis, cluster analysis, canonical correlation, path analysis, structural equation model, and Fisher's *Z* score)	523	51.9
Nonparametric statistics	Techniques for deriving logical conclusions about the characteristics of a population from corresponding characteristics of a sample or the population itself when the level of measurement of data is primarily nominal or ordinal scale (Chi-square, Mann-Whitney, Kruskal-Wallis, conjoint analysis, Wilcoxon's Sign Rank test, Fisher's Exact test, and Spearman rank correlation).	103	10.2
Total		1007	100.0

Source. Quarterman, J., Jackson, E. N., Kim, K., Yoo, E., Koo, G. Y., Prugger, B., & Han, K. (2006). Statistical data analysis techniques employed in the *Journal of Sport Management* January 1987 to October 2004. *International Journal of Sport Management,* 7(1), 13-30.

Note. The categories of statistical analysis were adapted from textbooks by Gall, Gall, & Borg, 2003; Howell, 1999; McMillan & Schumacher, 2001; Neuman, 2003.

Sport Marketing Quarterly, and *International Journal of Sport Management.* Among three statistical techniques (descriptive, parametric, nonparametric), parametric statistical techniques were most frequently used, accounting for slightly more than half of the techniques primary to the research purpose, questions, and/or hypotheses. In this study of Quarterman, Pitts, Jackson, Kim, & Kim (2005), a technique was considered primary when it was used to directly answer a study research purpose, questions, and/or hypotheses. Descriptive statistics were the second most frequently used techniques, accounting for more than one third, or 37.8%, of the techniques primary to the questions and hypotheses. Nonparametric statistical techniques were least frequently coded (10.2%) as primary to the research purpose, questions, and/or hypothesis.

Summary

The introduction of critical concepts and issues in statistics is the focal point of this chapter. The concepts of descriptive statistics and inferential statistics, parametric statistics and nonparametric statistics, and dependent variables and independent variables were reviewed and compared.

The nature and scope of hypothesis testing is another central topic. The two main types of hypotheses testing are the null hypothesis and the alternative hypothesis. The null hypothesis is a statement of no difference and it is tested statistically. On the other hand, the alternative hypothesis generally states that there will be a statistically significant difference between the dependent and independent variables, but it does not state the magnitude of the differences. There are two ways that alternative hypotheses can be formulated, either as a directional hypothesis or a nondirectional hypothesis.

The third topic of discussion in this chapter is the concept of statistical significance. There are two types of research questions written in relevance to inferential statistics: difference questions and relationship questions. Difference questions are typically used in the measurements for comparing differences between two or more groups, whereas relationship questions are raised in studies that assess the degree of relationship or association between two or more groups.

It is an important statistical procedure that tests operate under the assumption that the null hypothesis exactly describes the population and then test the sample's probability. This chapter also gives a brief overview of the Type I and Type II errors that could be committed in research. Sport management researchers carry the responsibility of deciding which error is more acceptable in their studies.

Study Questions

1. What does the term statistics mean?
2. What are the differences between descriptive statistics and inferential statistics?
3. What are the differences between parametric statistics and nonparametric statistics?
4. What are the differences between the dependent variable and the independent variable?
5. Properly state a null hypothesis.
6. What are the differences between the two types of alternative hypotheses?
7. Properly state an alternative hypothesis in a research situation.
8. What are the differences between the difference questions and the relationship questions?
9. Explain the concept of statistical significance.
10. What are the differences between Type I errors and Type II errors?
11. How can researchers reduce the probability of committing Type I errors and Type II errors?

Learning Activities

1. Identify an article published this year in one of the sport management related journals that used a parametric test procedure and give your justification of why the author used the parametric test procedure instead of a nonparametric one.
2. Identify a topic of interest in sport management in which you want to pursue a study.
   a. Draft a null hypothesis.
   b. Draft a directional alternative hypothesis.
   c. Draft a nondirectional alternative hypothesis.
   d. Draft a research question using the format of difference questioning.
   e. Draft a research question using the format of relationship questioning.

References

Bork, C. E. (1993). *Research in physical therapy.* Philadelphia, PA: J. B. Lippincott Company.

Buxton, B. A., Lankford, S. V., & Noda, J. E. (1992). Motivational congruency between certified athletic trainers and noncertified student athletic trainers in the State of Hawaii. *Journal of Athletic Training, 27*(4), 20, 22, 24.

Cohen, J. (1992). A power primer. *Psychological Bulletin, 112*, 115–159.

DePoy, E., & Gitlin, L. N. (2005). *Introduction to research; Understanding and applying multiple strategies* (3rd ed.). St. Louis, MO: Elsevier.

Gall M. D., Gall J. P., & Borg W. R. (2003). *Educational research: An introduction* (7th ed.). Boston, MA: Allyn and Bacon.

Howell, D. C. (1997). *Statistical methods for psychology* (4th ed.). Belmont, CA: Duxbury Press.

Keller, G., Warrack, B., Bartel, H. (1990). *Statistics for management and economics* (2nd ed.). Belmont, CA: Wadsworth.

Kerlinger, F. D., & Lee, H. B. (2000). *Foundations of behavioral research* (4th ed.). Fort Worth, TX: Harcourt.

Kiess, H. O. (1996). *Statistical concepts for the behavioral* sciences (2nd ed.). New York, NY: Pearson.

Kwon, H. H., & Armstrong, K. L. (2002). Factors influencing impulse buying of sport team licensed merchandise. *Sport Marketing Quarterly 11*(3), 151–162.

Leary, M. (1991). *Behavioral research methods.* NY: Wadsworth.

Leedy, P. D., & Ormrod, J. E. (2001). *Practical research; Planning and design* (7th ed.). Upper Saddle River, NJ: Prentice Hall, Inc.

Levine, D. M., Ramsey, P. R., & Berenson, M. L. (1995). *Business statistics for quality and productivity.* Englewood Cliffs, NJ: Prentice Hall Inc.

Lomax, R. (2007). *An introduction to statistical concepts* (2nd ed.). New York, NY: Routledge.

McMillan, J. H., & Schumacher, S. (2014). *Research in education; Evidence-based inquiry* (7th ed.). London, UK: Pearson.

Neuman, W. L. (2003). *Social research methods: Qualitative and quantitative approaches* (5th ed.). Boston, MA: Allyn and Bacon.

Quarterman, J., Pitts, B. G., Jackson, E. N., Kim, K., & Kim, J. (2005). Statistical data analysis techniques employed in the *Sport Marketing Quarterly* January 1982 to October 2004. *Sport Marketing Quarterly, 14*(1), 158–167.

Shiffler, R. E., & Adams, A. J. (1990). *Introductory business statistics with microcomputer applications.* Boston, MA: PWS-Kent Publishing Company.

Triola, M. F., & Franklin, L. A. (1994). *Business statistics.* Reading, MA: Addison-Wesley Publishing Co.

Wimmer, R. D., & Dominick, J. R. (2000). *Mass media research: An introduction* (6th ed.). Wadsworth Publishing Company: Belmont, CA.

19

Descriptive Statistics and Correlation

After completing the study of this chapter, you should be able to:

- Define the measures of central tendency (i.e., mean, median, mode), and measures of variability (i.e., range, variance, standard deviation), and measures of relative proportions (i.e., percentiles, percentile ranges, quartiles).
- Describe the characteristics and uses of frequency polygons, histograms and bar diagrams, and be able to interpret them.
- Recognize distributions by their shapes and name them.
- Define correlation.
- Interpret correlation coefficient.
- Determine when Pearson and Spearman correlation procedures are to be utilized.

The purpose of this chapter is to introduce the general nature of descriptive statistics and the ways in which such techniques are used to organize, describe, and summarize data. Specifically, this chapter covers relevant terms such as mean, median, and mode (measures of *central tendency*), range, variance, and standard deviation (measures of *variability*), and percentiles, percentile ranges, and quartiles (measures of *relative proportions*). The graphic presentation of descriptive data, including frequency polygons, histograms and bar diagrams, is also reviewed in this chapter.

Major Approaches for Presenting Descriptive Statistical Data

Contemporary scholars (DePoy & Gitlin, 1994; Gall, Gall, & Borg, 2003; Hunter & Brown, 1991; McMillian & Schumacher, 2014; Neuman, 2003) have identified six distinct categories of descriptive statistical data analysis techniques that are applicable to sport management research: (1) describing statistical data as *frequency distributions*, (2) describing statistical data by graphic presentations, (3) describing statistical data as measures of central tendency, (4) describing statistical data as measures of variability, (5) describing statistical data as measures of relative proportions, and (6) measuring statistical data by *correlations*.

A major reason for calculating statistics is to describe and summarize a set of data. A collection of numbers, the raw data, is not very informative, so the researcher needs to find ways of picking out and presenting the key information in a clear, concise, and comprehensible format. The following section introduces various methods used to present data and to summarize data.

Describing Statistical Data by Frequency Distributions

Researchers can make sense of a large amount of numerical data by formulating frequency distributions. A frequency distribution is used to group and display quantitative data. It is an initial way of arranging data so that we can know how often a particular score or unit of information occurs. Frequency distributions are usually presented as tables or graphs that present the number of times (frequency) different values of a variable occur in a group of observations or scores.

For an analysis of a large set of scores, such as those listed in Exhibit 19.1, there is a need to organize and summarize the data within tables and graphs. Let's assume that the table shows the list of ungrouped data for the ages of 100 high school girls' basketball coaches in the United States. The age obtained from each of the coaches has been recorded alphabetically. Let us assume the State High School Activities Association wants to analyze the demographic characteristics of its high school girls' basketball coaches. The sample of 100 coaches was selected and the age at nearest birthday of each coach was collected from the Board of Education office. This initial list of scores for each of the coaches is referred to as raw data.

As shown in Exhibit 19.1, it is difficult to determine and summarize the raw data as collected. How would you go about extracting meaningful information from the data for the Association? One simple way to gain some order is simply to list the scores in a descending or ascending order. This technique is called an *ordered array,* and the result is given in Exhibit 19.2. When the data is sorted into an ordered array, it is easier to locate the extremes. According to the ordered array, the oldest coach is 63 and the youngest coach is 23. This information is not easily seen in the raw data in Exhibit 19.1. Although this technique has helped us understand how the scores were distributed,

Exhibit 19.1. Ages of 100 Interscholastic Head Coaches of Female Basketball Teams

39	37	32	52	46	35	25	28	33	33
30	63	23	30	29	26	29	41	40	32
58	26	24	55	51	28	28	40	44	38
35	25	42	34	29	43	41	31	30	36
43	34	35	37	33	38	34	34	33	34
32	32	30	31	25	49	31	26	33	36
43	25	27	35	53	25	38	33	37	33
23	51	57	23	26	36	39	31	35	34
34	51	40	50	35	45	28	36	32	39
26	48	27	45	45	25	25	30	36	60

Exhibit 19.2. An Ordered Array of the Ages of 100 Interscholastic Head Coaches of Female Basketball Teams

63	50	43	38	36	34	33	30	28	25
60	49	42	38	35	34	32	30	28	25
58	48	41	38	35	34	32	30	27	25
57	46	41	37	35	34	32	30	27	25
55	45	40	37	35	33	32	30	26	25
53	45	40	37	35	33	32	29	26	25
52	45	40	36	35	33	31	29	26	24
51	44	39	36	34	33	31	29	26	23
51	43	39	36	34	33	31	28	26	23
51	43	39	36	34	33	31	28	25	23

it is still quite cumbersome because of the large number of scores. In this case, the most common way to consolidate the data is to create a frequency distribution table.

A frequency distribution table refers to a summary table in which the data are arranged into conveniently established numerical ordered class groupings or categories (Levine, Berensen, & Stephan, 1999). The ordered array of data (Exhibit 19.2) will be used to illustrate the construction of a frequency distribution table. To construct a frequency distribution table, a four-step process is recommended (Daniel & Terrell, 1995; Groebner, Shannon, Fry, & Smith, 2007):

1. Determine the number of groups or classes. There is no binding rule for grouping the number of class intervals. However, according to a rule of thumb, a frequency distribution table should have at least five class intervals and no more than 15. When selecting the number of intervals, the researcher must consider the fact that having too few intervals will result in the loss of too much specific information about the scores. On the other hand, having too many class intervals prevents us from condensing the original data enough to make clear the information they contain. In this example, nine classes were formulated.
2. Establish the width of each class grouping (class intervals). Given the class width equation of:

$$W = \frac{\text{Largest value} - \text{Smallest value}}{\text{Number of classes}}$$

   In applying this formula for the 100 coaches, we discover that the highest number is 63 and the lowest one is 23. Since there are nine classes, we then could generate nine classes that included five years each.
3. Determine the class boundaries for each class interval. The class boundary refers to the lower and upper values of each class interval. Class boundaries for the first interval begin with 20 and end with 24; for the last interval, they begin with 60 and end with 64.
4. Count the number of values in each of the class intervals. From the ordered array in Exhibit 19.2, or the raw data in Exhibit 19.1, at this point we count the number of coaches in each class. Shown in Exhibit 19.3 is a frequency table that uses the data already presented in Exhibit 19.2. When observing the frequency table, one can get a better feel for the distribution of the ages of the coaches.

Exhibit 19.3. A Frequency Distribution of the Ages of 100 Interscholastic Head Coaches of Female Basketball Teams

Age Intervals	Tally	Frequency
20–24	1111	4
25–29	111111111111111111111	21
30–34	1111111111111111111111111111	28
35–39	11111111111111111111	20
40–44	1111111111	10
45–49	111111	6
50–54	111111	6
55–59	111	3
60–64	11	2

Describing Data by Graphic Presentations

Graphic presentations are unique ways of providing a picture of the differences that exist in the statistical data being presented. At times it may be difficult to conceive the overall meaning of frequency distribution tables; however, a picture can better communicate the overall meaning of the data being presented. Typically, bar graphs, histograms, frequency polygons, and pie charts are the four types of graphic methods frequently used by researchers (Shao, 2002; Cozby, 2001). Histograms are simply extensions of the frequency distributions previously discussed and are used with continuous rather than categorical data; that is, usually they are used with interval and ratio-type data.

Bar Graph

A bar graph, also known as a bar chart, is used to give visual presentation of data when the scale of measurement is nominal. Bar graphs (which use horizontal or vertical rectangles) are often used to display categorical data where there is no emphasis on the percentage of a total represented by each category. The scale of measurement is nominal or ordinal (Aczel, 1995). The units of measure for the bar graphs are arranged from lowest to highest. The bar graph differs from the histogram in that it is the height of the bar that indicates the value, not the area of the bar. Bar charts or graphs are used for discrete variables. They can have either a vertical or horizontal orientation with a small space between the bars. The traditional form of a bar graph is a series of disjointed or separated rectangles; each rectangle represents a category of a qualitative variable, and the height of each rectangle corresponds to the frequency of such category (See Exhibits 19.4 and 19.5 for examples).

Dixon (2002) examined gender differences in perceptions and attitudes toward the Ladies Professional Golf Association (LPGA) and its tour professionals. As presented in Exhibits 19.4 and 19.5, she used a bar graph to summarize the perceptions of the LPGA Corporate sponsors between males and females.

Exhibit 19.4. Spectator Interest in the LPGA

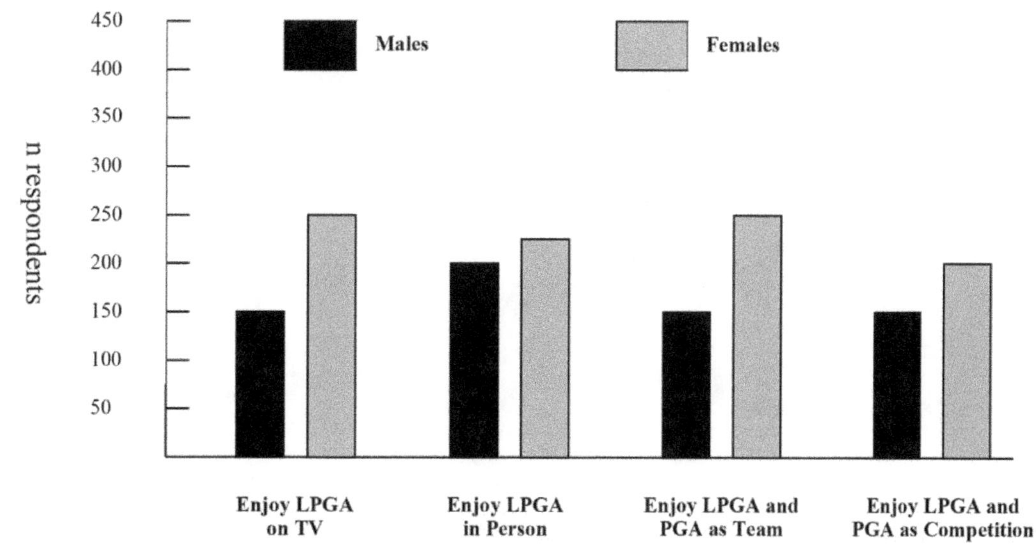

Exhibit 19.5. Perceptions of LPGA Corporate Sponsors

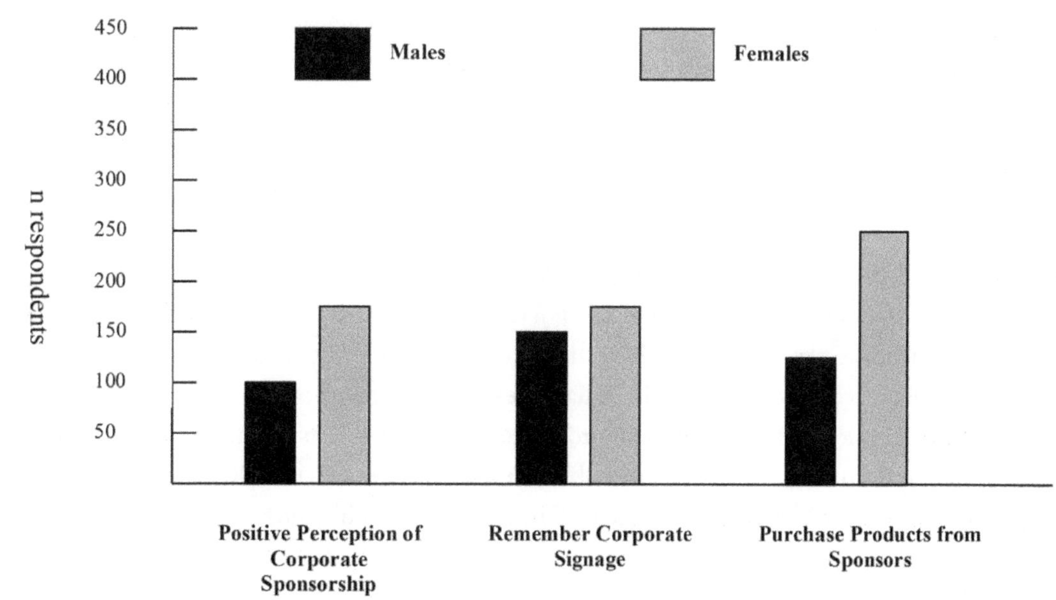

Histogram

A histogram is a graphical display of ordinal, interval, or ratio scales of measurement (Aczel, 1995). Unlike a bar graph, the bars in a histogram touch each other. When analyzing ordinal data, the bars of a histogram are all equal in width; however, when analyzing interval or ratio measurements, the bars may or may not be the same width. When grouped frequencies of unequal intervals are displayed, the bars must be constructed to show the different widths to correspond to the different sizes of the intervals.

Exhibit 19.6. Frequency Polygon with Ends of Polygon Open

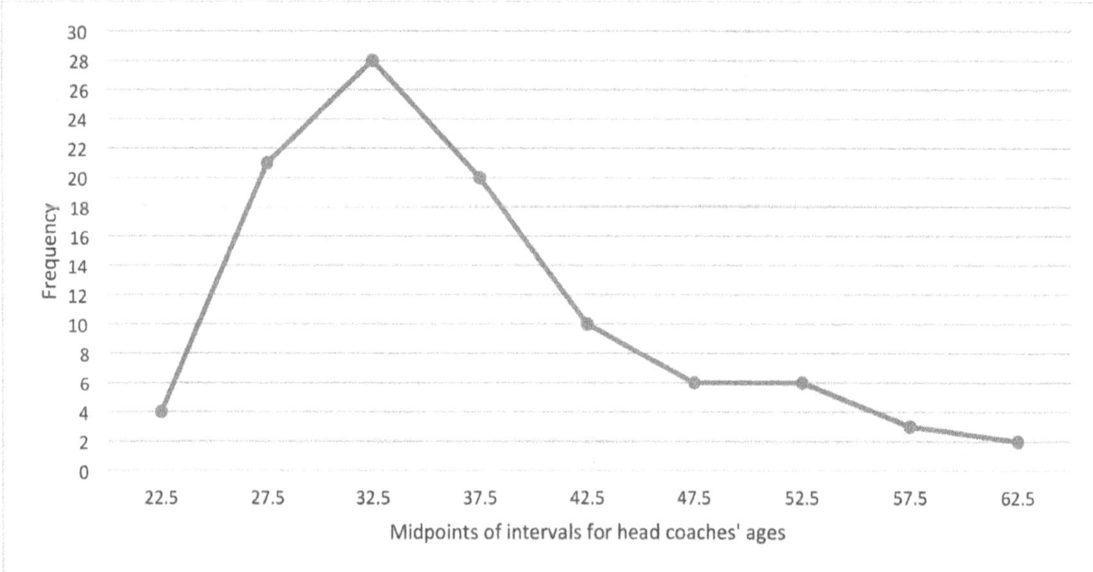

Frequency Polygon

The frequency polygon refers to a type of line graph that shows frequency distributions. It can be created with interval or ratio data (Aczel, 1995; Cozby, 1997). Using this procedure, the bars in a histogram are removed and a line is used to represent the frequencies (known as a line graph). The line is drawn at each end of the distribution of scores or values to connect it to the horizontal axis. Exhibit 19.6 is an illustration of a frequency polygon.

Pie Chart

A pie chart consists of a circle representing 100% of the total data, with each slice of pie being proportional to the percentage in a particular category. Thus, a pie chart can convey the information in a table at a glance. Pie charts are useful for displaying nominal scale information. It is recognized as the most illustrative way for displaying quantities as percentages of a given total (Aczel, 1995). Usually the raw frequencies are converted into percentages. Pie charts are useful as long as they are clearly labeled and limited to an easily readable number of slices (Shao, 2002). (For an example of the applications of a pie chart, see Exhibit 18.5 in the previous chapter.)

Describing Data by Measures of Central Tendency

The measures of central tendency refer to measures that describe the center of a frequency distribution of observations (Field, 2013). The most common measures of central tendency are the mean, median, and mode. Among these measures, the mean is most commonly known to the lay person and is the most preferred measure of central tendency because it includes all of the scores. However, when there are extreme scores in the set of observations, the mean score will be affected. In that case, the median would be preferred.

Mean

The mean is the most familiar and useful measure used to describe the central tendency or average of a distribution of scores for any set of data. It is calculated by adding the values of all scores in a dataset and then dividing by the number of scores in the set. The mean is distorted when there are extreme total values at either the high or low end of a scale: it is more accurate when the extremes are not widely separated.

Median

The median is the number that lies at the midpoint of the distribution of scores and divides the distribution of scores into two equal halves. Medians are the most representative measurement for ordinal data and skewed distributions. It is often referred to as the 50th percentile because it is the score that divides the group of scores in half (50% of the scores are below the median and 50% of the scores are above the median). Placing the scores in order is a more manageable way to locate the median. For an odd number of scores, the median is the middle score in the set of scores; for an even number of scores, the median is the mean of the two middle scores. Unlike the mean, the median is not distorted by extreme values. When presented in a journal article, median is usually illustrated by the word itself or by the symbol *Mdn*.

Mode

The mode is the most frequently occurring score(s) in a distribution of scores. There may be no mode if none of the scores appear more often than any other. There may also be bimodal (having two modes), trimodal (having three modes), or multimodal (having four or more modes).

Sport management researchers frequently use these three measures of central tendency to report their data. Armstrong-Doherty (1995) used the mean, standard deviation, median, and range to describe the profile of funding practices in Canadian interuniversity athletics and to examine any changes from previous reports, particularly in light of recent financial constraints.

Pitts and Ayers (2001) conducted an analysis of visitor spending in Amsterdam from the Gay Games V in 1998. They used the median to describe the average daily spending in US dollars per visitor and to describe the total spending by visiting participants and spectators.

Describing Data by Measures of Variability (also called Measures of Dispersion)

Unlike the measures of central tendency, measures of variability characterize the amount of spread in a distribution of scores. The three most common measures of variability include the range, variance, and standard deviation.

Range

The range of a set of scores is the difference between the highest and lowest scores in a distribution of scores. When presented in a journal article, it is usually illustrated by the word itself or by the symbol *R*. Parks and Bartley (1996) utilized range to examine "gender and salary ranges of the sport management professoriate" (p. 126).

Variance

The variance is equal to the square of the standard deviation. The variance is the arithmetic average of the squared differences between the values and the mean in the data collected.

Standard Deviation

The standard deviation indicates the average by which scores deviate from the mean. It is commonly symbolized as *SD*, σ or s. In general, the standard deviation for the population is denoted by σ whereas the standard deviation for the sample is denoted by s. Since the variance is the square of the standard deviation, the variance for the population is denoted by σ^2 whereas the variance for the sample is denoted by s^2. Because of its close links with the mean, standard deviation can be greatly affected if the mean gives a poor measure of central tendency.

Describing Data by Measures of Relative Proportions

The measures of relative proportions describe the spread of scores based on relative positions. Percentiles, percentile ranges, and quartiles are the most common ways.

Percentile

A percentile is defined as "a score below which a certain percentage of the distribution lies" (Lomax, p. 29). Percentile is a descriptive statistic that is used to tell the relative position of a given score in a distribution of scores. For example, if a student's test score was at the 80th percentile in a class, it means that 80% of the other students in the class had lower scores, or 20% of the students had higher scores. The study conducted by Colyer (2000) is an example from the sport management literature. In the study, the researcher investigated the organizational culture of a selected group of state sport organizations in Western Australia. The results were presented with the use of percentile scores to compare a gender ratio of sport employees and volunteers in the sport organizations.

Percentile Range

Percentile range is a descriptive statistic that is used to describe the interquartile range (IQR) of values extending from 25th to 75th percentile. The 50th percentile means a median. The 1st quartile is the middle value of all scores below median and the 3rd quartile is some distance above the mean. The IQR can be statistically calculated by getting the difference between the 75th and 25th percentiles.

Drain and Ashley (2000) utilized interquartile range (IQR) to determine the degree of consensus for critical issues facing intercollegiate athletics at the NCAA Division I-A level in the following 15 years that may lead to reforms in intercollegiate athletics. The Delphi Technique was used to derive the consensus of the panel.

> *The IQR was a vital statistic in determining consensus. Because the median was used as the measurement of central tendency, the IQR was a logical choice to determine consensus. The IQR can be calculated by determining the difference between the 75th and the 25th percentiles. It was deemed that an IQR score of one or less would show consensus, meaning that at the minimum, 50% of the answers were within one point of each other. (p. 80)*

The results showed that the 13 Division I-A athletic directors who participated in the panel forecasted 31 critical issues facing Division I-A athletics in the next 15 years. "These 31 issues fell into eight broad categories: funding issues, NCAA, bowl games versus play-offs, amateurism, academics, gender equity, facilities, and student-athlete issues" (Drain & Ashley, 2000, p. 80).

Measuring Relationships of Data by Correlational Statistical Techniques

The term correlation is used as a synonym for relationship and association. Unlike descriptive research, correlational statistical techniques are used to measure relationships between two variables. A correlation is a numerical measure of the strength of the relationship between two or more things. For example, a researcher may use this technique to determine how years of work experience and organizational loyalty among a group of workers in sport organizations are related. Researchers can also use the techniques to determine how well one variable can be used to predict the other variable. For instance, selection committee members use grade point averages and scores of standardized examinations, such as the Graduate Record Examination (GRE), to predict how successful a candidate will be in completing the academic requirements for earning a graduate degree in a sport management program.

A positive correlation means that high values of one variable are associated with high values of the other, and that low values of one variable are associated with low values of the other. A positive correlation exhibits a direct relationship between two variables. For example, if we measured the GRE scores of students entering a master's degree in sport management and found that the students graduating from the program with the highest grade point averages (GPAs) also had the higher GRE scores when entering the program, this would lead us to believe there is a positive correlation between high GRE scores and high GPAs for this group of students. In contrast, a negative correlation exhibits an inverse relationship between two variables. Considering the same example, if students entering the program with high GRE scores in fact graduated with lower GPAs, this would illustrate a systematic negative correlation since high values of one set of variables (high GRE scores) are paired with the other set of variables (low GPAs upon graduation). Zero correlation occurs when no systematic relationship exists between the two sets of variables (Cozby, 2001; Salkind 2014). If there is no relationship between two variables, then they are uncorrelated. For example, shoe sizes and IQ scores are probably uncorrelated.

Correlation Coefficients

A *correlation coefficient* is "a numerical index that provides information about the strength and direction of the relationship between two variables" (Johnson & Christense, 2014).

Exhibit 19.7. Five Graphs Representing Hypothetical Correlation Coefficient

High +0.90	Medium +0.50	Zero 0.0	Medium -0.50	High -0.90
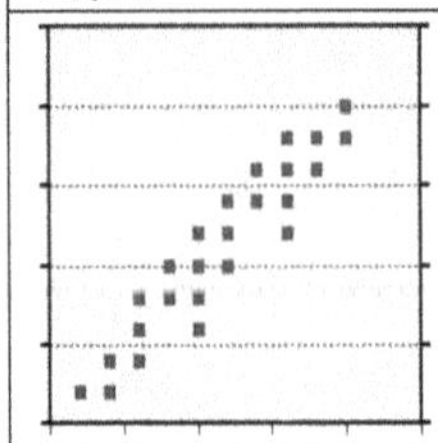	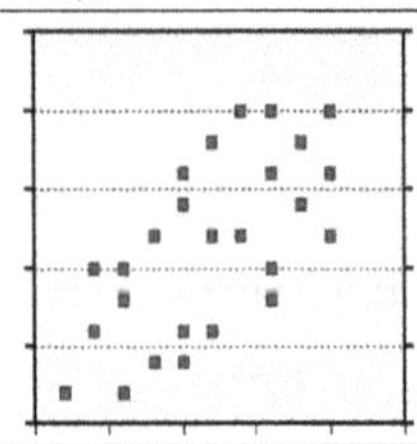	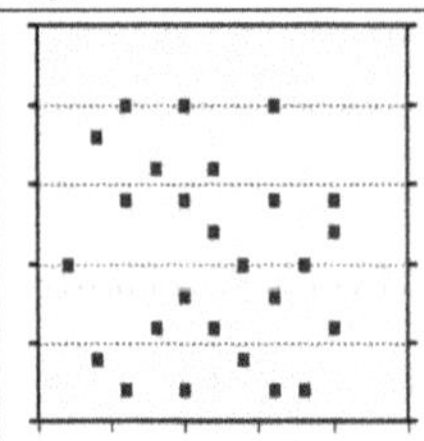	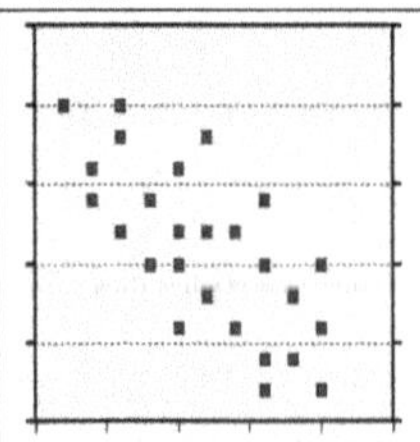	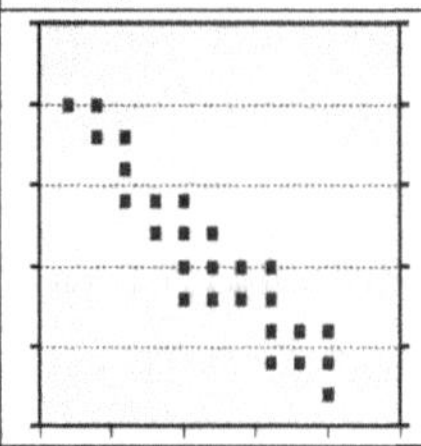
High positive correlation	Moderate positive correlation	No correlation; scores are totally unrelated	Moderate negative correlation	High negative correlation

As explained above, there are three types of relationships interpreted by correlation: positive correlations; negative correlations; and zero or no correlations (Huck, Cormier, & Bounds, 1974).

A correlation coefficient is expressed as a decimal fraction that is preceded by either a plus (+) or minus (-) sign. The degree of the relationships ranges along a continuum and is usually illustrated by three types of patterns. The patterns range between +1.00 (as positive correlation) through 0.0 (for no relationship) to -1.00 (as negative correlation). Very few variables are perfectly related to each other and, therefore, most correlation coefficients are not exact +1.00 or -1.00, but have more moderate values, such as +0.70 and -0.41. The closer the correlation coefficient is to +1.00 or -1.00, the stronger the correlation. The closer the coefficient is to zero, the lower or weaker the correlation (Huck, Cormier, & Bounds, 1974). Exhibit 19.7 shows five graphs representing hypothetical correlation coefficients.

Interpretation of Correlation Coefficients in Sport Management Research

There is no universal rule when it comes to interpretation of correlation coefficients. Therefore, several schemes have been proposed. The terminology for describing strength of the relationship varies by author and across different research investigations. The following represents the most common descriptors for interpreting relationships in sport management research.

0.00–0.24 extremely low relationship
0.25–0.49 low relationship
0.50–0.69 moderate relationship
0.70–0.89 high relationship
0.90–1.00 extremely high relationship

There is no simple answer to how high the reliability of a research instrument should be. According to Nunnally and Bernstein (1994), the judgment of how high the reliability should be based on the type of variable being measured and the stage of the development on the research instrument. A value of *r* of 0.8 is a high correlation. However, in order to state with confidence that there is a positive association between two sets of data, researchers need to refer to a statistical probability table that tells them whether the correlation has statistical significance at a defined confidence level. The level of confidence will depend on the size on the sample used. There are two common types of correlation coefficients to determine relationships of variables: the Pearson product-moment correlation and Spearman's rho.

Pearson Product-Moment Correlation Coefficient

The Pearson product-moment correlation coefficient can be obtained through a parametric statistical test if a researcher uses continuous data to analyze the strength and nature of a relationship between two variables.

Let us look at an example of sport management research that used the Pearson product-moment correlation coefficient. Branvold, Pan, and Gabert (1997) examined the effects of winning percentage and market size on attendance in Minor League Baseball (MiLB). The data were examined using correlation and hierarchical multiple regression procedures for winning percentage and host city population by competition levels on season attendance for a recent five-year period. The data were collected on 142 MiLB clubs from the National Association of Professional Baseball Leagues (NAPBL). The researchers examined the relationship between attendance and winning percentage, as well as market size, using correlation analysis. A significant but low relationship between attendance and winning percentage ($r = 0.1041$, $p < 0.01$) and a moderate relationship between attendance and market size ($r = 0.5152$, $p < 0.001$) were found in the overall analysis.

Spearman's Rank Order Correlation Coefficient (Rho)

Spearman's rho is a nonparametric statistical test used to analyze rank-ordered variables in terms of the strength and nature of a relationship between two variables. Armstrong-Doherty (1995) conducted a study to examine whether athletic departments are perceived to be controlled by the funding sources in their environment according to their relative resource dependence upon those sources. Spearman rank order calculations revealed the resource dependence-based, perceived control of the university central administration, corporate sponsors, and provincial/federal sport organizations and ministries ($p < 0.05$).

Shown in Exhibit 19.8 is a cross-tabulation of descriptive statistical techniques used by the journals between 2000 and 2004. Seven types of descriptive statistical techniques were identified that were directly related to answering the research purposes, questions, and/or hypotheses. More than a fourth of studies investigated percentages and frequencies. Percentage was the most dominant statistical technique to emerge, accounting for 28.37% of the techniques, followed by frequencies (26.96%) and the means (21.99%) used in the three sport management journals. The range, ranking, and median were the least used descriptive techniques; the mode was not used by any of the researchers during this time period.

Summary

This chapter explains the techniques that are essential for understanding descriptive statistical data analysis techniques. Statistics are a body of rules, procedures, and standards directed to extract units of information from a set of numerical data.

Descriptive statistics are often used to analyze data for descriptions about existing demographic variables such as age, sex, marital status, ethnicity, income, occupation, and religious status. The

Exhibit 19.8. A Summary of Descriptive Statistical Techniques Primary to the Research Questions and Hypotheses in *JSM, SMQ,* and *IJSM* by Journal: 1987–2006

	Journal							
	JSM (1987-2006)		SMQ (1992-2006)		IJSM (2000-2006)		Total	
Descriptive Statistics	N	%	N	%	N	%	N	%
Percentage	45	26.79	54	38.57	21	28.77	120	31.50
Frequency	39	23.21	40	28.57	22	30.14	101	26.51
Mean	36	21.43	21	15.00	15	20.55	72	18.90
Standard Deviation	35	20.83	5	3.57	12	16.44	52	13.65
Range	5	2.98	5	3.57	2	2.74	12	3.15
Ranking	4	2.38	10	7.14	0	0.00	14	3.67
Mode	0	0.00	2	1.43	0	0.00	2	0.52
Median	4	2.38	3	2.14	1	1.37	8	2.10
Column Total	168	100	140	100	73	100	381	100

Note. More than one statistical technique was used in some articles.
JSM=*Journal of Sport Management,* SMQ=*Sport Marketing Quarterly,* and IJSM: *International Journal of Sport Management.*

measures of central tendency include mean, median, and mode. Unlike the measures of central tendency, the measures of variability indicate a degree of *dispersion* or spread of a set of data. The three most common measures of variability are range, variance, and standard deviation. The measures of relative proportions describe the spread of scores based on relative positions.

The measures of relationships use coefficient of correlation to explore relationships between two or more variables. Charts in different forms are used for condensing and displaying statistical data in sport management research. Five of the most common charts used for describing statistics in graphical condensed forms are (1) pie charts, (2) bar graphs, (3) histograms, (4) frequency polygons, and (5) picture grams.

Study Questions

1. Explain how to use frequency polygons, histograms and bar diagrams, and interpret each of them.
2. Describe distributions by their shapes and name them.
3. Define measures of central tendency and dispersion.
4. What are the three measures of central tendency? Which one is most commonly used in sport management research?
5. What do the measures of variability mean?

6. What are the differences between the mean and the standard deviation?
7. Define correlation and interpret correlation coefficient.
8. Under what circumstances can the Pearson and Spearman correlation procedures be utilized?

Learning Activities

1. Shown in the following table are the salaries of head football coaches of NCAA Division IA member institutions (*USA Today* 2006, November 16). Construct a frequency table for this data using the procedures outlined in this chapter.

Compensation for NCAA Division I-A head football coaches (2006)

Coach	Total	Coach	Total	Coach	Total
F. DeBerry	755,370	K. Ferentz	2,840,000	J. Paterno	NA
J. Brookhart	207,500	D. McCarney	1,107,000	D. Wannstedt	NA
M. Shula	1,766,853	M. Mangino	1,501,241	J. Tiller	1,364,785
W. Brown	371,500	R. Prince	750,000	T. Graham	NA
M. Stoops	865,000	D. Martin	150,200	G. Schiano	911,000
D. Koetter	991,000	R. Brooks	729,165	C. Long	701,500
H. Nutt	1,049,644	L. Miles	1,450,000	D. Tomey	342,100
S. Roberts	189,655	J. Bicknell III	208,500	S. Spurrier	1,300,000
B. Ross	NA	R. Bustle	194,250	J. Leavitt	1,117,500
T. Tuberville	2,231,000	C. Weatherbie	130,000	P. Carroll	2,782,320
B. Hoke	159,220	B. Petrino	1,743,000	P. Bennett	495,602
G. Morriss	1,144,236	M. Snyder	390,603	J. Bower	349,983
C. Petersen	511,800	R. Friedgen	1,691,864	W. Harris	NA
T. O'Brien	733,626	T. West	925,000	G. Robinson	638,441
G. Brandon	150,000	L. Coker	1,800,000	A. Golden	NA
B. Mendenhall	NA	S. Montgomery	139,500	P. Fulmer	2,050,000
T. Gill	183,000	L. Carr	1,454,619	M. Brown	2,664,000
J. Tedford	1,505,300	J. Smith	1,545,834	D. Franchione	2,012,200
G. O'Leary	1,035,000	R. Stockstill	254,800	G. Patterson	952,162
B. Kelly	184,897	G. Mason	1,210,000	M. Leach	1,600,000
M. Dantonio	495,000	E. Orgeron	905,000	M. Price	543,000
T. Bowden	1,198,028	S. Croom	940,000	T. Amstutz	376,400
D. Hawkins	1,098,500	G. Pinkel	1,095,000	L. Blakeney	226,750
S. Lubick	529,900	P. Johnson	1,003,309	C. Scelfo	NA
R. Edsall	845,000	B. Callahan	1,690,317	S. Kragthorpe	438,533
T. Roof	370,200	C. Ault	360,000	K. Dorrell	881,000
S. Holtz	425,000	R. Long	431,175	M. Sanford	435,000

Coach	Total	Coach	Total	Coach	Total
J. Genyk	154,442	H. Mumme	283,800	K. Whittingham	672,000
U. Meyer	1,524,550	J. Bunting	311,200	B. Guy	241,900
H. Schnellenberger	353,147	C. Amato	995,000	B. Johnson	NA
D. Strock	282,100	D. Dickey	266,625	A. Groh	1,785,000
B. Bowden	1,691,900	J. Novak	212,496	F. Beamer	2,008,000
P. Hill	1,253,982	P. Fitzgerald	NA	J. Grobe	987,843
M. Richt	1,713,000	C. Weis	NA	T. Willingham	1,414,772
C. Gailey	1,009,191	F. Solich	262,172	B. Doba	559,042
J. Jones	820,008	J. Tressel	2,012,700	R. Rodriguea	1,100,000
A. Briles	550,000	B. Stoops	3,450,000	B. Cubit	188,000
D. Erickson	214,643	M. Gundy	800,000	B. Bielerna	761,600
R. Zook	1,241,750	M. Bellotti	1,160,000	J. Glenn	505,100
T. Hoeppner	600,000	M. Riley	835,000		

Source: USA Today (2006, November 16). Compensation for NCAA Division I-A head football coaches. *USA Today, 25*(45), 16A.

2. Shown in the following table is information on college football bowl game per-team payout in 2006–2007. Apply the central tendency techniques (mean, median, and mode) discussed in this chapter to analyze the payouts.

Bowl	Site	Per-Team Payout
Poinsettia	San Diego	$750,000
Pioneer Pure Vision Las Vegas	Las Vegas	$950,000
R+L Carriers New Orleans	New Orleans	$325,000
Papajohns.com	Birmingham, Ala.	$300,000
New Mexico	Albuquerque	$750,000
Bell Helicopter Armed Forces	Fort Worth	$600,000
Sheraton Hawaii	Honolulu	$398,000
Motor City	Detroit	$750,000
Emerald	San Francisco	$850,000
Petro Sun Independence	Shreveport, La.	$1,100,000
Texas (for Big East)	Houston	$500,000
Texas (for Big 12)	Houston	$750,000
Pacific Life Holiday	San Diego	$2,200,000
Gaylord Hotels Music City	Nashville	$1,600,000
Brut Sun	El Paso	$1,900,000
AutoZone Liberty	Memphis	$1,500,000
Insight	Tempe, Ariz.	$1,200,000
Champs Sports	Orlando	$2,250,000

Bowl	Site	Per-Team Payout
Meineke Car Care	Charlotte	$750,000
Alamo	San Antonio	$2,200,000
Chick-fil-A (for ACC)	Atlanta	$3,250,000
Chick-fil-A (for SEC)	Atlanta	$2,400,000
MPC Computers	Boise	$250,000
Outback	Tampa	$3,000,000
AT&T Cotton	Dallas	$3,000,000
Toyota Gator	Jacksonville	$2,250,000
Capital One	Orlando	$4,250,000
Rose	Pasadena, Calif.	$17,000,000
Tostitos Fiesta	Glendale, Ariz.	$17,000,000
FedEx Orange	Miami	$17,000,000
Allstate Sugar	New Orleans	$17,000,000
International	Toronto	$750,000
GMAC	Mobile, Ala.	$750,000
Tostitos BCS Championship	Glendale, Ariz.	$17,000,000
Total		$126,523,000

References

Aczel, A. (1995). *Statistics: Concepts and Applications*. Chicago, IL: Irwin.

Armstrong-Doherty, A. (1995). The structure of funding in Canadian interuniversity athletics. *Journal of Sport Management, 9*(1), 59–69.

Baird, K. (2004). Dominance in college football and the role of scholarship restrictions. *Journal of Sport Management, 18*(3), 217–235.

Branvold, S. E., Pan, D. W., & Gabert, T. E. (1997). Effects of winning percentage and market size on attendance in minor league baseball. *Sport Marketing Quarterly, 6*(4), 35–42.

Colyer, S. (2000). Organizational culture in selected western Australian sport organizations. *Journal of Sport Management, 14*(4), 321–341.

Cozby, P. C. (1997). *Methods in behavioral research* (6th ed.). Mountain View, CA: Mayfield Publishing Company.

Cozby, P. C. (2001). *Methods in behavioral research* (8th ed.). Boston, MA: McGraw- Hill.

Daniel, W. W., & Terrell, J. C. (1995). *Business statistics: Basic concepts and methodology* (7th ed.). Boston, MA: Houghton Mifflin Company.

DePoy, E., & Gitlin, L. N. (1994). *Introduction to research: Multiple strategies for health and human services.* Mosby-Year Book, Inc.: St. Louis, MO, p. 195.

Dixon, M. (2002). Gender differences in perceptions and attitudes toward the LPGA and its tour professionals: An empirical investigation. *Sport Marketing Quarterly. 11*(1), 44–54.

Drain, T. S., & Ashley, F. B. (2000). Intercollegiate athletics: Back to the future II: A comparison with branch and crow five years later. *Sport Marketing Quarterly, 9*(2), 77–84.

Field, A. (2013). *Discovering statistics using IBM SPSS statistics* (4th ed.). Thousand Oaks, CA: SAGE Publications Inc.

Gall, M. D., Gall, J. P., & Borg, W. R. (2003). *Educational research and introduction* (7th ed.). Boston, MA: Allyn and Bacon.
Groebner, D. F., Shannon, P. W., Fry, P. C., & Smith, K. D. (2007). *Business statistics: A decision-making approach* (7th ed.). Upper Saddle River, NJ: Prentice Hall.
Huck, S. W., Cormier, W. H., & Bounds, W. G., Jr. (1974). *Reading statistics and research.* Cherry Valley, CA: D & S Martin.
Hunter, R., & Brown, R. (1991). The application of inferential statistics with non-probability type samples. *Journal of Applied Recreation Research, 16*(3), 234–243.
Johnson, R. B., & Christensen, L. (2014). *Educational research: Quantitative, qualitative, and mixed approaches* (5th ed.). Thousand Oaks, CA: SAGE Publications, Inc.
Levine, D.M., Berenson, M .L., & Stephan, D. (1999). *Statistics for managers using Microsoft Excel* (2nd. ed.). Upper Saddle River, NJ: Prentice-Hall.
McMillan, J. H., & Schumacher, S. (2014). *Research in education: Evidence-based inquiry* (7th ed.). London, UK: Pearson.
Neuman, W. L. (2003). *Social research methods: Qualitative and quantitative approaches* (5th ed.). Boston, MA: Allyn and Bacon.
Nunnally, J. C., & Bernstein, I. (1994). *Psychometric theory* (3rd. ed.). New York: McGraw Hill.
Parks, J. B., & Bartley, M. E. (1996). Sport management scholarship: A professoriate in transition? *Journal of Sport Management, 10*(2), 119–130.
Pitts, B. G., & Ayers, E. K. (2001). An analysis of visitor spending and economic scale on Amsterdam from the Gay Games V, 1998. *International Journal of Sport Management, 2*(2), 134–151.
Salkind, N. J. (2014. *Exploring research* (4th ed.). New York, NY: Pearson.
Shao, A. T. (2002). *Marketing research: An aide to decision making* (2nd ed.). Cincinnati, OH: South-Western.

20

Statistical Testing of Differences

After completing the study of this chapter, you should be able to:

- Distinguish between parametric and nonparametric statistical techniques.
- Understand the circumstances under which type of *t*-test can be used.
- Explain the differences between the two types of analysis of variance.
- Explain the differences between a *t*-test and an analysis of variance.
- List reasons for utilizing nonparametric techniques versus parametric techniques.
- Identify the assumptions of the Chi-square test.
- Explain the difference between a *t*-test and a Mann-Whitney U-test.
- Explain the differences between an ANOVA and a Kruskal-Wallis test.
- Explain the differences between a Chi-square test and a Fisher's Exact test.

The purpose of this chapter is to provide students with an overview of the most commonly used techniques of statistical testing of differences in sport management research. The main intent is to present a conceptual rather than numerical review of inferential statistics and to describe the statistical procedures that investigators use to analyze data. As such, the numerical description of various statistical analysis procedures is purposefully omitted and deemphasized. Particularly, the *t-test* and the *analysis of variance* (ANOVA), two of the most commonly used parametric statistical techniques that assess differences between univariate means, will be reviewed. In the second half of this chapter, some nonparametric techniques, such as the *Chi-square test*, the *Mann-Whitney U-test*, the *Wilcoxon Sign Rank Test*, and the *Kruskal-Wallis test*, will be covered.

Parametric Statistical Techniques

There are two major dimensions of inferential statistics: parametric statistical techniques and non-parametric statistical techniques. For parametric tests, the fundamental assumptions should be met. These assumptions include assumption of normality (i.e., data taken from the populations

Exhibit 20.1. A Summary of Parametric Statistical Techniques

Assessing differences between univariate and multivariate means	*t*-tests Analysis of Variance (ANOVA) Analysis of Covariance (ANCOVA) Multiple Analysis of Variance (MANOVA) Multiple Analysis of Covariance (MANCOVA)
Discriminant Analysis	Discriminant Analysis
Assessing bivariate and multivariate means of observed variables	Canonical Correlation Analysis Path Analysis Cluster Analysis Trend Analysis
Regression analysis techniques	Simple Regression Analysis Multiple Regression Analysis Stepwise Regression Analysis Logistic Regression Analysis Hierarchical Regression Analysis
Parametric statistical techniques to determine degree of relationship of two sets of variables	Pearson Product-Moment Correlation Spearman Rho
Factor analysis	Exploratory Factor Analysis (EFA) Confirmatory Factory Analysis (CFA)
Structural equation modeling	Structural Equation Model (SEM)
Z-score	*Z*-score

are normally distributed), assumption of independence (i.e., the sample must be independent), and assumption of homogeneity of variance (i.e., data have equal variances on the variable(s) that is being measured).

Ways to Subdivide Parametric Statistical Techniques

There is no one way to subdivide the categories of parametric statistical techniques. For the purpose of this book, parametric statistical techniques are divided into eight categories. Such categories are shown in Exhibit 20.1. Exhibit 20.2 presents a summary of the parametric statistical analysis techniques used to address research questions and prove hypotheses in sport management studies published in such publications as *Journal of Sport Management*, *Sport Marketing Quarterly* and *International Journal of Sport Management* from 1987–2006. As shown in Exhibit 20.2, ANOVA, regression analysis, MANOVA, *t*-test, and factor analysis are the five parametric statistical analysis methods used most frequently by sport management researchers in this period.

Parametric statistics are used for making assumptions about the characteristics of a sample or population when the level of measurement of the data is primarily interval or ratio. They are the most powerful tools in testing questions and hypotheses about the differences of groups or a set of values via the use of interval and ratio scales, primarily by comparing means of different groups. Exhibit 20.3 illustrates the use of the level of measurement to select an appropriate parametric

Exhibit 20.2 A Summary of Parametric Statistical Techniques Used by Sport Management Researchers (as Reflected by the Manuscripts Published in *JSM*, *SMQ*, and *IJSM* by Journal: 1987–2006)

	Journal							
	JSM (1987–2006)		*SMQ* (1992–2006)		*IJSM* (2000–2006)		Total	
Parametric	N	%	N	%	N	%	N	%
ANOVA	50	21.19	37	21.14	26	23.21	113	21.61
Regression	46	19.49	33	18.86	21	18.75	100	19.12
MANOVA	32	13.56	23	13.14	12	10.71	67	12.81
T-test	28	11.86	20	11.43	16	14.29	64	12.24
Factor Analysis (EFA)	26	11.02	12	6.86	10	8.93	48	9.18
Pearson Product-Moment Correlation	14	5.93	15	8.57	9	8.04	38	7.27
Factor Analysis (CFA)	14	5.93	17	9.71	5	4.64	36	6.88
Discriminatory Analysis	5	2.19	4	2.29	2	1.79	11	2.10
Carnonical Correlation	5	2.19	0	0.00	1	0.89	6	1.15
Cluster Analysis	2	0.85	1	0.57	0	0.00	3	0.57
Structural Equation Model	5	2.19	8	4.57	7	6.25	20	3.82
Fishers *Z* score	3	1.27	3	1.71	0	0.00	6	1.15
MANCOVA	2	0.85	0	0.00	3	2.68	5	0.96
ANCOVA	1	0.42	1	0.57	0	0.00	2	0.38
Path Analysis	1	0.42	1	0.57	0	0.00	2	0.38
Trend Analysis	1	0.42	0	0.00	0	0.00	1	0.19
Hierarchical Linear Modeling (HLM)	1	0.42	0	0.00	0	0.00	1	0.19
Column Total	236	100	175	100	112	100	523	100

Note. More than one statistical technique was used in some articles.

JSM=Journal of Sport Management; SMQ=Sport Marketing Quarterly; and IJSM=International Journal of Sport Management.

statistical procedure for testing differences. For example, as shown in Exhibit 20.3, the *t*-test should be selected when the dependent variable is either interval or ratio in nature and is a nominal measurement, and the independent variable contains two categories.

Inferences about the Difference between Two Means

The *t*-test (also known as the *t*-statistics) is a group of related parametric tests used when we want to determine the statistical significance of a difference between the means of two groups. As the

Exhibit 20.3. Selection of an Appropriate Parametric Statistical Procedure to Test Differences

Parametric Statistical Procedures		
Test	**Dependent (Criterion) Variable**	**Independent (Predictor) Variable**
Independent *t*-test	Interval/ratio	Nominal (2 categories)
Paired *t*-test	Interval/ratio	(2 repeated measures)
One-way ANOVA	Interval/ratio	Nominal (3 or more categories)
Within subject one-way ANOVA	Interval/ratio	(2 or more repeated measures)
Pearson	Interval/ratio	Interval/ratio

Source. Weinbach, R. W., & Grinnell, R. M., Jr. (1995). *Statistics for social workers* (3rd ed.). White Plains, NY: Longman Publishers USA.

t-test compares the means of two groups, the dependent variable must have the characteristics that make calculation of a mean appropriate. In other words, the *t*-test is used when the dependent variable is at the interval or ratio level of measurement and the independent variable has two categories.

The *t*-test produces a *t* value that is compared with a standardized table to determine whether or not it was justifiable to reject the null hypothesis. There are two different forms of the *t*-test for comparing group means: the *t*-test for independent groups or variables and the *t* test for dependent groups or variables. Both forms of the *t*-test compare the two means of interval or ratio level data within a research sample with each other. They ask the question, "is the mean of one sample different enough from the mean of the other sample that it would be safe to conclude that there is a statistically significant difference between the two samples in relation to the dependent variable?" (Weinbach & Grinnell, 1995, p. 174).

McGehee, Yoon, and Cardenas (2003) examined relationships between recreational runners' involvement in travel to road races and behavioral characteristics. The subjects of the study were 444 members of three North Carolina running clubs. Based on the degree of sport involvement, they were divided into two groups: cluster 1 (low involvement group) and cluster 2 (high involvement group) based on nine involvement items. A number of *t*-tests were conducted to see if differences could be found in several dependent variables. For instance, the researchers examined whether the patterns of participation in road races, preparation for road races, travel behavior, and running-related expenditures are different according to the level of sport involvement. The study found statically significant differences between the high involvement group and low involvement group in terms of travel behavior and running-related expenditures, concluding that different levels of sport involvement should be considered by sport and tourism agencies when planning, marketing, and managing events targeted at traveling recreational runners.

Stoldt and Narasimhan (2005) examined how collegiate sports information professionals assessed their expertise levels relative to a variety of public relations tasks. Three hundred thirty-nine collegiate sports information professionals participated in the study. Results indicated collegiate sport information professionals assess themselves as having significantly greater levels of expertise in technical tasks than managerial tasks. They also deem technical tasks to be significantly more important to their organization's well-being than managerial tasks.

Analysis of Variance (ANOVA)

Analysis of variance (ANOVA) is a parametric test used when the researcher wants to determine the statistical significance of a difference between the means of three or more groups. ANOVA allows researchers to investigate the mean differences of one dependent variable on three or more independent variables. When the ANOVA technique is employed, the dependent variable must be at the interval or ratio level of measurement and the independent variable must be at the nominal or ordinal level of measurement with three or more categories. ANOVA is an extension of the *t*-test, but it avoids the error inherent in performing multiple *t*-tests. In general, there are two forms of ANOVA used by sport management researchers: one-way ANOVA and two-way ANOVA.

One-Way ANOVA

A one-way ANOVA is employed to analyze one independent variable when the dependent variable has more than two categories. It is an extension of the independent *t*-test. An *F* value is computed and compared with a standardized table to determine whether or not it is justifiable to reject the null hypothesis. To use this particular statistical procedure, a number of the following assumptions must be met:

- Assumption of normal distribution: the populations from which the samples were obtained must be normally or approximately normally distributed.
- Assumption of independence: the samples must be independent.
- Assumption of homogeneity of variance: the variances of the populations must be equal.

The study conducted by Mahony and Moorman (2000) is a good illustration of how the one-way ANOVA technique can be used by sport management professionals to meet their research needs. The authors employed one-way ANOVA to see if there were significant differences among the means of strong positive and strong negative attitudes on NFL and NBA viewing intentions of a pool of undergraduate and graduate students at a large Midwestern university. The NFL sample included 161 students and the NBA sample included 157 students. The independent variables were the teams that the students reported watching and the dependent variables were the self-reported intentions made by the students regarding their preferences for watching their favorite team. The major finding revealed that the students preferred to watch their favorite team. Their secondary choice was a team they disliked if this team threatened the success of their favorite team; otherwise, they would also watch a neutral game.

Two-Way ANOVA

A two-way ANOVA is used to analyze the mean differences of two independent variables. Each independent variable has more than two categories or levels. A two-way ANOVA is an extension of the independent *t*-test and the one-way ANOVA. In addition to the three assumptions that must be met before the one-way ANOVA can be employed, there is one additional assumption for the use of two-way ANOVA. The additional assumption is that the groups have the same sample size.

Sport management researchers use two-way ANOVA much less frequently than they use one-way ANOVA. Nevertheless, it is still worthwhile to briefly discuss this particular statistical technique using the following example. A study by London and Boucher (2000) attempted to determine whether or not there was any statistically significant difference among the athletic directors in Canadian universities in terms of their leadership style (i.e., transformational versus transactional leadership) with regard to the organizational effectiveness of their respective athletic programs. The results of the two-way ANOVA showed that transformational leadership has significant effects on organizational effectiveness. Interestingly, no significant effect resulted from transactional leadership on athletic program organizational effectiveness.

Post Hoc Tests of Statistical Significance for ANOVA

When an ANOVA yields a significant result (i.e., a significant value for the *F*-ratio, explained below), additional statistical analysis is necessary, known as post hoc tests. Because there are at least three groups of means, it is impossible to determine which means are statistically different from each other simply by visually comparing their relative magnitudes. To determine precisely the group means that are significantly different from the other group means, it is necessary to carry out a post hoc test of statistical significance (Crowl, 1996). Post hoc comparisons (also known as multiple comparison procedures) are computed in an ANOVA by what is called the *F* ratio (*F*-test). The computed *F* value is compared with a standardized table to determine whether or not it was justifiable to reject the null hypothesis. There are several commonly used post hoc tests that can be used to locate significant differences:

- Fisher LSD
- Newman-Keulis test
- Scheffe's test
- Tukey's test

Among these post hoc tests, Sheffe's test and Tukey's test are the most stringent methods and least likely to make a Type I error.

Z-Test

Z-test is used to express a score in terms of its distance from the mean and in terms of the normal curve. It is calculated by subtracting the mean score from a value of data and dividing by the standard deviation of a sample. For example, a person's relative standing on two or more measurements can be compared by converting the raw scores to *z*-scores. The *z*-distribution is used when samples are large and is used to determine the level of statistical significance of an observed difference between the groups (Gall, Gall, & Borg, 2003; Huck, 2000).

Nonparametric Statistical Techniques

Unlike parametric statistics, nonparametric statistics are used for deriving conclusions about the characteristics of a sample when the level of measurement of data is primarily nominal or ordinal. Nonparametric statistics do not make assumptions about the population (Gall, Gall, & Borg, 2003; Hollander & Wolf, 1999; Leedy & Ormrod, 2001; McMillan & Schumacher, 2014; Neuman, 2003).

Exhibit 20.4. A Summary of Nonparametric Statistical Techniques Used by Sport Management Researchers (as Reflected by the Manuscripts Published in *JSM, SMQ,* and *IJSM* by Journal: 1987–2006)

Nonparametric Statistical Techniques	*JSM* (1987–2006)		*SMQ* (1992–2006)		*IJSM* (2000–2006)		Total	
	N	%	N	%	N	%	N	%
Chi-Square	37	77.08	31	79.49	13	81.25	81	78.64
Mann-Whitney	1	2.08	2	5.13	1	6.25	4	3.88
Kruskal-Wallis	2	4.17	1	2.56	1	6.25	4	3.88
Spearman Rank Correlation	8	16.67	0	0.00	0	0.00	8	7.77
Fisher's Exact Test	0	0.00	1	2.56	0	0.00	1	0.97
Wilcoxon Signed Rank Test	0	0.00	2	5.13	0	0.00	2	1.94
Conjoint Analysis	0	0.00	2	5.13	1	6.25	3	2.91
Column Total	48	100	39	100	16	100	103	100

Note. More than one statistical technique was used in some articles.

JSM=*Journal of Sport Management*; SMQ=*Sport Marketing Quarterly*; and IJSM=*International Journal of Sport Management.*

Source. Quarterman, Pitts, Jackson, Hwang, Han, & Ahn (2007). Statistical data analysis techniques employed in the *Journal of Sport Management*, *Sport Marketing Quarterly*, and *International Journal of Sport Management*: 1987 to 2006. Unpublished manuscript.

Common Nonparametric Statistical Techniques

The nonparametric statistical procedures that sport management researchers may use include chi-square test, Spearman-rank order correlation, Mann-Whitney U-test, Kruskal-Wallis one-way ANOVA, Fisher's exact test, Wilcoxon's sign ranked test, conjoint analysis, Friedman's two-way ANOVA by rank, and Kendall's Tau. Presented in Exhibit 20.4 is a summary of the nonparametric tests employed to address research questions and hypotheses in sport management studies published in the *Journal of Sport Management*, *Sport Marketing Quarterly* and *International Journal of Sport Management* from 1987–2006. According to Exhibit 20.4, the Chi-square (χ^2) test is the most frequently used nonparametric statistical method by sport management researchers.

Exhibit 20.5 illustrates the use of the level of measurement to select an appropriate nonparametric statistical procedure for testing of differences. For example, to properly use the chi-square test, both the independent and dependent variables must be nominal in nature with two or more categories. In the use of the Mann-Whitney U-test, the dependent variable needs to be ordinal data, and the independent variable needs to be a nominal scale.

In sum, when the data takes the form of categories (male vs. female, undergraduate vs. graduate), ordinal groups (mild vs. moderate vs. severe), or rankings (academic standing in one's class), nonparametric statistics will be needed. It is also appropriate to use nonparametric statistics when numerical data are not normally distributed.

Exhibit 20.5. Selection of an Appropriate Nonparametric Statistical Procedure to Test Differences

Nonparametric Statistical Tests		
Test	**Dependent (Criterion) Variable**	**Independent (Predictor) Variable**
Chi-square test	Nominal (2 or more categories)	Nominal (2 or more categories)
Fisher's exact test	Nominal (2 categories)	Nominal (2 categories)
McNemar's test	Nominal (2 categories)	(2 repeated measures)
Mann-Whitney U-test	Ordinal	Nominal (2 categories)
Wilcoxon Sign test	Ordinal	(2 repeated measures)
Kruskal-Wallis test	Ordinal	Nominal (3 or more categories)
Friedman's test	Ordinal	(2 or more repeated measures)
Spearman rho test	Ordinal	Ordinal
Kendall's tau test	Ordinal	Ordinal

Source. Weinbach, R. W., & Grinnell, R. M., Jr. (1995). Statistics for social workers (3rd ed.). White Plains, NY: Longman Publishers USA.

Chi-Square Test

The chi-square test is used to test the differences in proportion in two or more independent groups of which the independent variable is in the nominal scale. It determines whether the two variables being examined are independent or related and whether the difference in frequencies of observed data exists with the comparison of the frequencies that could be expected to occur if the data categories were actually independent of each other.

The chi-square test is computed by summing differences between the observed frequencies in each cell and the expected frequencies—the frequencies that would be expected if there were no relationship between the two variables. For example, suppose a master's program in sport management had two concentrations: event management and fitness management. Of the 200 enrolled students, a total of 120 (60%) students chose event management, and a total of 80 (40%) students chose fitness management. A question that might be addressed would be whether the gender of the students is related to the choice of specialty area. The null hypothesis would be that there is no difference between males and females in relation to choice of a specialty area. In this case, we can use the chi-square test to determine whether a relationship between gender and choice of major exists and is statistically significant. The chi-square test compares the numbers we have observed with the numbers expected, if the two groups were equal.

Robinson and Carpenter (2002) conducted a study to determine if the choice of different day of the week is related to sociodemographic characteristics and consumption patterns of spectators attending a Ladies Professional Golf Association (LPGA) event. A chi-square test was utilized to determine if statistically significant differences existed on days attended related to the variables of spectators' gender, age, income, employment status, how they acquired their tickets, and frequency of play. Significant differences did exist for variables of age [χ^2 (33) = 60.04, p = 0.003]. The (33)

Exhibit 20.6. A Fictitious Specialty of Sport Management Programs by Gender of Student

Program	Gender: Female	Gender: Male	Row Totals
Event management	90 (56%)	30 (75%)	120 (60%) event management
Fitness management	70 (44%)	10 (25%)	80 (40%) fitness management
Column Total Sample, N = 200	160 females	40 males	Total

A statistical interpretation of the Chi-square (χ^2) test

$\chi^2 (33) = 60.04$; $p = .003$

means the test had 33 degrees of freedom. The 60.04 is the value of the test statistic. The test statistic represents the ratio of differences between the groups to measurement error. The p refers to the probability of making a Type I error, or the probability that the result occurred by chance. The 0.003 means that the probability of a Type I error was less than 0.03%.

Mann-Whitney U-Test

The Mann-Whitney U-test uses ordinal data (ranks) to compare two independent conditions. When you want to compare the distributions in two conditions and these conditions contain different entities, you can use the Mann-Whitney U test (Mann & Whitney, 1947). This test is the nonparametric equivalent of the independent *t*-test (Field, 2013). To perform the Mann-Whitney U-test, the basic requirement is that the two samples be independent, and that the observations be ordinal or continuous measurements. It is often presented as an alternative to a *t*-test when the data are not normally distributed. In general, the Mann-Whitney U-test technique is used when these three conditions are met: (1) when the researchers want to avoid the assumptions of the *t*-test, (2) when random selection has not occurred, (3) when the measurement variables generate categorical or ordinal data, and (4) when the sample sizes are small (Huck, Cormier, & Bounds, 1974; Witte & Witte, 2010).

Exhibit 20.7 is an illustration of the Mann-Whitney U-test to measure motivational congruency and discrepancy between certified athletic trainers and noncertified student athletic trainers in the state of Hawaii (Buxton, Lankford, & Noda, 1992).

As illustrated in Exhibit 20.7, significant differences ($p < 0.05$) were found in the hierarchical rankings of the motivational statements between the two groups of trainers. The noncertified student athletic trainers ranked job growth (item G) as their number one motivator. Additionally, differences ($p < 0.05$) were found between the two groups for rankings of freedom on the job (item C), job growth (item G), good benefits and wages (item I), being appreciated (item J), helping the organization to obtain its goals (item K), being an integral part of the work team (item M), job security (item N), and feedback on performance (item O).

Exhibit 20.7. Example of the Mann-Whitney U-Test

Motivational Statement	Certified Athletic Trainers (I) (N = 20)	Noncertified Athletic Trainers (II) (N = 16)
A. Designated as a leader	3.50 (± 0.76)	3.19 (± 0.83)
B. Importance of work	4.90 (± 0.31)	4.75 (± 0.45)
C. Freedom on the job	4.50 (± 0.61)	3.75 (± 0.86)*
D. Achieving work-related goals	4.70 (± 0.47)	4.44 (± 0.51)
E. Getting along with others	4.70 (± 0.57)	4.81 (± 0.40)
F. Opportunities for advancement	4.55 (± 0.61)	4.00 (± 0.82)*
G. Job growth	4.85 (± 0.37)	5.00 (± 0.00)
H. Good work condition	4.50 (± 0.51)	4.44 (± 0.51)
I. Good benefits and wages	4.75 (± 0.44)	3.75 (± 1.00)*
J. Being appreciated	4.25 (± 0.72)	4.50 (± 0.63)
K. Helping organization obtain goals	4.30 (± 0.47)	4.56 (± 0.51)
L. Receiving raises	3.95 (± 0.67)	3.44 (± 1.03)
M. Being an integral part of work team	4.55 (± 0.61)	4.75 (± 0.45)
N. Job security	4.75 (± 0.55)	3.94 (± 0.85)*
O. Feedback on performance	4.25 (± 0.55)	4.38 (± 0.62)
P. Role in decision making	4.65 (± 0.49)	4.13 (± 0.50)

Scale: 1 = Extremely Unimportant; 2 = Unimportant; 3 = Neutral; 4 = Important; 5= Extremely Important

* Significant $p < 0.05$

Source. Buxton, B. A., Lankford, S. V., & Noda, J. E. (1992). Motivational congruency between certified athletic trainers and noncertified student athletic trainers in the State of Hawaii. *Journal of Athletic Training, 27*(4) 20, 22, 24.

Another example of the use of the Mann-Whitney U-test in sport management is the study conducted by Zhang, Wall, and Smith (2000). The authors utilized the Mann-Whitney U-test to examine the relationship between gender and game attendance variables. Season home game ticket holders of a major Western Conference NBA team ($n = 924$) were the subjects of the study. Gender was the independent variable. Game attendance variables, number of home games, and number of playoff games in the previous season were the dependent variables. The results of the Mann-Whitney U-test revealed that gender was not significantly ($p > 0.05$) related to game attendance variables.

Westerbeek (2000) tested "the relationship between five place-specific factors of service quality in spectator sport settings (the dependent variable) and two critical demographic variables, frequency of attendance and age (the independent variables)" (p. 194). A Mann-Whitney U-test,

Exhibit 20.8. Rank Order of Motivational Statements by Groups with the Kruskall-Wallis Values

Motivational Statement	High School	Rank Order University	Clinic	K-W[a]
A. Designation as a leader	16	16	15	1.26
B. Importance of work	2	1	1	0.90
C. Freedom on the job	9	11	4	2.56
D. Achieving work-related goals	10	5	2	2.11
E. Getting along with others	8	2	6	0.95
F. Opportunities for advancement	7	10	11	0.99
G. Job growth	4	4	5	0.72
H. Good work conditions	5	6	7	0.51
I. Good benefits and wages	1	9	3	5.58
J. Being appreciated	14	12	10	0.58
K. Helping organization obtain goals	11	13	14	2.79
L. Receiving raises	15	14	13	0.84
M. Being an integral part of work team	13	3	12	8.34+
N. Job security	3	8	9	1.04
O. Feedback on performance	12	15	16	2.88
P. Role in decision making	6	7	8	0.56

[a] Value required for significance at the .05 level, $df=2$, is 5.99

\+ Significant $p < 0.05$; whereas High School differs from University (Mann-Whitney U)

Source. Buxton, B. A., Lankford, S. V., & Noda, J. E. (1992). Motivational congruency between certified athletic trainers and noncertified student athletic trainers in the State of Hawaii. *Journal of Athletic Training, 27*(4) 20, 22, 24.

performed to determine if older spectators differed from young spectators in relation to frequency of attendance ($Z = -0.757, p > 0.05$), delivered no significant results.

Kruskal-Wallis Test

The Kruskal-Wallis test compares differences between several independent groups similar to the parametric one-way ANOVA. Like the Mann-Whitney U-test, the Kruskal-Wallis analyzes ranked data but is an extension of the Mann-Whitney U-test in that it analyzes three or more groups. As a nonparametric technique, it does not assume a normal population and the population variabilities among groups do not have to be equal. So, the Kruskal-Wallis test can be applied to the situation where the ANOVA normality assumptions may not apply.

Exhibit 20.8 provides one example of the Kruskal-Wallis test. Buxton, Lankford, & Noda (1992) utilized the Kruskal-Wallis test to determine if any statistically significant differences existed between certified athletic trainers who worked in high schools and those who were employed by colleges and universities in the state of Hawaii in terms of their job motivations.

The study of Zhang, Wall, and Smith (2000) examined the relationship of selected variables to game attendance level of NBA season ticket holders in the following four areas: sociodemographics, decision making, special program, and customer service. The season home game ticket holders of a major Western Conference NBA team ($n = 924$) were the subject of the study. To conduct the Kruskal-Wallis test, the researchers used a number of sociodemographic variables, such as age, ethnicity, income, and education, as independent variables. Each of these variables had more than two categories or groups. The dependent variables examined were the number of home games attended and the number of playoff games in the previous season attended. The Kruskal-Wallis test revealed that age and household income were significantly ($p < 0.05$) related to attendance variables (i.e., number of home games attended, number of previous playoff games attended). Educational background was also found to be significantly ($p < 0.05$) related to the number of season home games that the respondents intended to attend next season.

Fisher's Exact Test

Fisher's exact test is a procedure that researchers can use for categorical (i.e., nominal) data in a two by two contingency table. There is one issue with the chi-square test, which is that the sampling distribution of the test statistic has an approximate chi-square distribution, resulting in it being affected by the size of samples. In small samples, the approximation may not be good enough, leading significance tests of the chi-square distribution to be inaccurate (Field, 2013).

An example of the use of the Fisher's exact test is provided by Cuneen and Spencer (2003), who analyzed the milk mustache ads of sport celebrities ($N = 34$) to ascertain if there were gender effects. Fisher's exact test was used in the study to measure relationships between sport celebrity gender and the units of analysis. The test was used in place of the regular test of relationship model because the small sample size rendered several under-filled cells (Agresti, 1996). Using Fisher's exact test, the researchers found that there was a significant relationship between model gender and activity type ($p = 0.0174$). Examples of their findings: (1) seven times more males ($N = 14$ or 88%) than females ($N = 2$ or 12%) represented the milk mustache ads; (2) two times more females ($N = 8$ or 62%) than males ($N = 4$ or 31%) represented the ads for individual sports; and (3) there were no females associated with extreme ($N = 2$ or 100%) and motor sports ($N = 2$ or 100%). All the sport celebrities pictured in these sports were males.

Wilcoxon Signed Ranks Test

As a nonparametric statistical procedure, the Wilcoxon Signed Ranks test is used to examine the differences between two related groups of participants for which the level of measurement for the dependent variables is of the ordinal level. Most specifically, it can be employed to measure the effect of a treatment on pre- and post-treatment observations, where the same subjects are measured twice (Zikmund, 2000, p. 542). The Wilcoxon Signed Rank test is a nonparametric alternative to the paired Student's t-test.

Brown, Nagel, McEvoy, and Rascher (2004) conducted a cross-sectional study to determine if a significant difference in net revenue change existed in NFL teams that moved into new facilities and to determine if there was a significant change in valuation for these franchises. The subjects of the study were NFL franchises that opened new stadia during the 1995–1999 timeframe. The Wilcoxon Signed Rank test was employed in data analysis. The variables compared included ticket sales, revenues from local TV and radio, loge boxes, concessions, advertising/parking/other, and total local revenue. The results of the study indicated that there was a significant difference between pre-stadium opening revenue and post-stadium opening revenue at the 0.05 level in the areas of ticket sales, loge box revenue, advertising/parking/other, and total local revenue.

Summary

The focus of this chapter was on testing differences. Commonly used procedures in both parametric and nonparametric statistics were reviewed. The *t*-test, the analysis of variance or ANOVA, and the *z* score are the three types of parametric statistical testing procedures discussed. The *t*-test is used when the dependent variable is at the interval or ratio level of measurement and the independent variable has two categories. There are two different forms of the *t*-test for comparing group means: the *t*-test for independent groups or variables and the *t*-test for dependent groups or variables. On the other hand, ANOVA is used when the researcher wants to determine the statistical significance of a difference between the means of three or more groups. One-way ANOVA and two-way ANOVA are the two forms of ANOVA that are used by sport management researchers for answering research questions and hypothesis testing. A number of post hoc comparison tests are typically run to determine which means are statistically different from each other when an ANOVA yields a statistically significant result. These tests include the Fisher LSD, the Newman-Keulis test, Tukey's test, and Scheffe's test.

Five types of nonparametric testing procedures are introduced in this chapter (i.e., the χ^2 test, the Mann Whitney U-test, the Kruskal-Wallis test, the Wilcoxon Signed Ranks test, and the Fisher Exact test).

The chi-square test is used to analyze nominal (categorical) data among groups. The Mann-Whitney U-test uses ordinal data (ranks) to determine whether two independent samples come from the same distribution. It is a distribution-free test for statistical significance. The Kruskal-Wallis test is analogous to the parametric one-way ANOVA and an extension of the Mann-Whitney U-test to three or more groups. Fisher's exact test is used for categorical (i.e., nominal) data in a two by two contingency table when the sample size is small and the application of the chi-square test becomes inappropriate. The Wilcoxon Signed Rank test is a nonparametric alternative to the paired Student's *t*-test for two related groups of subjects.

Study Questions

1. What are the differences between parametric and nonparametric statistical techniques?
2. Under what circumstances can the *t*-test be used?
3. What are the differences between the two types of analysis of variance?
4. What are the differences between a *t*-test and an analysis of variance?
5. What are the reasons for utilizing nonparametric techniques versus parametric techniques?

6. What are the assumptions of the chi-square test?
7. What are the differences between a *t*-test and a Mann-Whitney U-test?
8. What are the differences between an ANOVA and a Kruskal-Wallis test?
9. What are the differences between a chi-square test and a Fisher's Exact test?

Learning Activities

1. Identify an article published this year in one of the sport management related journals that used a parametric test procedure, and give your justification of why such a procedure was an appropriate one to use in that study.
2. Identify an article published this year in one of the sport management related journals that used a *t*-test in data analysis, and give your justification of why the *t*-test was an appropriate statistical procedure to use in that study.
3. Identify an article published this year in one of the sport management related journals that used ANOVA in data analysis, and give your justification of why ANOVA was an appropriate statistical procedure to use in that study.
4. Identify an article published this year in one of the sport management related journals that used one of the nonparametric test procedures to find out the justification given by the author in terms of why this type of nonparametric test was used.

References

Agresti, A. (1996). *An introduction to categorical data analysis.* New York: John Wiley & Sons.

Brown, M., Nagel, M., & McEvoy, C. (2004). Revenue and wealth maximization in the National Football League: The impact of stadia. *Sport Marketing Quarterly, 13*(4), 227–235.

Buxton, B. A., Lankford, S. V., & Noda, J. E. (1992). Motivational congruency between certified athletic trainers and noncertified student athletic trainers in the State of Hawaii. *Journal of Athletic Training, 27*(4) 20, 22, 24.

Crowl, T. K. (1996). *Fundamentals of education research* (2nd ed.). Boston, MA: McGraw Hill. p. 182–183.

Cuneen, J., & Spencer, N. (2003). Gender representations related to sport celebrity portrayals in the milk mustache advertising campaign. *Sport Marketing Quarterly, 12*(3), 140–150.

Field, A. (2013). *Discovering statistics using IBM SPSS statistics* (4th ed.). Thousand Oaks, CA: SAGE Publications Inc.

Gall, M. D., Gall, J. P., & Borg, W. R. (2003). *Educational research* (7th ed.). Boston, MA: Pearson Education, Inc.

Hollander, M., & Wolfe, D.A. (1999). Nonparametric statistical methods (2nd. ed.). New York: Wiley.

Huck, S. (2000). *Reading statistics and research* (3rd ed.). New York: Longman.

Huck, S. W., Cormier, W. H., Bounds, W. G., Jr. (1974). *Reading statistics and research.* New York: Harper & Row, pp. 68–69.

Leedy, P. D., & Ormrod, J. E. (2001). *Practical research: Planning and design.* (7th ed.). Upper Saddle River, NJ: Prentice Hall.

London, C., & Boucher, R. (2000). Leadership and organizational effectiveness in Canadian university athletics. *International Journal of Sport Management, 1*(1), 70–87.

Mahony, D. F., & Moorman, A. M. (2000). The relationship between the attitude of professional sport fans and their intentions to watch televised games. *Sport Marketing Quarterly, 9*(3), 131–139.

Mann, H. B., & Whitney, D. R. (1947). On a test of whether one of two random variables is stochastically larger than the other. *Annals of Mathematical Statistics, 18*, 50–60.

McGehee, N.G., Yoon, Y., & Cardenas, D. (2003). Involvement and travel for recreational runners in North Carolina. *Journal of Sport Management, 17*(3), 305–324.

McMillan, J. H., & Schumacher, S. (2014). *Research in education; Evidence-based inquiry* (7th ed.). New York, NY: Pearson.

Neuman, W. L. (2003). *Social research methods: Qualitative and quantitative approaches* (5th ed.). Boston, MA: Allyn and Bacon.

Quarterman, Pitts, Jackson, Hwang, Han, & Ahn (2007). Statistical data analysis techniques employed in the *Journal of Sport Management*, *Sport Marketing Quarterly*, and *International Journal of Sport Management*: 1987 to 2006. Unpublished manuscript.

Robinson, M. J., & Carpenter, J. R. (2002). The day of the week's impact on selected socio-demographic characteristics and consumption patterns of spectators at a Ladies Professional Golf Association event. *Sport Marketing Quarterly, 11*(4), 242–247.

Stoldt, G. C., & Narasimhan, V. (2005). Self assessments of collegiate sports information practitioners regarding their professional expertise. *International Sport Journal, 6*(3), 252–269.

Weinbach, R. W., & Grinnell, R. M., Jr. (1995). *Statistics for social workers* (3rd ed.). White Plains, NY: Longman Publishers USA.

Westerbeek, H. M. (2000). The influence of frequency of attendance and age on "place"-specific dimensions of service quality at Australian rules football matches. *Sport Marketing Quarterly, 9*(4), 194–202.

Witte, R. S., & Witte, J. S. (2010). *Statistics* (9th ed.). Hoboken, NJ: John Wiley & Sons, Inc.

Zhang, J. J., Wall, K. A., & Smith, D. W. (2000). To go or not? Relationship of selected variables to game attendance of professional basketball season ticket holders. *International Journal of Sport Management, 1*(3), 200–226.

Zikmund, W. G. (2000). *Business research methods* (6th ed.). Fort Worth, TX: Harcourt College Publishers.

21

Multivariate Data Analysis Techniques

After completing the study of this chapter, you should be able to:

- Understand the decision-making criteria used in the selection of a bivariate or multivariate analysis method.
- Understand how to classify multivariate analysis techniques.
- Understand the decision-making criteria in terms of how to select an appropriate technique of multivariate analysis.
- Explain the differences between simple regression analysis and multiple regression analysis.
- Explain under what circumstances a sport management researcher can use the discriminant analysis technique to solve his or her research problem.
- Explain the differences between an ANOVA and MANOVA.
- Understand the circumstances under which conjoint analysis is a preferred statistical analysis technique.
- Differentiate cluster analysis and factor analysis.
- Explain the differences between exploratory factor analysis and confirmatory factor analysis.

The statistical analysis of several groups using a set of criterion measures simultaneously is called *multivariate statistics* (Stevens, 2009). In this chapter, the commonly utilized *multivariate analysis* techniques by sport management researchers will be briefly reviewed. The chapter has three sections. The first section will focus on the classification of multivariate statistical analysis methods, followed by a section that examines the statistical procedures used to analyze data that are involved with only one dependent variable. The third section will deal with the analysis of interdependence of all the variables involved. The last section will provide a case in sport management research that exemplifies the application of a major statistical procedure of multivariate analysis.

Classification of Multivariate Analysis Techniques

To determine which multivariate analysis technique to use, a sport management researcher should ask this question: Does the research situation include one or more dependent variables? If the answer is yes, the researcher should choose a dependence analysis technique. Otherwise, an interdependence analysis technique, which can be applied to determine the interdependence of all the variables under investigation, should be chosen. Exhibit 21.1 shows how major multivariate analysis techniques are divided into these two classifications.

As Exhibit 21.1 demonstrates, sport management researchers will choose one of the dependence analysis techniques if the research involves at least one dependent variable. If there is only one dependent variable, the multivariate analysis options available to researchers include *multiple regression analysis* (if the dependent variable is interval-scaled) or *discriminant analysis* (if the data collected on the dependent variable is either nominal or ordinal in nature). When more than one dependent variable is involved, multivariate analysis of variance (MANOVA) should be chosen if the dependent variables are interval-scaled, otherwise *conjoint analysis* is the choice.

If the structural relationship among all the variables being invested is the interest of an investigation, researchers can select one of the interdependence analysis techniques. Again, depending on the nature of the data collected, either *factor analysis* or *cluster analysis* can be chosen if the variables are interval-scaled. If not, the researcher can use other special multivariate methods, such as multidimensional scaling with nominal or ordinal data.

Multivariate Dependence Techniques

Multiple Regression Analysis

Sport management researchers use *simple regression analysis* to test the relationships between two variables (one independent variable and one dependent) and use the independent variable to predict the dependent variable. The researchers, however, may have to employ multiple regression analysis to research situations in which there are several independent variables and only one dependent variable that is interval-scaled (Zikmund, 2000). Multiple regression analysis is a parametric statistical technique used to test the relationships between two or more independent (predictor) variables and one dependent (criterion or outcome) variable in order to predict a dependent variable from two or more independent variables.

$$Y = a + b_1X_1 + b_2X_2 + b_3X_3 \ldots b_nX_n$$

Ultimately, the goal is to be able to estimate the mean value of Y and/or predict particular values of Y to be observed in the future when the predictor variables assume specific values (Mendenhall & Beaver, 1995). The study conducted by Funk, Mahony, and Ridinger (2001), as shown in Exhibit 21.2, is an example that illustrates the use of multiple regression analysis in sport management research.

There are a number of different methods of entering variables into the multiple regression equations (George & Mallery, 2001, p. 185), such as the enter, forward, backward, stepwise, and remove methods. Several sport management researchers (e.g., Kwon & Armstrong, 2002; McDaniel,

Exhibit 21.1. Classification of Multivariate Analysis Techniques

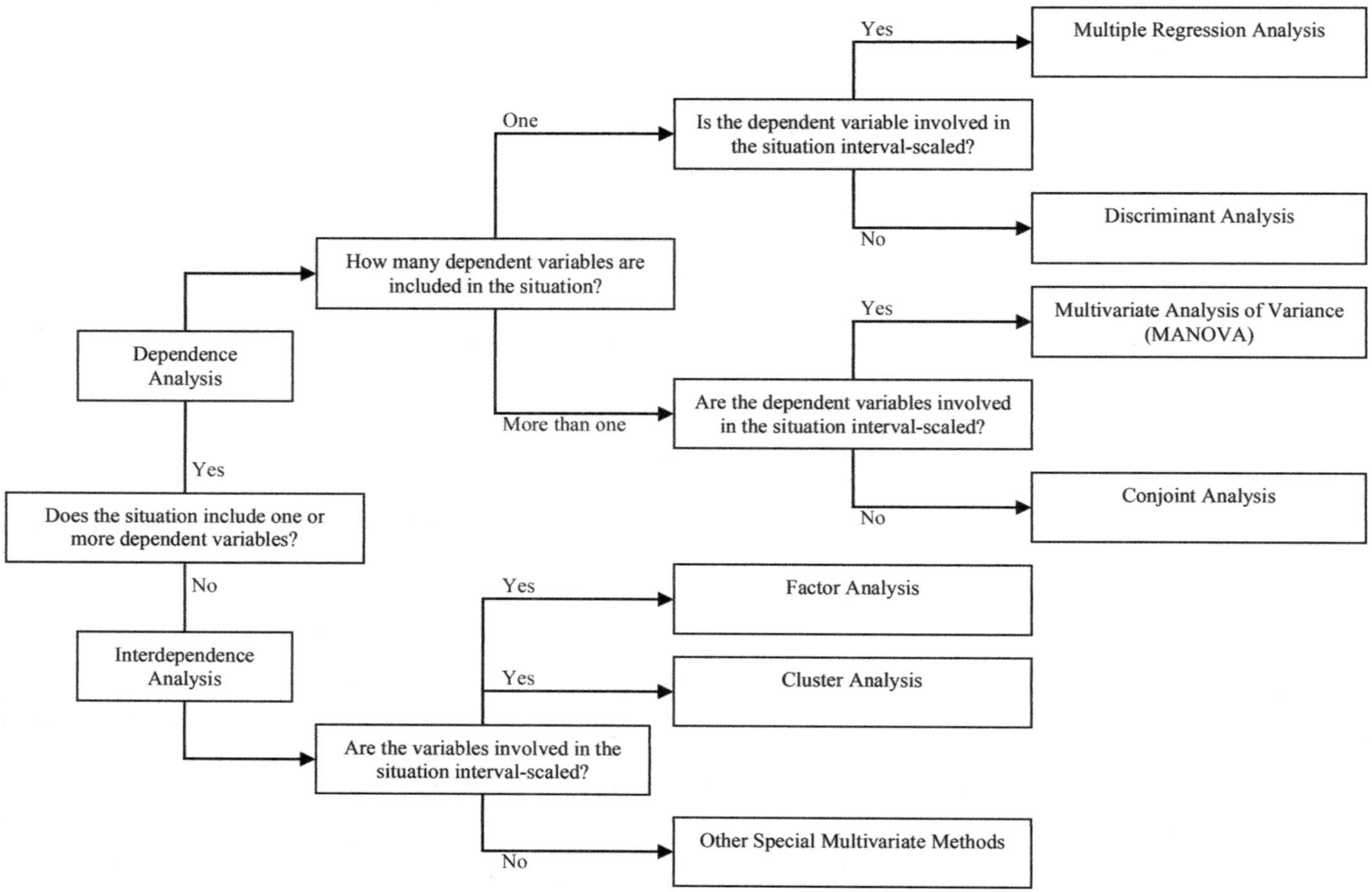

Exhibit 21.2. An Example of Multiple Regression Analysis in Sport Management Studies

Type	Source and Purpose of the Study	Sample	Independent Variables	Dependent Variables	Major Findings
Multiple Regression	Funk, Mahony, and Ridinger (2001) examined how individual difference factors could be used to explain various levels of consumer support for a specific sport property.	Spectators attending the 1999 US Nike Cup on October 10, 1999, in Louisville, Kentucky (*n*=580)	Fourteen Motives: Aesthetics, Family bonding, Drama, Excitement, Entertainment value, Interest in players, National pride, Role modeling, Interest in soccer, Social opportunities, Support for women's opportunities, Interest in team, Vicarious achievement, Wholesome environment	Spectator support level	Multiple linear regression analysis revealed that five motivational characteristics—sport interest, team interest, vicarious achievement, role modeling, and entertainment value—explained 54% of variance in level of spectator support for women's professional soccer. The results suggest that augmenting traditional spectator measures offers a better understanding of motivational characteristics in different sport situations and of the impact these motivations have on behavior. (p. 33)

2002; Zhang, Smith, Pease, & Jambor, 1997) have employed stepwise multiple regression. The enter method is used in a standard multiple regression analysis, in which the sport management researcher determines the number of predicting variables or predictors to enter the equation. All the predictors will enter the regression model simultaneously. In a stepwise multiple regression analysis, however, variables are entered one at a time based on a p value (typically less than 0.05) of each variable. When a new variable enters into the equation, the variance explained by the variables that have already been included in the equation may change. The stepwise method will eliminate those variables that become nonsignificant as other new variables are added.

Kwon and Armstrong (2002) conducted a study using the stepwise regression analysis technique to investigate impulsive tendencies to purchase sport team–licensed merchandise. Using a survey, they obtained data from 145 college students (48 males and 97 females) at a large Midwestern university. The impulse of buying sport team licensed merchandise was examined as a dependent variable. They wanted to see if any of the predicting variables, including shopping enjoyment, time availability, money availability, and sport team identification, are a significant predictor for the impulse of buying sport team licensed merchandise. The results of the stepwise multiple regression analysis revealed that the only significant antecedent to impulse buying of sport team licensed merchandise was the students' identification with the university's sport team. Sport team identification also influenced the amount of money spent on impulsive sport purchases. (p. 151)

Sport management researchers have also applied a number of advanced regression analysis procedures to meet their research needs (such as logistic regression analysis, hierarchical loglinear models) (Gladden & Milne, 1999; Harrison, Lee & Belcher, 1999).

Logistic Regression Analysis

Logistic regression analysis is a parametric statistical technique used to predict the probability of a dichotomous dependent variable (e.g., success/failure), typically using a combination of two or more independent variables to predict the probability of an event, to estimate relative risk (odds ratios). As an extension of multiple regression analysis (George & Mallery, 2001), logistic regression provides a way for studying the prediction of a criterion variable when the independent variables cannot be assumed to have a normal distribution. The independent variable can be continuous or categorical.

Peetz, Park, and Spencer (2004) used logistic regression analysis to examine the role of gender in the transfer of meaning from athlete endorsers to products and to purchase intentions of a select sample of undergraduate students. The overall description of the study is provided in Exhibit 21.3.

Hierarchical Linear Regression Analysis

Hierarchical loglinear regression analysis examines the predictive ability of several independent variables on one dependent variable when a given variable is selectively controlled. It provides the researcher with strategies for exploring causal relationships between independent and dependent variables. A number of studies using hierarchical linear regression analysis are found in sport management literature (e.g., Carptner, 2001; McDaniel & Chalip, 2002). The case described in Exhibit 21.4 exemplifies the use of such a statistical procedure.

Exhibit 21.3. An Example of Logistic Regression in Sport Management Studies

Type	Source and Purpose of the Study	Sample	Independent Variables	Dependent Variables	Major Findings
Logistic Regression	Peetz, Park, and Spencer (2004) examined the role of gender in the transfer of meaning from athlete endorsers to products and to purchase intentions of a select sample of undergraduate students.	Sport management and kinesiology students at a midsize university located in the Midwestern United States. (n=150; 102 males and 48 females)	Group by gender; male, female	Identification of endorsers, Athlete expertise, Intention to purchase	Male endorsers were more likely to be identified correctly and to influence participants' purchase intentions. Gender differences also existed in participants' perceptions of endorsers' expertise and influence on purchase intentions. These findings suggest that gender plays an important role in the transfer of meaning for the members of the Generation Y market segment.

Exhibit 21.4. An Example of Hierarchical Regression in Sport Management Studies

Type	Source and Purpose of the Study	Sample	Independent Variables	Dependent Variables	Major Findings
Hierarchical Regression	Matsuoka, Chelladurai, and Harada (2003) investigated the direct and interaction effects of team identification and satisfaction with facets of a game on intentions to attend future games.	1,256 spectators at seven J-League (Japan professional football/soccer league)	Team identification, satisfaction	Intention to attend games	Correlational and regression analyses showed that both team identification and facets of satisfaction were significantly correlated with intention to attend future games, with team identification correlating at a higher level. The significance of the interaction of identification and satisfaction indicated that the intentions of highly identified fans relative to low-identified fans were less influenced by any of the facets of satisfaction.

Discriminant Analysis

Discriminant analysis is a multiple analysis procedure used when the dependent variable is categorical (nominal or ordinal in nature) and the independent variables are continuous. It allows sport management researchers to study the difference between two or more groups of objects with respect to several variables simultaneously, determining whether a meaningful difference exists between the groups and identifying the discriminating power of each variable (Klecka, 1980). In addition, multiple discriminant analysis allows sport management researchers to "establish [a] model for classifying individuals or objects into groups on the basis of their values on the dependent variable" (McDaniel & Gates, 1991) or specifically, "to predict membership into two or more mutually exclusive groups" from a set of predicting variables (George & Mallery, 2001, p. 264). For example, Li and Burden (2004) examined such variables as institutional control, relationship with the local community, perceived product attractiveness, availability of in-house equipment, institutional support and facilities, and competency of athletic marketing staff to determine whether they affected athletic administrators' outsourcing decisions (outsourcing vs. in-house operations). Exhibit 21.5 provides a detailed description of this study.

MANOVA

Multivariate analysis of variance, also known as MANOVA, is a multivariate analysis technique that is used to test the group difference simultaneously across several interval-scaled dependent variables on one or more categorical (i.e., nominally-scaled) variables serving as independent variables. This technique overcomes the limitation and drawback of the *bivariate* analysis of variance, including a higher error due to separate tests or lack of accuracy (George & Mallery, 2001).

Sport management researchers have adopted the MANOVA technique for a number of research purposes, such as to determine divisional differences in the methods of employee evaluation used by collegiate athletic directors (Barber & Eckrich, 1998); to investigate the relationships among gender, type of sport, motives, and points of attachment to a team for spectators of selected intercollegiate sports (Robinson & Trail, 2005); to analyze differences in perception of leadership and climate related to division of competition in college athletics (Scott, 1999); to assess the efforts of gender, membership of college divisions, and mentor status on the member of promotion and salary (Weaver & Chelladurai, 2002). The study conducted by Robinson, Trail, Dick, and Gillentine (2005) is a good showcase of how MANOVA can be used in solving problems in sport management. Exhibit 21.6 provides a description of the study.

Conjoint Analysis

Conjoint analysis is one of the "many [statistical] techniques for dealing with a situation in which a decision maker has to choose among options that simultaneously vary across two or more attributes" (Green, Wind, & Rao, 1999, pp. 12). Conjoint analysis has been popular in market research to examine how consumers value different attributes of individual products and/or services (Daniel & Johnson, 2004; Green & Srinivsaan, 1990) and determine the price of a new product (Curry, 1996). For example, a tennis racquet has a number of attributes, such as color, price, stiffness, length, head size, weight, design, and reputation. Before purchasing a tennis racquet, consumers typically examine those attributes and then make trade-off judgments (e.g., picking reputation over pricing) to determine their final purchase choice. Conjoint analysis helps analyze these trade-off judgments

Exhibit 21.5. An Example of Discriminant Analysis in Sport Management Studies

Type	Source and Purpose of the Study	Sample	Independent Variables	Dependent Variables	Major Findings
Discriminant analysis	Li and Burden (2004) examined whether institutional control, relationship with the local community, perceived product attractiveness, availability of in-house equipment, institutional support and facilities, and competency of athletic marketing staff affect athletic administrators' outsourcing decisions.	Athletic directors from NCAA Division I-A, Division I-AA, and Division I-AAA athletic programs from the National Directory of Collegiate Directors (*n*=236)	Degree of control over the right to athletic properties, Perceived product attractiveness	Outsourcing, In-house operations	1. The degree of control over the rights to market the athletic properties did affect athletic administrator's decisions regarding outsourcing. 2. The relationship between the athletic department and its local business community did not have an effect on the athletic administrator's decisions regarding outsourcing. 3. The athletic administrators' perception of the attractiveness of the sports product was one of the factors affecting their decisions regarding outsourcing. 4. The competency of marketing staff was a factor influencing the athletic administrator's decisions regarding outsourcing.

Type	Discriminant Analysis
Source and purpose of the study:	Li and Burden (2004) examined whether institutional control, relationship with the local community, perceived product attractiveness, availability of in-house equipment, institutional support and facilities, and competency of athletic marketing staff affect athletic administrators' outsourcing decisions.
Sample:	Athletic directors from NCAA Division I-A, Division I-AA and Division I-AAA athletic programs from the National Directory of Collegiate Directors (n=236)
Independent variables:	Degree of control over the right to athletic properties, Perceived product attractiveness
Dependent variables:	Outsourcing, In-house operations
Major findings:	1. The degree of control over the rights to market the athletic properties did affect athletic administrator's decisions regarding outsourcing. 2. The relationship between the athletic department and its local business community did not have an effect on the athletic administrator's decisions regarding outsourcing. 3. The athletic administrators' perception of the attractiveness of the sports product was one of the factors affecting their decisions regarding outsourcing. 4. The competency of marketing staff was a factor influencing the athletic administrator's decisions regarding outsourcing.

Exhibit 21.6. An Example of Multivariate Analysis of Variance (MANOVA) in Sport Management Studies

Type	Source and Purpose of the Study	Sample	Independent Variables	Dependent Variables	Major Findings
MANOVA	Trail, Robinson, Gillentine, and Dick (2003) developed a model and conducted a study to determine how individuals who attend college football games should be classified at the four NCAA divisions of college football.	Data were collected from spectators at four intercollegiate football games during the 2001–2002 academic year. Out of 910 questionnaires distributed, 861 usable questionnaires were returned, for a return rate of 94%. The selection of the schools for the study was based on the geographic proximity to the researchers. The data were collected at one game in each of the divisions.	Collegiate Division levels	Overarching Motives, Spectator Motives, Vicarious Achievement Motive, Organizational Identification, Sport Identification, and Player Identification	Differences do exist among the divisions based on the motive variables and the points of attachment variables.

to determine the combination of attributes that will satisfy consumers the most. The results of the analysis can provide the tennis manufacturers with information about how they can present their consumers with a product that combines all the attributes that will be most satisfying to them.

A review of sport management related literature has found a few studies that applied the conjoint analysis technique (e.g., Chalip & McGuirty, 2004; Chang & Johnson, 1995; Daniel & Johnson, 2004). Daniel and Johnson (2004) employed conjoint analysis to determine whether or not it is viable to bundle benefits with different pricing structures in an Australian professional football team. The summary of this study is provided in Exhibit 21.7.

Multivariate Interdependence Techniques

Cluster Analysis

Cluster analysis is an exploratory data analysis technique used to identify objects or people with similar attributes and put similar objects or people into clusters according to well-defined similarity rules (McDaniel & Gates, 1991; Vermetten, 1999). However, it can only discover structures in data; it cannot explain why those structures exist. The identified structures are referred to as clusters. There are various methods to group objects or people into clusters. Different methods will result in different cluster patterns (George & Mallery, 2001; McDaniel & Gates, 1991; Nicol & Pexman, 1999). There are three generally used approaches to cluster analysis: hierarchical clustering, *k*-means clustering, and two-step clustering.

Hierarchical clustering is also referred to as agglomerative hierarchical clustering. It allows researchers to first define a distance criterion, and then use the criterion to form clusters, ultimately determining how many clusters will fit the data best. Hierarchical clustering is more appropriate for a study with a smaller sample. The hierarchical clustering method starts with the consideration of each object of interest as a cluster. Based on the criterion, two or more similar objects (i.e., those with the lowest distance or highest similarity) are combined into a cluster. Next, the object with the highest similarity to either of the first two is considered. If that third object is more similar to a fourth one than it is to either of the first two, the third and fourth objects will be clustered together. Otherwise, the third object is added to the first cluster. As the clustering process goes on, either more and more objects are added to the existing clusters or more and more new clusters are created or combined. If the process continues, all objects are joined together. When the process stops or the number of clusters is considered optimal depends on the purpose of the research. Exhibit 21.8 is an example of the hierarchical tree clustering plots.

K-Means Clustering

In *k*-means clustering, the number of clusters is first specified by the researcher. The researcher then decides how to assign objects into the clusters. *k* means clustering uses Euclidean distance as decision criterion. Euclidean distance is the most common distance measure. A given pair of objects is plotted on two variables, which form the x and y axes. The Euclidean distance is the square root of the sum of the square of the x difference plus the square of the y distance. The initial cluster centers are chosen randomly in a first pass of the data, then each additional iteration groups observations based on nearest Euclidean distance to the mean of the cluster. That is, the algorithm seeks to minimize within-cluster variance and maximize variability between clusters in an ANOVA-like

Exhibit 21.7. An Example of Conjoint Analysis in Sport Management Studies

Type	Source and Purpose of the Study	Sample	Variables Investigated	Major Findings
Conjoint Analysis	To investigate the viability of bundling benefits with different pricing structures so as to provide better service and encourage retention while increasing prices.	A questionnaire was sent to a random sample of 1500 members of an Australian professional football team, 500 for each of the three membership categories (standard members, full club members—concourse seating, and full club members—grandstand seating). About 35% of the surveyed responded.	The respondents were asked to rate their probability of renewal of their membership under each of the sixteen scenarios or hypothetical membership packages provided. The packages included one of the three price levels and various combinations of the five benefit attributes, such as grand final ticket preference, ticket resale program, an official glossy yearbook, discount on public transport, and entry to newly improved social club facilities.	1. The members in the standard membership group were the most price sensitive of the three groups surveyed. However, they were prepared to extend to the medium price level to full club members—concourse seating. 2. The full club members—concourse seating group found the access to grand final ticket option appealing and were interested in the newly improved social club facilities. This group seemed to be prepared for moving up to the next price level. 3. The full club members—grandstand seating group—was least sensitive to price increase. In fact, they were prepared to pay more for superior services.

Exhibit 21.8. An Example of Cluster Analysis

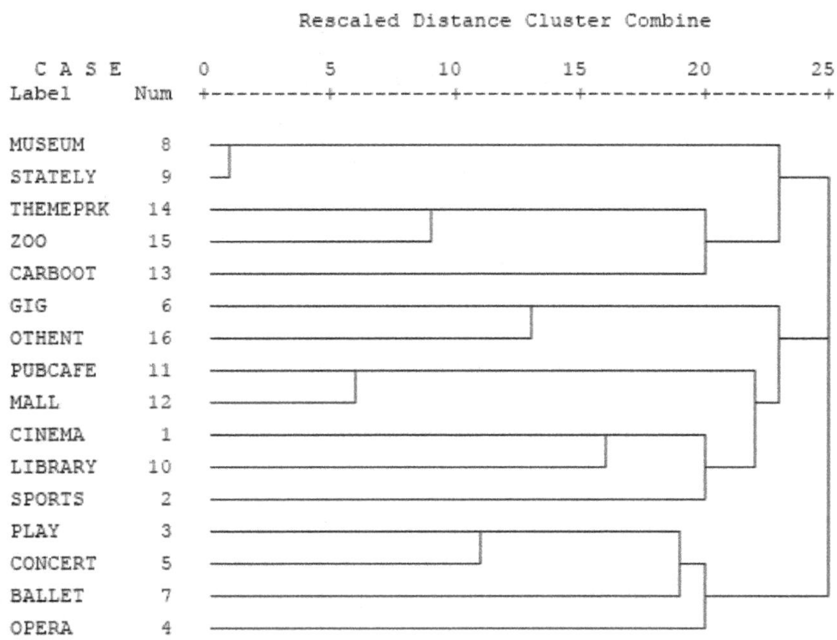

Source. Sturgis, P., & Jackson, J. (2003). Examining participation in sporting and cultural activities: Analysis of the UK 2000 Time Use Survey. London, UK: Department for Culture, Media and Sport.

fashion. Cluster centers change at each pass. The process continues until cluster means do not shift more than a given cut-off value or the iteration limit is reached. *k*-means clustering is much less computer-intensive and is therefore sometimes preferred when data sets are very large.

Two-Step Clustering

When using this method, the researcher first creates a number of pre-clusters. Standard hierarchical clustering is then applied to the pre-clusters in the second step. This is the method used when one or more of the variables are categorical (not interval or dichotomous). Also, because it is a method requiring neither a proximity table such as hierarchical classification nor an iterative process similar to *k*-means clustering, the two-step clustering method is recommended for very large data sets, as it is a one-pass-through-the-data-set method.

In sport management–related research, cluster analysis has been used in strategy studies to identify homogeneous groups of amateur sport organizations (Kikulis, Slack, Hinings, & Zimmerman, 1989), to classify recreational runners into mutually exclusive segmented groups based on their perceptions and attitudes on a list of involvement items (McGehee, Yoon, and Cardenas, 2003), and to identify segments of spectators based upon the brand associations held for a professional sport team (Ross, 2007).

The following is a brief description of the application of cluster analysis by Ross (2007) in identifying segments of spectators. Six hundred sixty-five season ticket holders in the 2004–2005 NBA season participated in the study. Data were collected on 11 measures of team brand associations for professional sport teams, including (1) non-player personnel, (2) team success, (3) team history, (4) stadium community, (5) team-play characteristics, (6) brand mark, (7) organizational attributes, (8) concessions, (9) social interaction, (10) rivalry, and (11) commitment. The hierarchical tree clustering method was applied to determine how many clusters of brand association were contained within the surveyed participants. This determination followed a number of diagnostic steps. Specifically, the clustering coefficients or the within-cluster sum of squares in the agglomeration schedule were examined. Homogeneous clusters with small coefficients were then merged. This process continued until the largest difference among clustering coefficients appeared, ultimately resulting in the emergence of two clusters.

Factor Analysis

Factor analysis is a multivariate analysis technique for reducing a large set of variables into a smaller set of variables with common characteristics or underlying dimensions. Factor analysis is used by researchers to develop underlying commonality or dimensions among variables. The idea is that some variables belong together, and they are referred to as factors. The primary goals of factor analysis are "to discover the factors that best explain a group of measures and describe the relation of each measure to the factor of underlying construct" (Thomas & Nelson, 1996, p. 184). In most instances, factor analysis is the starting point for examining the factor structure of a research instrument. This procedure identifies underlying patterns of procedures (Kline, 2016). It involves the application of complex statistical procedures; however, when used appropriately and effectively, it is an excellent starting point for the interested researcher.

There are two different types of factor analyses: *exploratory factor analysis* and *confirmatory factor analysis*. When employing factor analysis, researchers use a variety of techniques for reducing data to smaller and more manageable sets of underlying concepts. A number of criteria are used when making decisions about variables to be retained in a data set. As a rule of thumb, a typical factor loading cutoff criteria range of 0.30 to 0.55 is considered to be a strong factor loading coefficient (Stevens, 2009).

There are several important decisions for researchers to make when using factor analysis: (1) make clear the purpose of the instrument that is being developed, (2) test the construct validity of the variables for the proposed instrument, (3) make certain the sample size is adequate when testing the instrument, (4) select the most appropriate type of data reduction technique, (5) select the most appropriate type of rotation, and (6) stipulate how much loading a variable should have on each factor. Exhibit 21.9 is an example of factor and scale loadings in a factor analysis.

Exploratory Factor Analysis

The primary purpose of exploratory factor analysis (EFA) is to summarize all of the information that exist in a data set and present it in a more manageable form with minimal loss of variables. Typically, there is no theoretical ground in EFA about which variables should be quantitatively associated with which factors. The clusters of variables (or factors) that emerge are statistically rather than theoretically derived. Exploratory factor analysis is used because it describes and summarizes data by grouping together variables that are correlated (Tabachnick & Fidell, 2013). It is most often implemented to identify a small number of parts that may be used to represent relationships among sets of interrelated variables (George & Mallery, 2001).

Exhibit 21.10 provides an example of a study conducted by Ferreira and Armstrong (2004), who, using a student sample, employed EFA to explore the factors that may influence aspects of the purchase cycle regarding individual evaluations of sport products as a multi-attribute choice problem.

Exhibit 21.9. Oblique Factor Analysis of the Specialization Variables: Factor and Scale Loadings

Specialization Variables	Factor 1 (Investment)	Factor 2 (Experience)	Factor 3 (Lifestyle)
Amount of Equipment	.57	.07	-.04
Overall Investment in Related Expenses	.72	.13	.23
Subscription to Related Sport Magazines	.64	.29	.37
Self-Reported Skill Level	.17	.54	.17
Development of Skills	.08	.52	.21
Days Engaged in an Activity	.12	.62	.11
Percent/Leisure Time	.27	.18	.53
Importance of Activity	.43	.29	.48
Eigenvalue	3.43	1.52	1.25
% of Total Variance	38.80	10.90	6.70
Cumulative %	38.80	49.70	56.40

Note. Factor loadings greater than .45 are underlined.

Exhibit 21.10. An Example of Exploratory Factor Analysis (EFA) in Sport Management Studies

Type	Source and Purpose of the Study	Sample	Type of Analysis	Rotation	Major Findings
Exploratory Factor Analysis (EFA)	Ferreira and Armstrong (2004) explored the factors that may influence aspects of the purchase cycle regarding individual evaluations of sport products as a multi-attribute choice problem.	Students enrolled in three sport-related classes offered by the School of Physical Activity and Educational Services at a large Midwestern university (n=53)	Principal components	Orthogonal (Varimax)	The result of EFA suggested that some attributes may not be sport-specific and may be included in the individual's general evoked set for different sport events. (p. 207)

Exhibit 21.11. An Example of Confirmatory Factor Analysis (CFA) in Sport Management Studies

Type	Source and Purpose of the Study	Sample	Type of Analysis	Rotation
Confirmatory Factor Analysis (CFA)	James and Ross (2004) investigated research under two purposes of study: first, to extend understanding of sport consumers by identifying the motives that help explain an individual's interest in nonrevenue collegiate sports; second, to ascertain whether similar motives influence consumption across multiple nonrevenue sports.	People attending wrestling matches, baseball games, and softball games at a large Midwestern university (n=947; wrestling matches=292, baseball games=354, softball games=301)	Principal components	Orthogonal (Varimax)

Confirmatory Factor Analysis

Unlike exploratory factor analysis, the objective of confirmatory factor analysis (CFA) is to test a hypothesized factor or model and to assess its fit to the data (Bryant & Yarnold, 1995). The goal of CFA is to clearly identify whether a priori specified scale items are appropriately related to their specified factors (Kline, 2016). In most instances, it is used to test a proposed theory underlying an instrument or to confirm the model of an instrument used in a different context or population. The clusters of variables (or factors) that emerge are theoretically rather than statistically derived. Typically, it is based on a strong theoretical foundation that allows the researcher to specify the factor structure in advance. The major question is, are the correlations or covariances among the variables consistent with the hypothesized factor structure?

As a rule of thumb, a typical factor loading cutoff criteria range of 0.30 to 0.40 is considered to be a strong factor loading coefficient. CFA is quite useful for studying the factorial validity of data obtained with multiple-item and multiple-subscale instruments (Stevens, 2009).

As an interdependence analysis technique, CFA has often been adopted by sport management researchers in their determination as to whether the collected data fit the proposed conceptual framework (e.g., Gladden & Funk, 2002; Kwon & Trail, 2003). Exhibit 21.11 provides the description of study conducted by James and Ross (2004), who were interested in identifying the motives that help explain an individual's interest in nonrevenue collegiate sports and to ascertain whether similar motives influence consumption across multiple nonrevenue sports.

A Case Using Multivariate Analysis Techniques in Sport Management Research

Caro and Garcia (2007) conducted a study to measure consumer satisfaction in a popular long-distance race event and to examine the role of sport motivations on the "cognitive affective relationships

Exhibit 21.12. Summary of the Results of the Confirmatory Factor Analysis

	DIS	PLE	ARO	SAT	LOY
Disconfirmation (DIS)	0.747	0.174	0.213	0.341	0.334
Pleasure (PLE)	0.441	0.747	0.736	0.323	0.209
Arousal (ARO)	0.462	0.858	0.514	0.508	0.318
Satisfaction (SAT)	0.584	0.568	0.713	0.616	0.593
Loyalty (LOY)	0.578	0.457	0.564	0.770	0.611
Composite reliability	0.855	0.897	0.759	0.865	0.823

Note. Intercorrelations are presented in the lower triangle of the matrix. The average variance extracted (AVE) is depicted on the diagonal. Shared variances are given in the upper triangle of the matrix.

Source. Caro, L. M., & García, J. A. M. (2007),.Measuring perceived service quality in urgent transport service. *Journal of Retailing and Consumer Services, 14*(1), 60–72.

that drive satisfaction evaluation" (p. 70). The study employed a number of multivariate analysis techniques, including EFA, CFA, cluster analysis, and MANOVA.

A survey was used for data collection; out of 352 questionnaires sent, 137 were returned for a response rate of approximately 38%. Two conceptual models of satisfaction were proposed. Both the models contained the same constructs, but each had different structural relationships among the constructs. The constructs in the models, which included disconfirmation (two items), pleasure (six items), arousal (four items), overall satisfaction, loyalty (five items), and sport motivations (eight items), were first measured. The researchers employed EFA to evaluate the items included in the proposed model. To improve the quality of the psychometric properties of the measures, a number of the original items were deleted from the model based on the results of EFA. Then, a CFA was followed. Exhibit 21.12 summarizes the results of the CFA.

Cluster analysis was used in the study to analyze the moderating effect of sport motivation. A two-cluster solution was derived. In other words, all the respondents were different enough to be put into two groups or clusters based on their scores on all the motivation factors. A MANOVA was then applied, with the cluster membership as the independent variable, and the eight motivation measures as the dependent variables, to verify that differences indeed existed among members of the two clusters. Exhibit 21.13 shows the results of the MANOVA.

Summary

There are a number of multivariate analysis techniques that are commonly used in sport management research. These techniques include, among others, multiple regression analysis, discriminant analysis, MANOVA, conjoint analysis, factor analysis, and cluster analysis. To use each of the techniques, sport management researchers should understand the circumstances or prerequisites under which they should be used.

In general, all the multivariate analysis techniques can be categorized as either methods dealing with the explanation or prediction of the dependent variable (referred to as methods of dependence analysis), or methods used to depict the structure of all the variables under investigation when

Exhibit 21.13. Results of the MANOVA Used to Determine if Differences Exist Among Members of the Two Clusters in Each of the Eight Motivation Factors

	Mean					
	High Motivated	Low Motivated	Brown Porsythe Statistic	P Value	Effect Size[1]	Power
Factors positive affect	4.23	3.70	6.69	<.05	.24	.75
Social contact	3.53	2.25	56.07	<.001	.57	1.00
Body image	3.44	2.98	6.67	<.05	.23	.71
Competition	3.47	2.25	33.36	<.001	.47	1.00
Challenge	4.57	3.81	20.93	<.001	.40	.99
Physical exercise	4.65	4.59	.24	n.s.	.04	.07
Entertainment	4.44	3.40	32.82	<.001	.48	1.00
Coping strategy	3.52	2.07	48.62	<.001	.54	1.00

[1] Effect size conventions for the F test (Cohen, 1998); small: .10; medium: .25; large: .40.

Source. Caro, L. M., & García, J. A. M. (2007). Measuring perceived service quality in urgent transport service. *Journal of Retailing and Consumer Services, 14*(1), 60–72.

there is no dependent variable involved (referred to as methods of interdependence analysis). In this chapter, multiple regression analysis, discriminant analysis, MANOVA, and conjoint analysis are some of the methods of dependence analysis that have been reviewed, while factor analysis and cluster analysis are two techniques of interdependence analysis that have also been reviewed.

Study Questions

1. What are the decision-making criteria used in the selection of a bivariate or multivariate analysis method?
2. How can multivariate analysis techniques be classified?
3. What are the decision-making criteria in terms of how to select an appropriate technique of multivariate analysis?
4. What are the differences between simple regression analysis and multiple regression analysis?
5. Under what circumstances can sport management researchers use the discriminant analysis technique to solve their research problems?
6. What are the differences between an ANOVA and MANOVA?
7. Under what circumstances is conjoint analysis a preferred statistical analysis technique?
8. What are the differences between cluster analysis and factor analysis?
9. What are the differences between exploratory factor analysis and confirmatory factor analysis?

Learning Activities

1. Identify a research article in one of the sport management related journals published this year that employed the multiple analysis of variance (MANOVA) technique and describe the variables involved in the study.

2. Design a research project that will use multiple regression analysis to predict the attendance of college football games at your school with a number of independent variables. What are those variables?
3. Design a study that involves the use of the cluster analysis technique.

References

Barber, H., & Eckrich, J. (1998). Methods and criteria employed in the evaluation of intercollegiate coaches. *Journal of Sport Management, 12*(4), 301–322.

Bryant, F. B., & Yarnold, P. R. (1995). Principal-components analysis and exploratory and confirmatory factor analysis. In L. C. Grimm & P. R. Yarnold (Eds.), *Reading and understanding multivariate statistics* (pp. 99–106). Washington, DC: American Psychological Association.

Caro, L. M., & García, J. A. M. (2007). Measuring perceived service quality in urgent transport service. *Journal of Retailing and Consumer Services, 14*(1), 60–72.

Chalip, L., & McGuirty, J. (2004). Bundling sport events with the host destination. *Journal of Sport Tourism, 9*(3), 267–282.

Chang, M. G. S. J., & Johnson, L. W. (1995). Segmenting the triathlon association membership market: An Australian example. *Sport Marketing Quarterly, 4*(4), 25–28.

Curry, J. (1996). *Understanding conjoint analysis in 15 minutes.* Technical report. Evanston, IL: Sawtooth Technologies, Inc.

Daniel, K., & Johnson, L. W. (2004). Pricing a sporting club membership package. *Sport Marketing Quarterly, 13*, 113–116.

Ferreira, M., & Armstrong, K. L. (2004). An exploratory examination of attributes influencing students' decisions to attend college sport events. *Sport Marketing Quarterly, 13*(4), 194–208.

Funk, D. C., Mahony, D. F., & Ridinger, L. L. (2001). Characterizing consumer motivation as individual difference factors: Augmenting the Sport Interest Inventory (SII) to explain level of spectator support. *Sport Marketing Quarterly, 11*(1), 33–43.

George, D., & Mallery, P. (2001). *SPSS for Windows step by step: A simple guide and reference 10.0 update.* Boston, MA: Allyn and Bacon.

Gladden, J. M., & Funk, D. C. (2002). Developing an understanding of brand associations in team sport: Empirical evidence from consumers of professional sport. *Journal of Sport Management, 16*(1), 54–81.

Gladden, J. M., & Milne, G. R. (1999). Examining the importance of brand equity in professional sport. *Sport Marketing Quarterly, 8*(1), 21–29.

Green, P. E., & Srinivasan. (1990). Conjoing analysis in marketing: New developments with implications for research and practice. *Journal of Marketing, 54*(4), 3–19.

Green, P. E., Wind, J., & Rao, V. R. (1999). Conjoint analysis: Methods and applications. In R. C. Darf (Ed.). *Technology management handbook.* Boca Raton, FL: CRC Press.

Harrison, L., Jr., Lee, A., & Belcher, D. (1999). Self-schemata for specific sports and physical activities: The influence of race and gender. *Journal of Sport and Social Issues, 23*(3), 287–307.

James, J. D., & Ross, S. D. (2004). Comparing sport consumer motivations across multiple sports. *Sport Marketing Quarterly, 13*(1), 17–25.

Kikulis, L.M., Slack, T., Hinings, B., & Zimmerman, A. (1989). A structural taxonomy of amateur sport organizations. *Journal of Sport Management, 3*(2), 129–150.

Klecka, W. R. (1980). *Discriminant analysis.* Thousand Oaks, CA: Sage Publications.

Kline, R. (2016). *Principles and practice of structural equation modeling* (4th ed.). New York, NY: Guilford Press.

Kwon, H. H., & Armstrong, K. L. (2002). Factors influencing impulse buying of sport team licensed merchandise. *Sport Marketing Quarterly, 11*(3), 151–162.

Kwon, H. H., & Trail, G .T. (2003). A reexamination of the construct and concurrent validity of the Psychological Commitment to Team Scale. *Sport Marketing Quarterly, 12*(3), 88-93.

Li, M., & Burden, W. (2004). Institutional control, perceived product attractiveness, and other related variables in affecting athletic administrators' outsourcing decisions. *International Journal of Sport Management, 5*(4), 295–305.

McDaniel, C., & Gates, R. (1991). *Contemporary marketing research.* St. Paul, MN: West Publishing Company.

McDaniel, S. R. (2002). An exploration of audience demographics, personal values, and lifestyle: Influences on viewing network coverage of the 1996 summer Olympic Games. *Journal of Sport Management, 16*(2), 117–131.

McDaniel, S. R., & Chalip, L. (2002). Effects of commercialism and nationalism on enjoyment of an event telecast: Lessons from the Atlanta Olympics. *European Sport Management Quarterly, 2*(1), 3–22.

McGehee, N. G., Yoon, Y., & Cardenas, D. (2003). Involvement and travel for recreational runners in North Carolina. *Journal of Sport Management, 17*(3), 305–324.

Mendenhall, W., & Beaver, R. J. (1995). *A brief course in business statistics.* Belmont, CA: Wadsworth Publishing Company.

Nicol, A. A. M., & Pexman, P. M. (1999). *Presenting your findings: A practical guide for creating tables.* Washington, DC: The American Psychological Association.

Peetz, T. B., Parks, J. B., & Spencer, N. E. (2004). Sport heroes as sport product endorser: The role of gender in the transfer of meaning process for selected undergraduate students. *Sport Marketing Quarterly, 13*(3), 141–150.

Quarterman, J., Jackson, E. N., Kim, K., Yoo, E., Koo, G. Y., Prugger, B.. & Han, K. (2006). Statistical data analysis techniques employed in the *Journal of Sport Management* January 1987 to October 2004. *International Journal of Sport Management, 7*(1), 13–30.

Robinson, M. J., & Trail, G. T. (2005). Relationships among spectator gender, motives, points of attachment, and sport preference. *Journal of Sport Management, 19*, 58–80.

Robinson, M. J., Trail, G. T., Dick, R. J., & Gillentine, A. J. (2005). Fans vs. spectators: An analysis of those who attend intercollegiate football games. *Sport Marketing Quarterly, 14*, 43–53.

Ross, S. D. (2007). Segmenting sport fans using brand associations: A cluster analysis. *Sport Marketing Quarterly, 16*, 15–24.

Scott, D. K. (1999). A multiframe perspective of leadership and organizational climate in intercollegiate athletics. *Journal of Sport Management, 13*(4), 298–316.

Stevens, J. (2009). *Applied multivariate statistics for the social sciences* (5th ed.). New York, NY: Routledge.

Sturgis & Jackson (2003). *Examining participation in sporting and cultural activities: Analysis of the UK 2000 Time Use Survey.* London, UK: Department for Culture, Media and Sport.

Tabachnick, B. G., & Fidel, L. S. (2013). *Using multivariate statistics* (6th ed.). Upper Saddle River, NJ: Pearson.

Thomas, J. R., & Nelson, J. K. (1996). *Research methods in physical activity* (3rd ed.). Champaign, IL: Human Kinetics.

Vermetten, Y. J. M. (1999). Consistency and variability of student learning in higher education. Unpublished doctoral dissertation, Tilburg University, Netherlands.

Weaver, M. A., & Chelladurai, P. (2002). Mentoring in intercollegiate athletic administration. *Journal of Sport Management, 16*(2), 96–116.

Zhang, J. J., Smith, D. W., Pease, D. G., & Jambor, E. A. (1997). Negative influence of market competitors on the attendance of professional sport game: The case of a Minor League Hockey team. *Sport Marketing Quarterly, 6*(3), 31–40.

Zikmund, W. G. (2000). *Business research methods* (6th ed.). Fort Worth, TX: Harcourt College Publishers.

Index

D

E

F

G

H

I

J

K

L

M

N

S

T

U

V

W

Y

About the Authors

Brenda G. Pitts, EdD, Professor at Georgia State University, is well known nationally and globally in sport marketing and sport business management. She is among the first research fellows of both the North American Society for Sport Management and the Sport Marketing Association, and is the honored recipient of the 2016 Diversity Award, the prestigious Dr. Earle F. Zeigler Scholar Award, Distinguished Sport Management Educator, and the Dr. Garth Paton Distinguished Service Award. She has delivered over 250 presentations at conferences around the world, and is published widely, including 19 books (some in different languages) and over 200 papers in such scholarly journals as *Sport Management Education Journal*, *Journal of Sport Management*, *Sport Marketing Quarterly*, *Journal of Vacation Marketing*, *International Journal of Sports Marketing and Sponsorship*, *Women in Sport and Physical Activity*, *International Journal of Sport Management*, *Journal of Legal Aspects of Sport*, *Sport Management International Journal*, *Sport Management Review*, and *Journal of Hospitality, Leisure, Sport & Tourism Education*. She loves and plays all kinds of sports. She played in the first Women's Professional Basketball League (WBL), was one of the first scholarship basketball player at the University of Alabama, and is an inductee in the Huntsville Sports Hall of Fame, the Women's Basketball Hall of Fame, and a nominee for the Alabama Sports Hall of Fame. Putting research into action, her study on the first women's basketball for the WBL was used by the WBCA in their decision to adopt the basketball nationally. A documentary film about this can be found on YouTube (https://www.youtube.com/watch?v=FTklCekY72s). Dr. Pitts also enjoys spending time with her wife, Melita, and playing with their furry kids, Jazz the Corgi, Tucker the Terrier, and Blu the cat.

Ming Li, EdD, is professor and dean of the College of Education and Human Development at Western Michigan University. He received his bachelor's degree in education from Guangzhou Sport University, his master's degree in education from Hangzhou University, and his doctor of education degree in physical education (sport administration) from the University of Kansas. His research interests are in financial and economic aspects of sport and in management of sport business in a global context. He has published more than 30 articles in refereed journals, numerous book chapters and three books: *Economics of Sport*,

Research Methods in Sport Management, and *International Sport Management*. He has made numerous refereed presentations at state, national, and international conferences. Currently, Dr. Li serves as President of the Chinese American Deans and Presidents (CCADP). Previously, he served as President of the Michigan Council for Deans of Public Colleges of Education, President of the Alliance for Sport Business (ASB), Chair of the Board of Commissioners of the Commission on Sport Management Accreditation (COSMA), and President of the North American Society for Sport Management (NASSM). He also served on the editorial boards of several professional journals, including *Journal of Sport Management*, *Sport Marketing Quarterly*, and *International Journal of Sport Management*. Dr. Li is an honorary guest professor of a number of institutions in China, including Sun Yat-sen University, Central University of Finance and Economics, Beijing Sport University, Tianjin University of Sport, and Guangzhou Sport University. He was the recipient of the Taylor Dodson Award given by the Southern District American Alliance of Health, Physical Education, Recreation and Dance (AAHPERD), both the Garth Paton Distinguished Service Award and the Diversity Award bestowed by NASSM, the Founding President Award presented by ASB, the Distinguished Alumni Award of the Guangzhou Sport University, and the 2017 Giving Back Diversity Leadership Award recognized by the Insight into Diversity Magazine. Dr. Li worked for the Atlanta Committee for the Olympic Games (ACOG) as an Olympic Envoy in 1996. In November 2011, he served as a consultant for the sports department of the Organizing Committee for the Asian Games held in Guangzhou, China. Professor Li's photo appears courtesy of Western Michigan University.

Dr. Amy Chan Hyung Kim is an Assistant Professor for the Department of Sport Management at Florida State University. Her research inquiry lies in promoting community well-being and health through community sport programs from a social epidemiological perspective. She also has been interested in exploring the evolution of knowledge structure in the field of Sport Management. She has taught research methods in sport management, strategic management in sport, facility and event management in sport, human resource management in sport, and introduction to sport management for undergraduate and graduate students at Florida State University. If she is not in her office, you may find her on the tennis court.

Crystal Southall, PhD is currently an Assistant Professor of Exercise and Sport Science (ESS) at Western State Colorado University (Western) in Gunnison, CO where she teaches numerous ESS courses, primarily within the Sport and Fitness Management Emphasis. Dr. Southall teaches courses in sociology of sport, sport marketing, sport ethics, facility and event management, risk management/legal issues, and lifetime wellness. Dr. Southall earned a B.A. in History from the University of Colorado at Boulder, a Master of Science in human movement science with an emphasis in sport and leisure commerce from the University of Memphis, and a Ph.D. in exercise and sport science with an emphasis in Sport Administration from the University of Northern Colorado. She is also a certified yoga instructor who has taught classes for various organizations in Colorado, Nebraska, and Massachusetts. Dr. Southall's research focuses primarily on the socio-cultural aspects of sport, with a primary focus on gender issues (e.g., masculine hegemony) and issues in college sport (e.g., institutional logics and hypercommercialization) and has given numerous refereed and invited presentations at national conferences. She also serves as a member of the editorial review board for the *Journal of Issues in Intercollegiate Athletics* (JIIA). Dr. Southall is also a contributing author to the textbook *Introduction to Sport Management: Theory and Practice* (2nd ed.).

Dr. Richard M. Southall is a Professor in the Department of Sport and Entertainment Management at University of South Carolina. He is also Director of the College Sport Research Institute (CSRI) at the University of South Carolina. CSRI sponsors the annual CSRI Conference on College Sport and the *Journal of Issues in Intercollegiate Athletics*. Dr. Southall has disseminated his research, which examines the National Collegiate Athletic Association's (NCAA) Collegiate Model of Athletics, in more than 100 commentaries, book chapters, and journal articles in such publications as *The Chronicle of Higher Education, College Athletics Clips, Ethnic and Racial Studies, International Sports Law Journal, Journal of Issues in Intercollegiate Athletics, Journal of the Legal Aspects of Sport, Journal of Sport Management, Journal of Sport and Social Issues, Sociology of Sport Journal,* and *Sport Marketing Quarterly*. Dr. Southall has given more than 150 national and international presentations. He is co-author of *Introduction to Sport Management: Theory and Practice* (2nd ed.), and *Sport Facility Management: Organizing Events and Mitigating Risks* (3rd ed.). A recognized expert on NCAA "big-time" college-sport, he has been quoted in such news outlets as *The Chronicle of Higher Education, insidehighered.com, The New York Times, USA Today, The Washington Post, ESPN Outside the Lines, The Wall Street Journal, Time* and *CNN*. Dr. Southall was a consultant for the plaintiffs in the 2009–2016 *O'Bannon v. NCAA* litigation. In July, 2014 he was invited to testify before the U.S. Senate Committee on Commerce, Science, & Transportation during a hearing on the safety and well being of college athletes.